THE INTERNATIONAL SERIES OF

Professor S. V. Ley, FRS
Professor T. Mukaiyama
Professor J. P. Simons, FRS

THE INTERNATIONAL SERIES OF MONOGRAPHS ON CHEMISTRY

---

1 J. D. Lambert: *Vibrational and rotational relaxation in gases*
2 N. G. Parsonage and L. A. K. Staveley: *Disorder in crystals*
3 G. C. Maitland, M. Rigby, E. B. Smith, and W. A. Wakeham: *Intermolecular forces: their origin and determination*
4 W. G. Richards, H. P. Trivedi, and D. L. Cooper: *Spin-orbit coupling in molecules*
5 C. F. Cullis and M. M. Hirschler: *The combustion of organic polymers*
6 R. T. Bailey, A. M. North, and R. A. Pethrick: *Molecular motion in high polymers*
7 Atta-ur-Rahman and A. Basha: *Biosynthesis of indole alkaloids*
8 J. S. Rowlinson and B. Widom: *Molecular theory of capillarity*
9 C. G. Gray and K. E. Gubbins: *Theory of molecular fluids. Volume 1: Fundamentals*
10 C. G. Gray and K. E. Gubbins: *Theory of molecular fluids. Volume 2: Applications* (in preparation)
11 S. Wilson: *Electron correlations in molecules*
12 E. Haslam: *Metabolites and metabolism: a commentary on secondary metabolism*
13 G. R. Fleming: *Chemical applications of ultrafast spectroscopy*
14 R. R. Ernst, G. Bodenhausen, and A. Wokaun: *Principles of nuclear magnetic resonance in one and two dimensions*
15 M. Goldman: *Quantum description of high-resolution NMR in liquids*
16 R. G. Parr and W. Yang: *Density-functional theory of atoms and molecules*
17 J. C. Vickerman, A. Brown, and N. M. Reed (editors): *Secondary ion mass spectrometry: principles and applications*
18 F. R. McCourt, J. Beenakker, W. E. Köhler, and I. Kuščer: *Nonequilibrium phenomena in polyatomic gases. Volume 1: Dilute gases*
19 F. R. McCourt, J. Beenakker, W. E. Köhler, and I. Kuščer: *Nonequilibrium phenomena in polyatomic gases. Volume 2: Cross-sections, scattering, and rarefied gases*
20 T. Mukaiyama: *Challenges in synthetic organic chemistry*
21 P. Gray and S. K. Scott: *Chemical oscillations and instabilities: non-linear chemical kinetics*
22 R. F. W. Bader: *Atoms in molecules: a quantum theory*
23 J. H. Jones: *The chemical synthesis of peptides*
24 S. K. Scott: *Chemical chaos*
25 M. S. Child: *Semiclassical mechanics with molecular applications*
26 D. T. Sawyer: *Oxygen chemistry*
27 P. A. Cox: *Transition metal oxides: an introduction to their electronic structure and properties*
28 B. R. Brown: *The organic chemistry of aliphatic nitrogen compounds*
29 Y. Yamaguchi, Y. Osamura, J. D. Goddard, and H. F. Schaeffer: *A new dimension to quantum chemistry: analytic derivative methods in* ab initio *molecular electronic structure theory*
30 P. W. Fowler and D. E. Manolopoulos: *An atlas of fullerenes*
31 T. Baer and W. L. Hase: *Unimolecular reaction dynamics*
32 A. J. Stone: *The theory of intermolecular forces*

The Theory of Intermolecular Forces

A. J. STONE

Department of Chemistry
University of Cambridge

CLARENDON PRESS • OXFORD

This book has been printed digitally in order to ensure its continuing availability

OXFORD
UNIVERSITY PRESS

Great Clarendon Street, Oxford OX2 6DP

Oxford University Press is a department of the University of Oxford.
It furthers the University's objective of excellence in research, scholarship,
and education by publishing worldwide in

Oxford New York

Auckland Bangkok Buenos Aires Cape Town Chennai
Dar es Salaam Delhi Hong Kong Istanbul Karachi Kolkata
Kuala Lumpur Madrid Melbourne Mexico City Mumbai Nairobi
São Paulo Shanghai Singapore Taipei Tokyo Toronto

with an associated company in Berlin

Oxford is a registered trade mark of Oxford University Press
in the UK and in certain other countries

Published in the United States
by Oxford University Press Inc., New York

First published 1996
First published in paperback 1997
Reprinted (with corrections) 2000, 2002

A catalogue record for this book is available from the British Library

Library of Congress Cataloging in Publication Data
(Data available)

ISBN 0-19-855883-X

PREFACE

There have been many developments in the theory of intermolecular forces over the last twenty years or so that have not so far been collected together in book form. After many years of effort, practical methods have emerged for the accurate calculation of intermolecular interactions by perturbation theory. Increasing computer power is making it possible to pension off the Lennard-Jones potential after 75 years of valuable service and to use more realistic models; similarly point-charge models are giving way to more sophisticated multipole descriptions. This has come at a time when experimental methods for determining intermolecular potentials are becoming more and more powerful and reliable, and when interest in accurate models of intermolecular interactions is greater than ever before, not least because of the increasing use of computer modelling for simulating systems of biological interest and for designing biologically active molecules for use as drugs and pesticides and the like.

However these more elaborate descriptions call for corresponding developments in the formalism. The mathematical expressions for interaction energies involving electrostatic multipoles, for instance, are daunting for the beginner, but they can be expressed in a relatively simple form with appropriate mathematical techniques. Forces and torques arising from these interactions can also be manipulated much more easily than is often supposed. I have been interested in these techniques for some time, and I have attempted in this book to set them out in a coherent fashion. The calculation of intermolecular interactions *ab initio* and the construction of accurate potential models are closely related topics that I have described in some detail. The determination of intermolecular potentials from experiment, on the other hand, is not an issue that I have attempted to cover in detail, and I have given only an outline of this aspect of the subject to put the theory in context, together with references to fuller accounts.

It is a pleasure to acknowledge the contributions of my research students and postdoctoral research assistants to the ideas described here, especially those of Sally Price and David Wales, with whom I have had many stimulating discussions and rewarding collaborations in recent years. I thank all my colleagues in the Theoretical Chemistry group at Cambridge for their support and encouragement; many of them have read parts of the text in draft and made many helpful suggestions. David Buckingham, Nicholas Handy, Ruth Lynden-Bell, David Clary and Roger Amos have all been most agreeable and stimulating people to work with. It is a particular pleasure, as he approaches retirement, to acknowledge the enormous contribution made by David Buckingham to the study of molecular properties and intermolecular forces. I am grateful also to Paul Hoang and Claude Girardet for their hospitality at the Université de Franche Comté in Besançon, where I was Visiting Professor for a time in 1993 and where the first draft of this book was written. I thank Bill Meath for much valuable advice and for permission to reproduce Fig. 6.3. Many others have contributed in various ways; it would be invidious to name a few and impossible to list everyone, but I am grateful to all of them. It goes without saying, but it must be said, that all errors that remain are my responsibility alone.

This book was prepared using LaTeX 2ε with the mathptm package for Times and Symbol fonts, BibTeX with the harvard bibliography style, makeindex and dvips. I am glad to acknowledge the generosity of the TeX community worldwide in making their programs freely available to all.

Cambridge A. J. S.
December 1995

The original hardback edition contained a number of errors which were corrected in the paperback reprint of 1997. Some further errors have been found since then, and have been corrected in this reprint. I am indebted to those readers, especially Yuthana Tantirungrotechai, who have told me about errors that they have found.

For details of the corrections, go to http://fandango.ch.cam.ac.uk/timf.html and follow the links. If any further corrections come to light they will be listed there also.

Cambridge A. J. S.
August 2000

CONTENTS

To Sybil

1

INTRODUCTION

1.1 The evidence for intermolecular forces

The idea that matter is made up of atoms and molecules was known to the Greeks, though the evidence for it did not become persuasive until the eighteenth and nineteenth centuries, when the ideal gas laws, the kinetic theory of gases, Faraday's laws of electrolysis, the stoichiometry of most chemical reactions, and a variety of other indications combined to decide the matter beyond doubt. In the twentieth century, further techniques such as X-ray diffraction and, more recently, high-resolution microscopy, have provided more evidence.

Given the idea that matter consists of molecules, however, the notion that there must be forces between them rests on much simpler evidence. The very existence of condensed phases of matter is conclusive evidence of attractive forces between molecules, for in the absence of attractive forces, the molecules in a glass of water would have no reason to stay confined to the glass. Furthermore, the fact that water has a definite density, and cannot easily be compressed to a smaller volume, shows that at short range the forces between the molecules become repulsive.

It follows that the energy of interaction U between two molecules, as a function of the distance R between them, must take the form shown in Fig. 1.1. That is, it must have an attractive region at long range, where the force $-\partial U/\partial R$ is negative, and a steeply repulsive region at close range to account for the low compressibility of condensed materials. There is a separation R_m where the energy is a minimum, and a closer distance σ where the energy of interaction goes through zero before climbing steeply. These are all conventional notations, as is the symbol ε for the depth of the attractive well. The precise form of the function $U(R)$ will depend on the particular molecules concerned, but these

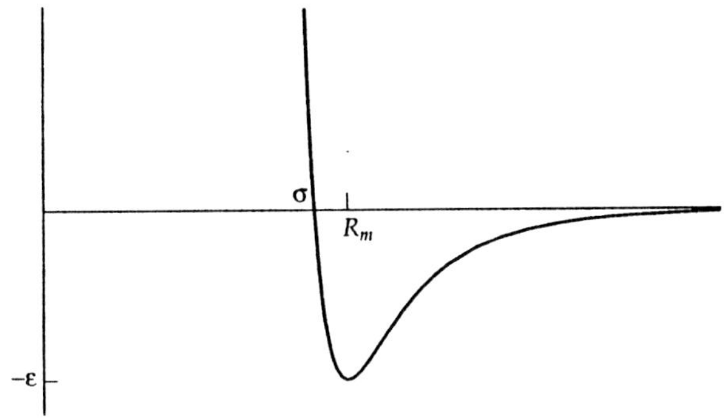

FIG. 1.1. A typical intermolecular potential energy function

general features will be universal.*

Van der Waals was the first to take these ideas into account in describing the departure of real gases from ideality. He suggested that the volume V occupied by a gas included a volume b that was occupied by the incompressible molecules, so that only $V-b$ remained for the free movement of the molecules; and that the attractive forces between the molecules had the effect of reducing the pressure exerted by a gas on its container, by an amount proportional to the square of the density. Thus the 'true' pressure in the sense of the gas laws is not the measured pressure P but $P+a/V^2$. The gas law $PV=RT$ then took the form $(P+a/V^2)(V-b)=RT$, where P and V are now the measured pressure and volume rather than the ideal values.

This simple equation gave a remarkably good account of the condensation of gases into liquids, and the values of the constants a and b correspond tolerably well with the properties of molecules as we now understand them. Although this approach has been superseded, the forces of attraction and repulsion between molecules are still often called 'Van der Waals' forces.

1.1.1 *Magnitudes*

The value of the Van der Waals parameter b has a clear interpretation in terms of molecular size; but more direct methods, such as X-ray crystallography, provide accurate values for the equilibrium separation in the solid. This is not quite the same as R_m, because there are attractive forces between more distant molecules that compress the nearest-neighbour distance slightly. The depth ε of the attractive well can be estimated *via* calorimetric data. One simple order-of-magnitude method is based on *Trouton's rule*, an empirical observation which states that the enthalpy of vaporization is approximately related to the boiling point at atmospheric pressure T_b by $\Delta H_{vap} \approx 10RT_b$. Although this is an empirical rule, it can be understood quite easily in a general way: the change in Gibbs free energy $G=H-TS$ between liquid and vapour is zero at the boiling point, so $\Delta H_{vap}=T_b\Delta S_{vap}$. We approximate ΔS in terms of the change of volume between liquid and gas, $\Delta S = R\ln(V_g/V_l)$, as if the liquid were just a highly compressed gas. Typically $(V_g/V_l)\approx 10^3$, which gives $\Delta S\approx 7R$; the remaining $3R$ or so can be accounted for by the fact that liquids are more structured than gases, and so have a lower entropy than is assumed in this simple picture.

Now the energy required to separate a condensed liquid into its constituent molecules is approximately ε for each pair of molecules that are close together. If each molecule has n neighbours in the liquid, the total energy for N molecules is $\frac{1}{2}Nn\varepsilon$ (where the factor of $\frac{1}{2}$ is needed to avoid counting each interaction twice). This is approximately the latent heat of evaporation; we should really correct for the zero-point energy of vibration of the molecules in the liquid, but if we assume that the intermolecular vibrations are classical the energy in these vibrations will be the same as the translational energy in the gas. Thus we find that $\frac{1}{2}N_A n\varepsilon \approx 10RT_b$, or $\varepsilon/k_B \approx 20T_b/n$.

*We shall see later that this is not quite true. In certain orientations the function $U(R)$ may be repulsive at all distances. However, unless the molecules are both ions with the same sign of charge, there will always be an orientation where the long-distance interaction is attractive.

Table 1.1 *Pair-potential well-depths.*

	T_b/K	n	$(20T_b/n)/K$	$(\varepsilon_{exp}/k_B)/K$
He	4.2	12	7	11
Ar	87	12	145	142
Xe	166	12	277	281
CH_4	111.5	12	186	180–300
H_2O	373.2	4	1866	2400 approx.

Table 1.1 shows the values that result for a few atoms and molecules. The value of the coordination number n has been assumed here to be the same as in the solid; this is probably a slight overestimate in most cases. The results show that the predictions are quite good, and certainly adequate for estimating the order of magnitude of ε.

We see, then, that the strength of intermolecular interactions between small molecules, as measured by the well-depth ε, is typically in the region 1–25 kJ mol^{-1} or 100–2000 K. This is considerably weaker than a typical chemical bond, which has a dissociation energy upwards of 200 kJ mol^{-1}, and this is why intermolecular interactions can be broken easily by simple physical means such as moderate heat, while chemical bonds can only be broken by more vigorous procedures.

In this analysis, we have tacitly assumed that the energy of an assembly of molecules, as in a liquid, can be treated as a sum of pairwise interactions; that is, that the total energy W of the assembly can be expressed in the form

$$\begin{aligned} W &= \sum_i W_i + \sum_{i=2}^{N} \sum_{j=1}^{i-1} U_{ij} \\ &= \sum_i W_i + \sum_{i>j} U_{ij} \\ &= \sum_i W_i + \tfrac{1}{2} \sum_j \sum_{i \neq j} U_{ij}, \end{aligned} \tag{1.1.1}$$

where W_i is the energy of isolated molecule i and U_{ij} is the energy of interaction between i and j. We have to be careful to count each pair of molecules only once, not twice, and the sum over pairs can be written in several ways, as shown, the second form being a conventional abbreviation of the first.

This assumption of *pairwise additivity* is only a first approximation. Eqn (1.1.1) is just the start of a series which should include three-body terms, four-body terms, and so on:

$$\begin{aligned} U &= W - \sum_i W_i \\ &= \sum_{i>j} U_{ij} + \sum_{i>j>k} U_{ijk} + \sum_{i>j>k>l} U_{ijkl} + \cdots \\ &= W_{2body} + W_{3body} + W_{4body} + \cdots. \end{aligned} \tag{1.1.2}$$

For instance, if there are three molecules A, B and C, the pairwise approximation to the total energy is $U_{AB}+U_{BC}+U_{AC}$, where U_{AB} is evaluated as if molecule C was not present, and so on. However the presence of molecule C will modify the interaction between the other two molecules, so there is a three-body correction: $U = U_{AB}+U_{BC}+U_{AC}+U_{ABC}$. When there are four molecules, we have to include a three-body correction for each set of three molecules, and the error that remains is the four-body correction. We hope that the total contribution $W_{3\text{body}}$ of the three-body corrections will be small, and the four-body term smaller still. In most circumstances this is a reasonably good approximation, so most effort is concerned with getting the pair interactions right. However the many-body terms are not small enough to neglect altogether, and we shall need to consider them later.

1.2 Classification of intermolecular forces

We can identify a number of physical phenomena that are responsible for attraction and repulsion between molecules. Here we give an overview of the main contributions to forces between molecules. All of the important ones arise ultimately from the electrostatic interaction between the particles comprising the two molecules. They can be separated into two main types: 'long-range', where the energy of interaction behaves as some inverse power of R, and 'short-range', where the energy decreases exponentially with distance. This apparently arbitrary distinction has a clear foundation in theory, as we shall see.

The long-range effects are of three kinds: *electrostatic*, *induction* and *dispersion*. The electrostatic effects are the simplest to understand in general terms: they arise from the straightforward classical interaction between the static charge distributions of the two molecules. They are strictly pairwise additive and may be either attractive or repulsive. Induction effects arise from the distortion of a particular molecule in the electric field of all its neighbours, and are always attractive. Because the fields of several neighbouring molecules may reinforce each other or cancel out, induction is strongly non-additive. Dispersion is an effect that cannot easily be understood in classical terms, but it arises because the charge distributions of the molecules are constantly fluctuating as the electrons move. The motions of the electrons in the two molecules become correlated, in such a way that lower-energy configurations are favoured and higher-energy ones disfavoured. The average effect is a lowering of the energy, and since the correlation effect becomes stronger as the molecules approach each other, the result is an attraction.

We shall discuss all these effects in much more detail later. For the moment they are summarized in Table 1.2. Two other effects that can arise at long range are included in the Table: these are resonance and magnetic effects. Resonance interactions occur either when at least one of the molecules is in a degenerate state—usually an excited state—or when the molecules are identical and one is in an excited state. Consequently they do not occur between ordinary closed-shell molecules in their ground states. They will be discussed in Chapter 10. Magnetic interactions involving the electrons can occur only when both molecules have unpaired spins, but in any case are very small. Magnetic interactions involving nuclei can occur whenever there are nuclei with non-zero spin, which is quite common, but the energies are several orders of magnitude smaller still, and are never of any significance in the context of intermolecular forces.

Table 1.2 *Contributions to the energy of interaction between molecules.*

Contribution	Additive?	Sign	Comment
Long-range ($U \sim R^{-n}$)			
Electrostatic	Yes	$\pm$	Strong orientation dependence
Induction	No	$-$	
Dispersion	approx.	$-$	Always present
Resonance	No	$\pm$	Degenerate states only
Magnetic	Yes	$\pm$	Very small
Short-range ($U \sim e^{-\alpha R}$)			
Exchange	No	$-$	
Repulsion	No	$+$	Dominates at very short range
Charge Transfer	No	$-$	Donor-acceptor interaction
Penetration	Yes	$-$	Can be repulsive at very short range
Damping	approx.	$+$	Modification of dispersion and induction

Further contributions to the energy arise at short range—that is, at distances where the molecular wavefunctions overlap significantly. The most important of these are exchange and repulsion, often taken together and described as exchange-repulsion or just exchange. The remaining effects listed in Table 1.2, namely penetration, charge transfer and damping, are modifications of the long-range terms arising from the overlap of the wavefunctions; the charge-transfer interaction is often viewed as a separate effect but is better thought of as a part of the induction energy. These terms will be discussed in more detail in Chapter 6.

1.3 Potential energy surfaces

Only in the case of atoms is a single distance sufficient to describe the relative geometry. In all other cases further coordinates are required, and instead of contemplating a potential energy curve like Fig. 1.1, we need to think about the 'potential energy surface', which is a function of all the coordinates describing the relative position of the molecules. In the case of two non-linear molecules, there are six of these coordinates (we discuss them below) and the potential energy surface becomes very difficult to visualize. For many purposes, however, we can think in three dimensions: a vertical dimension for the energy, and two horizontal dimensions which are representative of the six or more coordinates that we should really be using. This simplified picture is then like a landscape, with hills and valleys. There are energy minima on the potential energy surface that correspond to depressions in the landscape. In the real landscape such depressions are usually filled with water, to give lakes or tarns; there may be many of them, at different heights. In the quantum-mechanical landscape of the potential-energy surface, there may also be many minima, but there is always one 'global minimum' that is lower than any of the others. (Sometimes there are several global minima of equal energy, related to each other by symmetry.) The rest are 'local minima'; the energy at a local minimum is lower than the energy of any point in its immediate neighbourhood, but if we move further away we can find other points that are lower still.

Just as the depressions in the landscape are filled with water, we might think of the minima in the potential-energy surface as being filled with zero-point energy; and just as lakes may merge into each other, so the zero-point energy may permit molecular systems to move freely between adjacent minima if the barriers between them are not too high. Given enough energy, the molecular system can rearrange, passing from one energy minimum to another via a 'saddle point' or 'transition state' which is the analogue of the mountain pass. The molecular system however can also tunnel between minima, even if the energy of the system is not high enough to overcome the barrier in the classical picture. A well-known example is the ammonia molecule, which can invert via a planar transition structure from one pyramidal minimum to the other. The height of the barrier in this case is 2020 cm^{-1} or about 24 kJ mol^{-1}. If tunnelling is ignored, the lowest-energy stationary states for the inversion vibration have an energy of 886 cm^{-1}, and there are two of them, one confined to each minimum. When tunnelling is taken into account, these two states can mix. If they combine in phase the resulting state has a slightly lower energy; if out of phase, the energy is slightly higher. The difference in energy—the 'tunnelling splitting'—is 0.79 cm^{-1} in this case, and can be measured very accurately by spectroscopic methods. Such tunnelling splittings are a useful source of information about potential energy surfaces.

Along with the minima, then, the barriers constitute the other important features of the surface, corresponding to the mountain passes of the real landscape. To understand the nature of the potential energy surface, we need to characterize the minima and the transition states. In mathematical terms, these are both 'stationary points', where all the first derivatives of the energy with respect to geometrical coordinates are zero. To distinguish between them, we need to consider the second derivatives. If we describe the energy U in the neighbourhood of a stationary point by a Taylor expansion in the geometrical coordinates q_i, it takes the form

$$U = U(0) + \tfrac{1}{2}\sum_{ij} q_i q_j \left.\frac{\partial^2 U}{\partial q_i \partial q_j}\right|_0 + \cdots; \tag{1.3.1}$$

the first derivatives are all zero at a stationary point, and we can ignore the higher derivatives if the q_i are small. The matrix of second derivatives is called the 'Hessian'; its components are

$$H_{ij} = \frac{\partial^2 U}{\partial q_i \partial q_j}. \tag{1.3.2}$$

The eigenvalues of the Hessian are all positive at a minimum; in this case the Hessian is said to be 'positive definite'. Whatever direction we walk away from a minimum, we find ourselves going uphill. Such a point is said to have 'Hessian index zero': that is, the Hessian has no negative eigenvalues there. At a saddle-point or mountain pass, precisely *one* of the eigenvalues is negative—the Hessian index is 1. If we walk along the eigenvector corresponding to that eigenvalue, in either direction, we find ourselves going downhill, but the other eigenvectors take us uphill. Just as in the real landscape, the saddle-points provide the routes from one valley or minimum to another. If two or more of the eigenvalues are negative, there are two orthogonal directions that will take us downhill. In this case (in the real landscape) we are at the top of a hill, and all directions lead downhill; in the many-dimensional system there may be other directions that lead uphill. However it is never necessary to go over the top of the hill to get to the other side; there is always a lower route round the side of the hill (Murrell and Laidler 1968). If we are interested in characterizing the potential-energy surface, then, the most important stationary points are the minima, with Hessian index 0, and the saddle points, with Hessian index 1. Stationary points with higher values of the Hessian index are much less important.

Often there are several equivalent minima with the same energy, related to each other by the symmetry of the system. In the HF dimer, for example, the equilibrium structure is hydrogen-bonded, with only one of the two H atoms in the hydrogen bond. There are two such structures, one with the H atom of molecule 1 forming the hydrogen bond, the other with the H atom of molecule 2 in that rôle. For reasons of symmetry they have the same energy; they are distinguishable only if we can label the atoms in some way. The two cases are said to be different 'versions' of the same structure. In the same way there may be several versions of a transition state, related to each other by symmetry (Bone *et al.* 1991).

1.4 Coordinate systems

In this book, we shall be particularly concerned with the dependence of intermolecular forces on the orientation of the molecules—indeed, we shall see that the anisotropy of the interaction is a key feature, vital for understanding many aspects of structure and properties—and we shall need to use a variety of coordinate systems to describe the orientations. We summarize the principles here, but the details can be skipped on a first reading.

It is helpful to establish first the idea of 'local' and 'global' coordinate systems, often called 'molecule-fixed' and 'space-fixed' respectively. The 'global' coordinate system is defined by a set of cartesian axes, fixed in space, which we may call X, Y and Z. Their directions may be defined by reference to macroscopic features such as an external electric field, or the directions of the incident and scattered beams in Raman spectroscopy, or the directions of the molecular beam and the laser beam in molecular beam spectroscopy, but often they just provide an abstract frame of reference. Usually it is only the direction of the global axes that matters, and the position of the origin is unimportant.

The 'local' or 'molecule-fixed' axis system is tied to the molecule, and moves with it. Usually the origin is at the centre of mass, but it may be more convenient to put it at one of the nuclei—at the oxygen nucleus in the water molecule, for example. The directions of the coordinate axes are then fixed by reference to atom positions, so in water we might take the z axis along the C_2 symmetry axis, the x axis perpendicular to the plane, and the y axis to form a right-handed orthogonal set. Provided that we treat the molecules as rigid, the definition of local axes is straightforward. Important subtleties arise if we contemplate vibrating molecules, and there are further complications in molecules with large-scale internal motions, such as the torsions around the backbone bonds in a polypeptide. We shall for the most part ignore such issues, but note that the definition of molecule-fixed axes in vibrating molecules is usually achieved by imposing the 'Eckart

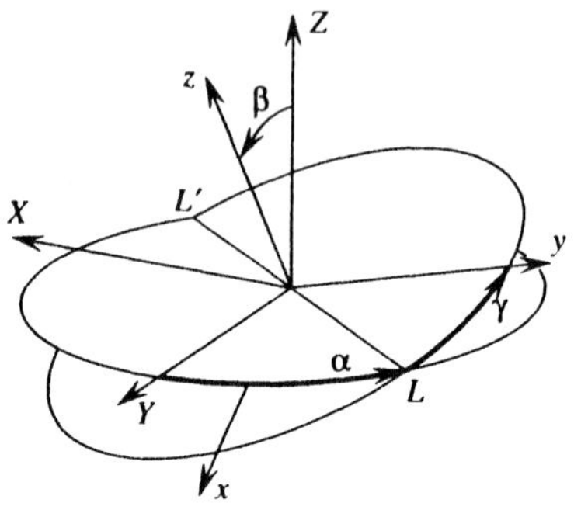

FIG. 1.2. Euler angles α, β and γ defining the orientation of an axis system x, y, z relative to X, Y, Z.

conditions', which are nearly, but not quite, equivalent to the requirement that the angular momentum of the molecule relative to its local axis system is zero. A detailed discussion can be found in the classic text by Wilson *et al.* (1955, Chapter 11). We return to the question of molecules with large-amplitude internal motions in Chapter 11.

The relationship between global and local axes is described by a rotation and a translation. Starting from a configuration in which the two sets of axes coincide, we first rotate the molecule into the desired orientation, and then translate it to the desired position. The rotation is usually described in terms of Euler angles α, β and γ which are defined as follows.

- Start from an orientation in which the local or molecule-fixed axes x, y and z are parallel to the X, Y and Z axes.
- Rotate through γ about the Z axis (or the z axis—they are parallel initially).
- Rotate through β about the Y axis.
- Rotate through α about the Z axis.

In each case a positive rotation is such as to advance a corkscrew outwards along the specified axis. Figure 1.2 shows the resulting configuration. The angle between the Z and z axes is β. The other angles, α and γ, can be defined by reference to the *line of nodes*, which is the line of intersection of the XY and xy planes (LL' in Fig. 1.2). α is the angle from the Y axis to the line of nodes, measured corkscrew-wise around the Z axis, while γ is the angle from the line of nodes to the y axis, measured corkscrew-wise around the z axis. If β is zero or π, the xy and XY planes coincide and the line of nodes is undefined, so that α and γ are not individually defined. When $\beta = 0$, their sum, $\alpha + \gamma$, is the angle from the Y axis to the y axis, while if $\beta = \pi$, $\alpha - \gamma$ is the angle from the Y axis to the y axis, measured around the Z axis.

In an alternative definition of the Euler angles, precisely equivalent to the one above, the rotations are taken around the molecule-fixed axes instead. Starting with the axes parallel as before, the rotations are then

- Rotate through α about the z axis (initially parallel to the Z axis).

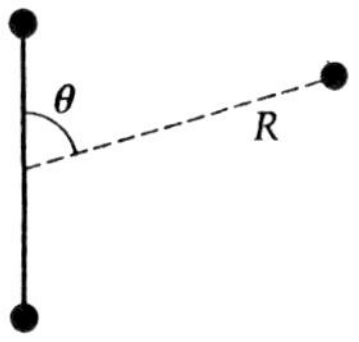

FIG. 1.3. The position of an atom relative to a linear molecule can be described by a distance R and an angle θ.

- Rotate through β about the y axis (which at this stage is directed along the line of nodes).
- Rotate through γ about the z axis.

Note that the angles are taken in the reverse order in this case. The equivalence of the two definitions can be verified by reference to Fig. 1.2; see Zare (1988, p. 78) for a more formal proof.

Some systems can be described more simply. For linear molecules, we take the local z axis along the molecular axis, and then the configuration is completely described by α and β, and γ is not needed. In this case, the angles α and β are more usually called φ and θ respectively: they are the spherical polar angles that describe the direction of the molecular axis relative to the global frame.

If we are concerned only to describe the interactions of the molecules with each other, rather than with some external field, the 'global' coordinate system may be more conveniently defined by reference to the molecules themselves. Thus for the interaction of an atom with a molecule, we can use a global coordinate system that coincides with the local coordinate system for the molecule, and then use spherical polar coordinates to specify the position of the atom relative to this axis system. If the molecule is linear the energy does not depend on the angle φ, and one distance and one angle suffice to determine the geometry (Fig. 1.3). For an atom interacting with a non-linear molecule we would need φ as well.

In the case of two linear molecules, it is often convenient to use a coordinate system in which one of the molecules is at the origin, and the other is situated on the positive Z axis, at $(0,0,R)$. Here we are defining the global axis system by reference to the positions of the two molecules. The direction of each molecular axis (its local z axis) is then defined by two polar angles, θ and φ. Notice that there is an asymmetry between θ_1 and θ_2: when $\theta_1 = 0$ the z_1 axis points towards molecule 2, but when $\theta_2 = 0$ the z_2 axis points away from molecule 1. This asymmetry in the coordinate system leads to an asymmetry in the formulae for the interaction energy. It is not unduly troublesome, but it means that care must be taken in defining which molecule is which.

A rotation of the entire system about the Z axis leaves the energy unchanged (in the absence of external fields). This has the effect of changing both φ_1 and φ_2 by the same amount, leaving $\varphi_1 - \varphi_2$ unchanged but altering $\varphi_1 + \varphi_2$. It follows that the energy may depend on $\varphi = \varphi_1 - \varphi_2$ but not on $\varphi_1 + \varphi_2$. The relative geometry is described by the four coordinates R, θ_1, θ_2 and $\varphi = \varphi_1 - \varphi_2$.

If either molecule is non-linear, the interaction energy will vary as it is rotated about

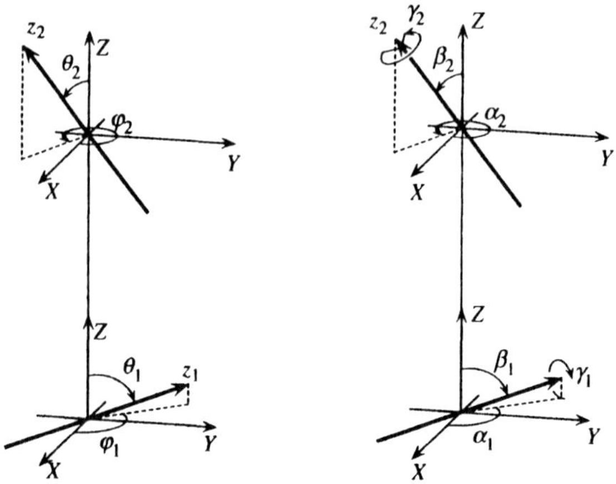

FIG. 1.4. (*Left*) A coordinate system for two linear molecules; (*Right*) A coordinate system for two non-linear molecules.

its molecular axis, and the third Euler angle becomes necessary. Here it is usual to use α, β and γ. If both molecules are non-linear, the description of the geometry requires six angles, but the energy is independent of $\alpha_1 + \alpha_2$. Thus we require the distance R and the five angles $\alpha_1 - \alpha_2$, β_1, β_2, γ_1 and γ_2 to describe the configuration.

Sometimes it is more natural to use a genuinely 'global' coordinate system, rather than a coordinate system tied to the molecules under discussion. The vector from molecule 1 to molecule 2 is now defined by three coordinates, R, θ and φ, and the orientation of each molecule is defined as before by three Euler angles. In this case we have nine variables in all, but only six of them are independent, in the absence of external fields, because the energy is unaffected by any rotation of the entire system. Here, however, it is a much more complicated matter to impose this energy invariance. Nevertheless, in spite of its more complicated appearance, such a description is often useful, and the simpler descriptions can be recovered from it by rotating the whole system so that $\theta = \varphi = 0$, i.e., so that the intermolecular vector lies along the global Z axis.

When there are more than two molecules in the system the number of intermolecular degrees of freedom increases further. For a system comprising N molecules, of which L are linear, and A atoms, there are $6(N-1) - L + 3A$ intermolecular degrees of freedom, plus one more if the system as a whole is linear. The simplest description is then to specify the position of each atom and molecule and the orientation of each molecule. This provides $6N - L + 3A$ coordinates, but the energy is unaffected by translation or rotation of the entire system.

1.4.1 *Internal coordinates*

Intermolecular forces are also affected by the vibrational coordinates of the molecules, which modify the molecular properties on which the intermolecular forces depend. These effects lead to a coupling between intermolecular and intramolecular motions. For the most part we shall not refer explicitly to such effects, since the fundamental theory of the interaction is the same whatever the molecular geometry may be, and many phenomena are not critically affected by the dependence of the intermolecular forces on the internal coordinates. However the coupling between internal coordinates and intermolecular coordinates may lead to changes in equilibrium bond lengths and bond angles, and to modification of vibrational frequencies. An example is the well-known vibrational red-shift caused by hydrogen bonding, which cannot be addressed without considering the relationship between the vibrational motion and the intermolecular forces. Even where the internal motions do not affect the phenomena in a fundamental way, however, they will modify them to some degree, so they should be taken into account in principle. Properties on which the intermolecular forces depend, such as multipole moments, should be regarded as functions of the internal coordinates, or at least averaged over the zero-point motion of the molecules. In practice, it is more usual in *ab initio* calculations to evaluate them at the equilibrium geometry of the isolated molecules.

2

MOLECULES IN ELECTROSTATIC FIELDS

2.1 Molecular properties: multipole moments

Of the contributions to the interaction energy listed in Table 1.2, the electrostatic interaction is often the most important, as we shall see in later chapters. However all of the contributions to the interaction energy between molecules, except for the unimportant magnetic terms, derive ultimately from the Coulombic interactions between their particles. In order to develop the theory of all these effects, we need to be able to describe the way in which the charge is distributed in a molecule. For most purposes, this is done most simply and compactly by specifying its *multipole moments*, and while this description has its limitations it provides an essential part of the language that we use to discuss intermolecular forces.

Multipole moments can be defined in two ways. One uses the mathematical language of cartesian tensors, while the other, the spherical-tensor formulation, is based on the spherical harmonics. The two descriptions are very closely related, and in many applications it is possible to use either. For more advanced work, the spherical-tensor approach is more flexible and powerful, but we begin with the cartesian approach because it is somewhat easier to understand. Later we shall use the spherical-tensor and cartesian tensor definitions more or less interchangeably, using whichever is more convenient.

2.1.1 *Cartesian definition*

The simplest multipole moment is the total charge: $q = \sum_a e_a$, where e_a is the charge on particle a and the sum is taken over all the electrons and nuclei. The next is the dipole moment, so called because the simplest example consists of two charges of equal magnitude q separated by a distance d, in which case the magnitude of the dipole moment vector is qd and its direction is from the negative charge to the positive one. In the more general case of an assembly of charges it takes the form

$$\hat{\mu}_z = \sum_a e_a a_z,$$

and similarly for $\hat{\mu}_x$ and $\hat{\mu}_y$, where we are using the vector $\mathbf{a}$ to describe the position of particle a. We can write the three components in one equation using 'tensor notation':

$$\hat{\mu}_\alpha = \sum_a e_a a_\alpha,$$

where α may stand for x, y or z. The caret over the symbol μ is to remind ourselves that this is an operator. If we require the value of the dipole moment in state $|n\rangle$, we construct it by taking the expectation value of this operator:

$$\mu_\alpha = \langle n|\hat{\mu}_\alpha|n\rangle,$$
$$= \int \rho_n(\mathbf{r}) r_\alpha \, \mathrm{d}^3\mathbf{r},$$

where the second form is expressed in terms of the molecular charge density $\rho(\mathbf{r})$, and again α may stand for x, y or z. If we integrate over electronic coordinates only (but include the contribution of the nuclear charge) we get the dipole moment for a fixed nuclear configuration. We have to integrate over nuclear coordinates too (vibrational average) to get a value that corresponds to the one that is measured experimentally. For example, the dipole moment of HF in its ground state has a magnitude of $0.72\,ea_0$* and is directed from the F atom to the H.†

The most accurate experimental technique for the measurement of dipole moments is the Stark effect in microwave spectroscopy. Unfortunately this does not provide the sign of the moment, and although the sign can usually be determined unambiguously on theoretical grounds, this is not always possible. A famous example is carbon monoxide, for which the dipole moment is in the direction $^-CO^+$, contrary to elementary chemical intuition and to *ab initio* calculations at the SCF level, though more accurate calculations allowing for the effects of electron correlation give the correct sign. Experimental determination of the sign requires a separate experiment; one method involves the determination of the effect of isotopic substitution on the rotational magnetic moment (Townes *et al.* 1955).

The next of the multipole moments is the *quadrupole moment*, so called because a quadrupolar charge distribution using charges of equal magnitude needs four of them, two positive and two negative‡. We define the operator

$$\hat{\Theta}_{zz} = \sum_a e_a\left(\tfrac{3}{2}a_z^2 - \tfrac{1}{2}a^2\right),$$
$$= \sum_a e_a a^2\left(\tfrac{3}{2}\cos^2\theta - \tfrac{1}{2}\right), \qquad (2.1.1)$$

where in the last line the vector **a** has been expressed in spherical polar coordinates. The expectation value of this operator for a particular state has the form

*The quantity ea_0 is the atomic unit of dipole moment, where e is the elementary charge (the charge on a proton) and a_0 is the Bohr radius. Other common units are the Debye, D= 10^{-18} esu, and the SI unit, the coulomb metre. 1 atomic unit = 2.5418 D = 8.478×10^{-30} C m. See Appendix D for other conversion factors.

†There is the possibility of confusion about the direction of a dipole moment. As defined here and in virtually all of the recent literature, the direction is from negative to positive charge. However, Debye used a crossed arrow to represent the dipole moment vector, thus: +——→, the arrow pointing from positive to negative charge, and this convention for the direction of the dipole moment may still occasionally be encountered. It causes much confusion and should be avoided.

‡There is some diversity of spelling of 'quadrupole' in the literature. The Latin prefix for 4 is 'quadri-' (as in quadrilateral, for example) so 'quadripole' would be acceptable. However, 'quadri-' usually becomes 'quadru-' before the letter 'p' (as in 'quadruped') so 'quadrupole' is correct and is the usual spelling. 'Quadrapole' is sometimes seen, but this is definitely incorrect; it is presumably formed by analogy with words like 'quadrangle', but there the 'a' belongs to 'angle' and the prefix has lost its final vowel. 'Octopole' too occurs with alternative spellings. In this case there is no clearcut rule: 'octopole' is more usual, but 'octapole' is also acceptable.

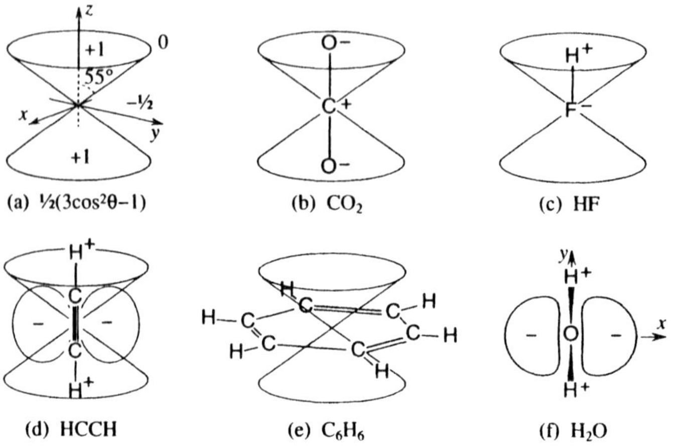

FIG. 2.1. Quadrupole moments.

$$\Theta_{zz} = \int \rho(\mathbf{r}) r^2 (\tfrac{3}{2}\cos^2\theta - \tfrac{1}{2})\, d^3\mathbf{r}, \qquad (2.1.2)$$

The angular factor in parentheses is positive when the angle θ is less than about 54° or more than 126°, and is negative between these values, in the region of the xy plane (see Fig. 2.1a). To estimate the quadrupole moment for a particular molecule, we can superimpose the angular function on the molecular charge distribution and carry out the integration schematically. For CO_2 (Fig. 2.1b) we see that the negatively charged oxygen atoms are in regions where the angular factor is positive, while the positively charged carbon atom occupies the region near the origin (small r) and contributes little to the quadrupole moment. Accordingly we expect a negative quadrupole moment. The experimental value is $-3.3\, ea_0^2$.* For HF (Fig. 2.1c) the H atom is positively charged, so with the origin at the centre of mass, as shown, we expect a positive quadrupole moment, in agreement with the experimental value of $+1.76\, ea_0^2$. In this case, however, the result depends on the choice of origin; see Section 2.7 for a detailed discussion.

Other interesting examples are acetylene and benzene. Acetylene (Fig. 2.1d) has a small charge separation between C and H, with the H atoms carrying small positive charges. These charges are in regions where $(3\cos^2\theta - 1)$ is large and positive, and are at relatively large r. There is also a substantial region of negative charge arising from the π-bonding orbitals in the region where $(3\cos^2\theta - 1)$ is negative. Consequently the quadrupole moment is quite large and positive; its value is $5.6\, ea_0^2$—larger than for CO_2. A similar situation arises in benzene (Fig. 2.1e), but here the H atoms are in the xy plane, where $(3\cos^2\theta - 1)$ is negative, and the π electrons are in the region where $(3\cos^2\theta - 1)$ is positive or small in magnitude. In this case, therefore, the quadrupole moment is negative; its value is $-6.7\, ea_0^2$.

*Once again, we are using atomic units, here ea_0^2. The SI unit is $\mathrm{C\,m^2}$, and the electrostatic unit is also commonly used. $ea_0^2 = 1.3450 \times 10^{-26}\,\mathrm{esu} = 4.487 \times 10^{-40}\,\mathrm{C\,m^2}$.

The quadrupole moment has other components besides Θ_{zz}, though they are all either zero or related to Θ_{zz} in the molecules discussed so far. They are

$$\Theta_{xx} = \sum_a e_a\left(\tfrac{3}{2}a_x^2 - \tfrac{1}{2}a^2\right),$$
$$\Theta_{yy} = \sum_a e_a\left(\tfrac{3}{2}a_y^2 - \tfrac{1}{2}a^2\right),$$
$$\Theta_{xy} = \sum_a e_a \tfrac{3}{2}a_x a_y,$$
$$\Theta_{xz} = \sum_a e_a \tfrac{3}{2}a_x a_z,$$
$$\Theta_{yz} = \sum_a e_a \tfrac{3}{2}a_y a_z.$$

Notice that $\Theta_{xx} + \Theta_{yy} + \Theta_{zz} = 0$, as a direct consequence of the definition. For linear and axially symmetric molecules, $\Theta_{xx} = \Theta_{yy}$, so that both are equal to $-\frac{1}{2}\Theta_{zz}$.

The water molecule has the charge distribution shown in Fig. 2.1f. The H atoms carry positive charges and are in the yz plane, while the lone pairs are directed in the xz plane and contain substantial amounts of negative charge. These facts fit well with the observed quadrupole moments, which are $\Theta_{xx} = -1.86\,ea_0^2$ and $\Theta_{yy} = +1.96\,ea_0^2$. Θ_{zz}, on the other hand, is small, with a value of only $-0.10\,ea_0^2$. Note again that $\Theta_{xx} + \Theta_{yy} + \Theta_{zz} = 0$.

It is convenient to summarize all these definitions in 'tensor notation':

$$\Theta_{\alpha\beta} = \sum_a e_a\left(\tfrac{3}{2}a_\alpha a_\beta - \tfrac{1}{2}a^2\delta_{\alpha\beta}\right). \tag{2.1.3}$$

The subscripts α and β can each be x, y or z. The quantity $\delta_{\alpha\beta}$ is the 'Kronecker delta': $\delta_{\alpha\beta} = 1$ if $\alpha = \beta$ and 0 if $\alpha \neq \beta$. This provides a much more compact way to write down the definitions, and also provides a powerful notation for manipulating them. Appendix A gives a brief summary of tensor notation, which we shall use extensively.

In the same notation, the octopole moment operator is defined by

$$\hat{\Omega}_{\alpha\beta\gamma} = \sum_a e_a\left[\tfrac{5}{2}a_\alpha a_\beta a_\gamma - \tfrac{1}{2}a^2(a_\alpha\delta_{\beta\gamma} + a_\beta\delta_{\alpha\gamma} + a_\gamma\delta_{\alpha\beta})\right], \tag{2.1.4}$$

Here there are 27 components in all, though they are not all independent: permutation of the subscripts does not change the value, and the definition guarantees that $\hat{\Omega}_{\alpha\beta\beta} = \hat{\Omega}_{\beta\alpha\beta} = \hat{\Omega}_{\beta\beta\alpha} = 0$. (The repeated Greek suffix in each of these expressions implies a summation over that suffix, so that, for example, $\hat{\Omega}_{\alpha\beta\beta} \equiv \sum_\beta \hat{\Omega}_{\alpha\beta\beta}$. This convention was first adopted by Einstein to express the equations of general relativity in a compact form, and is known as the Einstein summation convention.) When these relationships are taken into account, we are left with only 7 independent components.

As an example, consider methane (Fig. 2.2). The hydrogen atoms carry small net positive charges, while the regions between are negative, so it is not difficult to conclude that Ω_{xyz} is positive when the axes are chosen as shown, with a hydrogen atom in the octant $x > 0$, $y > 0$, $z > 0$. Its value is $+3\,ea_0^3$.* The symmetry of the molecule ensures

* $ea_0^3 = 0.7118 \times 10^{-34}$ esu $= 2.374 \times 10^{-50}$ C m^3.

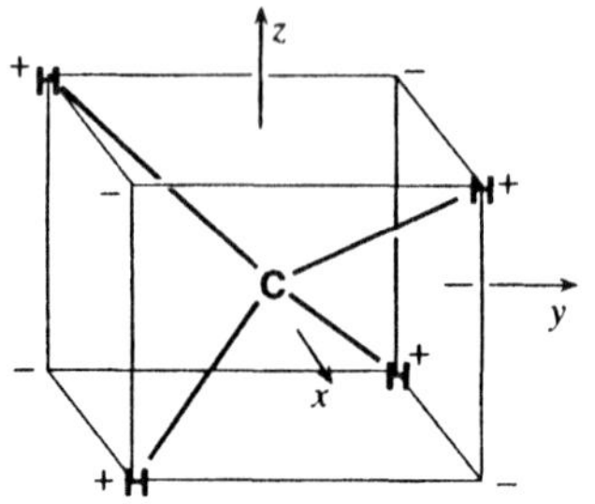

FIG. 2.2. Octopole moment of methane.

that the remaining components, except for those obtained by permuting the suffixes of Ω_{xyz}, are all zero.

These definitions can be generalized to any rank. The multipole moment operator of rank n has n suffixes, and takes the form

$$\hat{\xi}^{(n)}_{\alpha\beta\ldots\nu} = \frac{(-1)^n}{n!}\sum_a e_a a^{2n+1}\frac{\partial}{\partial a_\nu}\cdots\frac{\partial}{\partial a_\beta}\frac{\partial}{\partial a_\alpha}\left(\frac{1}{a}\right).$$

It follows at once from this definition that $\hat{\xi}^{(n)}_{\alpha\beta\ldots\nu}$ is symmetric with respect to permutation of its suffixes. Also, because

$$\frac{\partial}{\partial a_\alpha}\frac{\partial}{\partial a_\alpha}\left(\frac{1}{r}\right) = \nabla^2\frac{1}{r} = 0, \qquad (2.1.5)$$

it is traceless with respect to any pair of suffixes: $\hat{\xi}^{(n)}_{\alpha\alpha\gamma\ldots\nu} = 0$. We can use these properties to determine the number of independent components. All the components with m x suffixes and $(k-m)$ y suffixes are the same, and can be written as $\xi^{(n)}_{xx\ldots xy\ldots yz\ldots z}$. The value of k (which is the number of x and y suffixes) can range from 0 to n. For each of these values, the number of x suffixes can range from 0 to k, so there are $k+1$ distinct components for each value of k. The total number of distinct components is therefore $\sum_{k=0}^{n}(k+1) = \frac{1}{2}(n+1)(n+2)$.

This however ignores the trace conditions. Because of the invariance with respect to permutation of suffixes, we can write all the trace conditions in the form

$$\xi^{(n)}_{\alpha\alpha x\ldots xy\ldots yz\ldots z} = 0.$$

By the same argument as before, there are exactly $\frac{1}{2}(n-1)n$ of these conditions, one for each selection of the $(n-2)$ x, y and z suffixes. The total number of independent components is therefore reduced from $\frac{1}{2}(n+1)(n+2)$ to

$$\tfrac{1}{2}(n+1)(n+2) - \tfrac{1}{2}(n-1)n = 2n+1.$$

The multipole moment of rank n is sometimes called the '2^n-pole' moment. A dipole requires two charges arranged along a line, and a quadrupole requires four charges arranged in a two-dimensional array. Figure 2.2 shows that an octopole can be constructed from eight charges, four positive and four negative, arranged in a three-dimensional array. For rank 4 we would need $2^4 = 16$ charges, arranged in a four-dimensional array, which is not possible in three-dimensional space, but the rank-4 moment is nevertheless called the 2^4-pole or hexadecapole moment.

2.1.2 *Spherical tensor definition*

It is no accident that the number of independent components of the rank n multipole moment is $2n+1$, the same as the number of spherical harmonics of that rank. The components of the quadrupole moment are clearly proportional to the spherical harmonics of rank 2 (this is easier to see in the case of Θ_{xx} and Θ_{yy} if we consider $\Theta_{xx} - \Theta_{yy}$ rather than the separate quantities) and the same is true of the higher moments. Many advanced applications are easier if the multipole moments are defined at the outset in terms of the spherical harmonics, especially when quadrupoles or moments of higher rank are involved. Unfortunately some of the formulae are more difficult to derive, involving some advanced angular momentum algebra, but they are often much easier to apply than the corresponding cartesian formulae, and it is worth making the effort to master the theory, though perhaps not at a first reading. In the basic treatment, both cartesian and spherical tensor derivations will be given, but later we shall use whichever treatment is most convenient for the purpose. For this reason, it is useful not only to be able to handle both the cartesian and spherical tensor formalisms, but to be able to switch readily from one to the other. A summary of the basic ideas of the spherical tensor formalism is given in Appendix B, and some familiarity with those ideas is needed in what follows.

We now define the multipole moments in terms of the regular spherical harmonics defined in eqn (B.1.4) on p. 213:

$$\hat{Q}_{lm} = \sum_a e_a R_{lm}(\mathbf{a}),$$

or

$$Q_{lm} = \int \rho(r) R_{lm}(\mathbf{r})\, \mathrm{d}^3\mathbf{r}.$$

In practice it is usually more convenient to use the real forms:

$$\hat{Q}_{l\kappa} = \sum_a e_a R_{l\kappa}(\mathbf{a}),$$

or

$$Q_{l\kappa} = \int \rho(r) R_{l\kappa}(\mathbf{r})\, \mathrm{d}^3\mathbf{r}.$$

Here and subsequently we use the label κ to denote a member of the series 0, $1c$, $1s$, $2c$, $2s$, ... Where there is a sum over κ in a quantity labelled by $l\kappa$, the sum runs over the values 0, $1c$, $1s$, ..., lc, ls. The relationship between these and the earlier, cartesian, definitions is shown in Table E.1.* The inverse relationships can be derived without much

*With this definition of the Q_{lm} we have $Q_{l0} = \xi^{(l)}_{zz\ldots z}$ for all l. Some authorities, including the otherwise excellent text by Gray and Gubbins (1984), define the spherical components in terms of the Y_{lm} rather than the C_{lm}. This makes for a much less satisfactory relationship between the cartesian and spherical tensor definitions, and makes the spherical-tensor formalism even more forbidding for most people.

difficulty, but it is necessary to use the trace conditions, e.g., $\Theta_{xx}+\Theta_{yy}+\Theta_{zz}=0$, as well as the relationships in Table E.1. In this way, one finds, for example, that

$$\begin{aligned}\Theta_{xx} &= -\tfrac{1}{2}Q_{20}+\tfrac{1}{2}\sqrt{3}Q_{22c},\\ \Theta_{yy} &= -\tfrac{1}{2}Q_{20}-\tfrac{1}{2}\sqrt{3}Q_{22c}.\end{aligned} \tag{2.1.6}$$

The octopole and hexadecapole components can be obtained in the same way, and are listed in Table E.2.

2.2 The energy of a molecule in a non-uniform electric field

Consider a molecule in an external potential $V(\mathbf{r})$. Associated with this potential is an electric field $F_\alpha = -\partial V/\partial r_\alpha = -\nabla_\alpha V$. There may be non-zero higher derivatives such as the *field gradient* $F_{\alpha\beta} = -\partial^2 V/\partial r_\alpha \partial r_\beta = -\nabla_\alpha\nabla_\beta V$. We shall use the notation V_α for $\partial V/\partial r_\alpha$, $V_{\alpha\beta}$ for $\partial^2 V/\partial r_\alpha\partial r_\beta$, and so on. We shall work mainly in terms of derivatives of the potential, but the corresponding field or its derivative is obtained merely by changing the sign; that is, $F_\alpha = -V_\alpha$, $F_{\alpha\beta} = -V_{\alpha\beta}$, and so on.

We choose a suitable origin and set of coordinate axes, and expand the potential in a Taylor series:

$$V(\mathbf{r}) = V(0) + r_\alpha V_\alpha(0) + \tfrac{1}{2} r_\alpha r_\beta V_{\alpha\beta}(0) + \frac{1}{3!} r_\alpha r_\beta r_\gamma V_{\alpha\beta\gamma}(0) + \cdots.$$

Once again we are using the Einstein summation convention: a repeated suffix implies summation over the three values x, y and z of that suffix. We note that we should enquire about the convergence of this expansion, but we leave that problem until later. We shall be interested in the energy of our molecule in the presence of this potential. The operator describing this energy is

$$\mathcal{H}' = \sum_a e_a \hat{V}(\mathbf{a}),$$

where the sum is taken over all the nuclei and electrons in the molecule as before; particle a is at position $\mathbf{a}$ and carries charge e_a. Then

$$\mathcal{H}' = V(0)\sum_a e_a + V_\alpha(0)\sum_a e_a a_\alpha + \tfrac{1}{2}V_{\alpha\beta}(0)\sum_a e_a a_\alpha a_\beta + \cdots,$$

which we write as

$$\mathcal{H}' = \hat{M}V + \hat{M}_\alpha V_\alpha + \tfrac{1}{2}V_{\alpha\beta}\hat{M}_{\alpha\beta} + \cdots, \tag{2.2.1}$$

abbreviating $V_\alpha(0)$ to V_α, etc, and introducing the zeroth moment M, the first moment M_α, the second moment $M_{\alpha\beta}$, and so on. We can immediately identify the zeroth moment $\hat{M} = \sum_a e_a$ with the total charge q, and the first moment $\hat{M}_\alpha = \sum_a e_a a_\alpha$ with the dipole moment $\hat{\mu}_\alpha$.

The second moments are a little more complicated. We are interested only in the energy (2.2.1) of the interaction with the field. We define a new quantity $\hat{M}'_{\alpha\beta} = \hat{M}_{\alpha\beta} - k\delta_{\alpha\beta}$, where k is some constant and $\delta_{\alpha\beta}$ is the Kronecker tensor. Then

$$\begin{aligned}\tfrac{1}{2}V_{\alpha\beta}\hat{M}'_{\alpha\beta} &= \tfrac{1}{2}V_{\alpha\beta}\hat{M}_{\alpha\beta} - \tfrac{1}{2}k\delta_{\alpha\beta}V_{\alpha\beta}\\ &= \tfrac{1}{2}V_{\alpha\beta}\hat{M}_{\alpha\beta} - \tfrac{1}{2}kV_{\alpha\alpha}\\ &= \tfrac{1}{2}V_{\alpha\beta}\hat{M}_{\alpha\beta},\end{aligned}$$

where the last line follows from Laplace's equation:

$$V_{\alpha\alpha} = \sum_{\alpha}\frac{\partial^2 V}{\partial a_\alpha^2} = \nabla^2 V = 0.$$

This is true for any value of k. We now choose k so that $\hat{M}'_{\alpha\beta}$ becomes *traceless*: $\hat{M}'_{\alpha\alpha} \equiv \hat{M}'_{xx} + \hat{M}'_{yy} + \hat{M}'_{zz} = 0$. Then $\hat{M}_{\alpha\alpha} - k\delta_{\alpha\alpha} = 0$, or $k = \frac{1}{3}\hat{M}_{\alpha\alpha} = \frac{1}{3}\sum_a e_a a^2$. (Remember that $\delta_{\alpha\alpha} = 3$.) Then we have

$$\hat{M}'_{\alpha\beta} = \sum_a e_a(a_\alpha a_\beta - \tfrac{1}{3}a^2\delta_{\alpha\beta}) = \tfrac{2}{3}\hat{\Theta}_{\alpha\beta}.$$

So by subtracting away the trace of $\hat{M}_{\alpha\beta}$, which does not contribute to the electrostatic energy, we arrive at the quadrupole moment in the form given previously, except for a numerical factor.

The higher moments are manipulated in a similar way. When we modify $\hat{M}_{\alpha\beta\gamma}$ so as to remove the trace terms that do not contribute to the electrostatic energy, we arrive at the octopole moment $\hat{\Omega}_{\alpha\beta\gamma}$ in the form given in eqn (2.1.4) on p. 15, except for the numerical factor $\frac{5}{2}$, while $\hat{M}_{\alpha\beta\gamma\delta}$ leads to the hexadecapole moment $\hat{\Phi}_{\alpha\beta\gamma\delta}$, and so on. The operator describing the interaction becomes

$$\begin{aligned}\mathcal{H}' &= qV + \hat{\mu}_\alpha V_\alpha + \tfrac{1}{3}\hat{\Theta}_{\alpha\beta}V_{\alpha\beta} + \frac{1}{3.5}\hat{\Omega}_{\alpha\beta\gamma}V_{\alpha\beta\gamma} + \frac{1}{3.5.7}\hat{\Phi}_{\alpha\beta\gamma\delta}V_{\alpha\beta\gamma\delta} + \cdots\\ &\quad + \frac{1}{(2n-1)!!}\hat{\xi}^{(n)}_{\alpha\beta\ldots\nu}V_{\alpha\beta\ldots\nu} + \cdots,\end{aligned} \tag{2.2.2}$$

where $(2n-1)!! \equiv (2n-1)(2n-3)\ldots 5.3.1$ for integer n.

In terms of the spherical-tensor form of the multipole moments, the same expression can be written as

$$\mathcal{H}' = \sum_{lm}(-1)^m \hat{Q}_{l,-m}V_{lm}, \tag{2.2.3}$$

where $V_{lm} = [(2l-1)!!]^{-1}R_{lm}(\nabla)V\big|_{r=0}$. By $R_{lm}(\nabla)$ we mean the regular spherical harmonic whose argument is the vector gradient operator ∇, so for example $R_{10}(\nabla) = \nabla_z = \partial/\partial z$. The proof of this result is a little indirect. The general term in eqn (2.2.2) is a scalar (independent of the choice of coordinate axes) and it is the *only* non-zero scalar that can be constructed from the n-th rank tensors $\hat{\xi}^{(n)}_{\alpha\beta\ldots\nu}$ and $V_{\alpha\beta\ldots\nu}$. Similarly, the $l = n$ term in eqn (2.2.3), namely $\sum_{nm}(-1)^m\hat{Q}_{n,-m}V_{nm}$ is a scalar, and is the *only* scalar that can be constructed from the $\hat{Q}_{nm}$ and the V_{nm} (Brink and Satchler 1993, Zare 1988). Since the V_{nm} are linear combinations of the $V_{\alpha\beta\ldots\nu}$, and the $\hat{Q}_{nm}$ are linear combinations of the $\hat{\xi}^{(n)}_{\alpha\beta\ldots\nu}$, these scalars must be the same to within a multiplicative constant. Their identity is confirmed by examining the coefficient of the $\hat{\xi}^{(n)}_{zzz\ldots z}V_{zzz\ldots z}$ term.

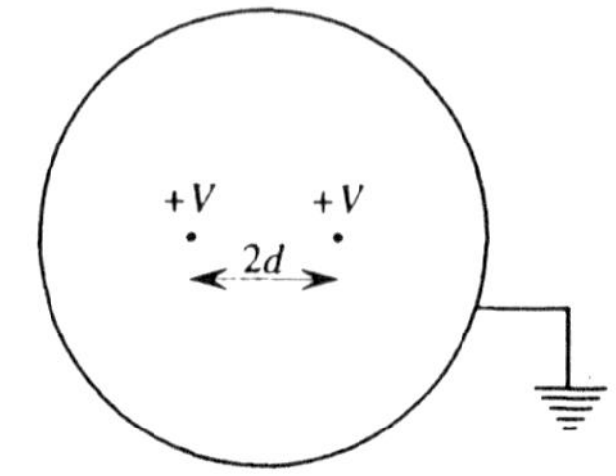

FIG. 2.3. Two-wire cell for quadrupole moment measurements.

A spherical-tensor 'scalar product' of the form $A \cdot B = \sum_m (-1)^m A_{l,-m} B_{lm}$, as in eqn (2.2.3), can be replaced by the scalar product in terms of the real components (see eqn (B.4.3)). That is,

$$\sum_m (-1)^m A_{l,-m} B_{lm} = \sum_\kappa A_{l\kappa} B_{l\kappa}. \tag{2.2.4}$$

so we can write the interaction with the field as

$$\mathcal{H}' = \sum_{l\kappa} \hat{Q}_{l\kappa} V_{l\kappa}. \tag{2.2.5}$$

This is a much more convenient form than (2.2.3) for most purposes.

We can now work out the energy of the molecule using perturbation theory.* To first order this is trivial: we just replace $\mathcal{H}'$ by its expectation value for the state of interest, usually the ground state $|0\rangle$, to obtain

$$\begin{aligned} W' &= \langle 0|\mathcal{H}'|0\rangle \\ &= qV + \mu_\alpha V_\alpha + \frac{1}{3}\Theta_{\alpha\beta} V_{\alpha\beta} + \frac{1}{3.5}\Omega_{\alpha\beta\gamma} V_{\alpha\beta\gamma} + \frac{1}{3.5.7}\Phi_{\alpha\beta\gamma\delta} V_{\alpha\beta\gamma\delta} + \cdots \\ &= \sum_{l\kappa} Q_{l\kappa} V_{l\kappa}, \end{aligned}$$

where $\mu_\alpha = \langle 0|\hat{\mu}_\alpha|0\rangle$, $\Theta_{\alpha\beta} = \langle 0|\hat{\Theta}_{\alpha\beta}|0\rangle$ and so on. This equation shows that to first order the energy of a dipole depends only on the electric field $F_\alpha = -V_\alpha$, the energy of a quadrupole involves only the field gradient $F_{\alpha\beta} = -V_{\alpha\beta}$, and so on. So the response to a uniform field (as for example in the Stark effect) provides information only about the molecular dipole. To study the molecular quadrupole, we require a potential for which the field is zero but the field gradient is non-zero. Buckingham and his colleagues (Buckingham *et al.* 1983, and references therein) have performed many experiments to determine molecular quadrupole moments and other properties using a cell in which two parallel wires charged to potential V are surrounded by an earthed cylinder (Fig. 2.3). The potential at a distance r from a wire charged to potential V is proportional to $-V \ln r$, so if there are wires at $(x, y) = (\pm d, 0)$, and the enclosing cylinder is far enough away, the potential at a point with coordinates $(x, y) = (\xi d, \eta d)$ close to the origin is

*See Appendix C for a summary of the principles of perturbation theory

$$\begin{aligned}V_{2\text{wire}} &\propto -V\left(\ln\sqrt{(d-x)^2+y^2}+\ln\sqrt{(d+x)^2+y^2}\right)\\ &= -V\ln d - \tfrac{1}{2}V\left(\ln[1-2\xi+\xi^2+\eta^2]+\ln[1+2\xi+\xi^2+\eta^2]\right.\\ &= -V\ln d - V(\xi^2-\eta^2)+O((\xi^2+\eta^2)^2)\\ &= -V\ln d - V(x^2-y^2)/d^2+O((x^2+y^2)^2/d^4).\end{aligned}$$

So at the origin, the electric field is zero, but the field gradient has non-zero components $F_{xx} = -F_{yy} \propto 2V/d^2$. In practice, the calculation of the field gradient is more complicated, because the size of the wires and the effect of the earthed cylinder must be taken into account (Buckingham and Disch 1963). This arrangement is used in measurements of the quadrupole moment, discussed further in Chapter 12.

2.3 Polarizabilities

The second-order energy of the ground state is given by the standard Rayleigh–Schrödinger sum over states:

$$\begin{aligned}W'' &= -\sum_n{}' \frac{\langle 0|\mathcal{H}'|n\rangle\langle n|\mathcal{H}'|0\rangle}{W_n - W_0}\\ &= -\sum_n{}' \frac{\langle 0|\hat{\mu}_\alpha V_\alpha + \frac{1}{3}\hat{\Theta}_{\alpha\beta}V_{\alpha\beta}+\cdots|n\rangle\langle n|\hat{\mu}_{\alpha'}V_{\alpha'} + \frac{1}{3}\hat{\Theta}_{\alpha'\beta'}V_{\alpha'\beta'}+\cdots|0\rangle}{W_n - W_0}\\ &= -V_\alpha V_{\alpha'}\sum_n{}' \frac{\langle 0|\hat{\mu}_\alpha|n\rangle\langle n|\hat{\mu}_{\alpha'}|0\rangle}{W_n - W_0}\\ &\quad -\frac{1}{3}V_\alpha V_{\alpha'\beta'}\sum_n{}' \frac{\langle 0|\hat{\mu}_\alpha|n\rangle\langle n|\hat{\Theta}_{\alpha'\beta'}|0\rangle}{W_n - W_0} - \frac{1}{3}V_{\alpha\beta}V'_{\alpha}\sum_n{}' \frac{\langle 0|\hat{\Theta}_{\alpha\beta}|n\rangle\langle n|\hat{\mu}_{\alpha'}|0\rangle}{W_n - W_0}\\ &\quad -\frac{1}{9}V_{\alpha\beta}V_{\alpha'\beta'}\sum_n{}' \frac{\langle 0|\hat{\Theta}_{\alpha\beta}|n\rangle\langle n|\hat{\Theta}_{\alpha'\beta'}|0\rangle}{W_n - W_0} - \cdots.\end{aligned} \tag{2.3.1}$$

As is customary, the prime on the summation indicates that the state $|0\rangle$ is to be omitted from the sum. Notice that the term qV can be dropped from the perturbation, even if the molecule is charged, because it is a constant and its matrix elements between different eigenstates are zero by orthogonality.

Now we define a set of polarizabilities by

$$W'' = -\tfrac{1}{2}\alpha_{\alpha\beta}V_\alpha V_\beta - \tfrac{1}{3}A_{\alpha,\beta\gamma}V_\alpha V_{\beta\gamma} - \tfrac{1}{6}C_{\alpha\beta,\gamma\delta}V_{\alpha\beta}V_{\gamma\delta} - \cdots,$$

and by comparing coefficients and relabelling the subscripts we find

$$\alpha_{\alpha\beta} = \sum_n{}' \frac{\langle 0|\hat{\mu}_\alpha|n\rangle\langle n|\hat{\mu}_\beta|0\rangle + \langle 0|\hat{\mu}_\beta|n\rangle\langle n|\hat{\mu}_\alpha|0\rangle}{W_n - W_0}, \tag{2.3.2}$$

$$A_{\alpha,\beta\gamma} = \sum_n{}' \frac{\langle 0|\hat{\mu}_\alpha|n\rangle\langle n|\hat{\Theta}_{\beta\gamma}|0\rangle + \langle 0|\hat{\Theta}_{\beta\gamma}|n\rangle\langle n|\hat{\mu}_\alpha|0\rangle}{W_n - W_0}, \tag{2.3.3}$$

$$C_{\alpha\beta,\gamma\delta} = \frac{1}{3}\sum_n{}' \frac{\langle 0|\hat{\Theta}_{\alpha\beta}|n\rangle\langle n|\hat{\Theta}_{\gamma\delta}|0\rangle + \langle 0|\hat{\Theta}_{\gamma\delta}|n\rangle\langle n|\hat{\Theta}_{\alpha\beta}|0\rangle}{W_n - W_0}. \tag{2.3.4}$$

To understand the significance of these quantities, it is helpful to write down the complete expression for the energy (for a neutral molecule, $q = 0$):

$$\begin{aligned} W &= W^0 + W' + W'' + \cdots \\ &= W^0 + \mu_\alpha V_\alpha + \tfrac{1}{3}\Theta_{\alpha\beta}V_{\alpha\beta} + \frac{1}{3.5}\Omega_{\alpha\beta\gamma}V_{\alpha\beta\gamma} + \cdots \\ &\quad - \tfrac{1}{2}\alpha_{\alpha\beta}V_\alpha V_\beta - \tfrac{1}{3}A_{\alpha,\beta\gamma}V_\alpha V_{\beta\gamma} - \tfrac{1}{6}C_{\alpha\beta,\gamma\delta}V_{\alpha\beta}V_{\gamma\delta} - \cdots \end{aligned} \tag{2.3.5}$$

Now

$$\begin{aligned} \frac{\partial W}{\partial V_\xi} &= \mu_\alpha\delta_{\alpha\xi} - \tfrac{1}{2}\alpha_{\alpha\beta}(V_\alpha\delta_{\beta\xi} + V_\beta\delta_{\alpha\xi}) - \tfrac{1}{3}A_{\alpha,\beta\gamma}\delta_{\alpha\xi}V_{\beta\gamma} - \cdots \\ &= \mu_\xi - \alpha_{\xi\beta}V_\beta - \tfrac{1}{3}A_{\xi,\beta\gamma}V_{\beta\gamma} - \cdots \\ &= \mu_\xi + \alpha_{\xi\beta}F_\beta + \tfrac{1}{3}A_{\xi,\beta\gamma}F_{\beta\gamma} + \cdots. \end{aligned} \tag{2.3.6}$$

So $(\partial W/\partial V_\xi)_{V\to 0}$ is the static dipole moment μ_ξ, and indeed this is commonly used as its definition, as well as the basis for some methods of calculating it. $\alpha_{\xi\beta}$ describes the additional dipole induced by an applied electric field F_β, and $A_{\xi,\beta\gamma}$ the dipole induced by an applied field gradient $F_{\beta\gamma}$.

In the same way,

$$3\frac{\partial W}{\partial V_{\xi\eta}} = \Theta_{\xi\eta} + A_{\alpha,\xi\eta}F_\alpha + C_{\alpha\beta,\xi\eta}F_{\alpha\beta} + \cdots, \tag{2.3.7}$$

so that A also describes the quadrupole induced by an electric field, and C describes the quadrupole induced by a field gradient.

In the spherical-tensor treatment, there is a generic polarizability $\alpha_{l\kappa,l'\kappa'}$:

$$\alpha_{l\kappa,l'\kappa'} = \sum_n{}' \frac{\langle 0|\hat{Q}_{l\kappa}|n\rangle\langle n|\hat{Q}_{l'\kappa'}|0\rangle + \langle 0|\hat{Q}_{l'\kappa'}|n\rangle\langle n|\hat{Q}_{l\kappa}|0\rangle}{W_n - W_0}, \tag{2.3.8}$$

and the energy takes the form

$$W = W^0 + \sum_{l\kappa} Q_{l\kappa}V_{l\kappa} - \frac{1}{2}\sum_{l\kappa l'\kappa'} \alpha_{l\kappa,l'\kappa'}V_{l\kappa}V_{l'\kappa'} + \cdots. \tag{2.3.9}$$

The polarizability $\alpha_{l\kappa,l'\kappa'}$ describes the moment $Q_{l\kappa}$ induced by the field $V_{l'\kappa'}$, and also the moment $Q_{l'\kappa'}$ induced by the field $V_{l\kappa}$. To see this, differentiate with respect to $V_{l\kappa}$:

$$\frac{\partial W}{\partial V_{l\kappa}} = Q_{l\kappa} - \sum_{l'\kappa'} \alpha_{l\kappa,l'\kappa'}V_{l'\kappa'} + \cdots. \tag{2.3.10}$$

The components of $\alpha_{1\kappa,1\kappa'}$ correspond precisely with those of the cartesian dipole–dipole polarizability $\alpha_{\alpha\beta}$, with the usual identification $10 \equiv z$, $11c \equiv x$ and $11s \equiv y$. $\alpha_{1\kappa,2\kappa'}$ is the dipole–quadrupole polarizability and corresponds to $A_{\alpha,\beta\gamma}$, and $\alpha_{2\kappa,2\kappa'}$ is the quadrupole–quadrupole polarizability and corresponds to $C_{\alpha\beta,\gamma\delta}$. Notice that one minor advantage of the spherical tensor formalism is that we do not need a new symbol for each rank.

To express the spherical-tensor components of the higher-rank polarizabilities in terms of the cartesian ones, we can start with eqn (2.3.8) and use Table E.1 to replace the spherical-tensor operators by their cartesian equivalents. Finally we refer to eqns (2.3.2)–(2.3.4) to identify the cartesian polarizabilities that result. For example, $\alpha_{20,22c}$ involves the operators $\hat{Q}_{20}$ and $\hat{Q}_{22c}$. In cartesian form,

$$\hat{Q}_{20} = \hat{\Theta}_{zz} \quad \text{and} \quad \hat{Q}_{22c} = \sqrt{\tfrac{1}{3}}(\hat{\Theta}_{xx} - \hat{\Theta}_{yy}).$$

Using eqn (2.3.4) we can then write down immediately that

$$\alpha_{20,22c} = \sqrt{3}(C_{zz,xx} - C_{zz,yy}).$$

Obtaining the cartesian components in terms of the spherical ones is a little more tricky. Where the dipole operators are involved, we can translate directly: $10 \leftrightarrow z$, $11c \leftrightarrow x$ and $11s \leftrightarrow y$. Most of the quadrupole operators similarly translate one to one (see Table E.1) but the tracelessness condition must be invoked to express polarizabilities involving the operators Θ_{xx} and Θ_{yy} in spherical form (see eqn (2.1.6)). In this way we find, for example,

$$\begin{aligned} A_{z,xx} &= -\tfrac{1}{2}\alpha_{10,20} + \tfrac{1}{2}\sqrt{3}\alpha_{10,22c}, \\ A_{z,yy} &= -\tfrac{1}{2}\alpha_{10,20} - \tfrac{1}{2}\sqrt{3}\alpha_{10,22c}. \end{aligned} \qquad (2.3.11)$$

A similar procedure can be used to express the cartesian components of the quadrupole–quadrupole polarizability in terms of the spherical ones. Full lists of cartesian components of the dipole–quadrupole and quadrupole–quadrupole polarizabilities are given in Tables E.3 and E.5 respectively.

Readers familar with spherical tensors may note that it is possible to use the complex components of the multipole moments in the definition of the polarizability. In this case the perturbation takes the form (2.2.3), and the energy to second order is

$$W = W^0 + \sum_{lm}(-1)^m Q_{lm}V_{l,-m} - \frac{1}{2}\sum_{lml'm'}(-1)^{m+m'}\alpha_{lm,l'm'}V_{l,-m}V_{l',-m'} + \cdots, \qquad (2.3.12)$$

with the polarizability given by

$$\alpha_{lm,l'm'} = {\sum_n}' \frac{\langle 0|\hat{Q}_{lm}|n\rangle\langle n|\hat{Q}_{l'm'}|0\rangle + \langle 0|\hat{Q}_{l'm'}|n\rangle\langle n|\hat{Q}_{lm}|0\rangle}{W_n - W_0}. \qquad (2.3.13)$$

In this formulation, $\alpha_{lm,l'm'}$ describes the moment $Q^*_{lm} = (-1)^m Q_{l,-m}$ induced by the field $V^*_{l',m'} = (-1)^{m'}V_{l',-m'}$, and the moment $Q^*_{l',m'} = (-1)^{m'}Q_{l',-m'}$ induced by the field $V^*_{lm} = (-1)^m V_{l,-m}$.

Since $\alpha_{lm,l'm'}$ is defined in terms of the operators Q_{lm} and $Q_{l'm'}$, it transforms under rotations like the product of the spherical harmonics C_{lm} and $C_{l'm'}$. For some purposes it is convenient to recouple the underlying angular momenta represented by the labels $lm, l'm'$. That is, we can construct from the $\alpha_{lm,l'm'}$ a new quantity that transforms according to angular momentum labels kq:

$$\alpha_{ll':kq} = \sum_{mm'} \langle ll'mm'|kq\rangle \alpha_{lm,l'm'}, \tag{2.3.14}$$

where $\langle ll'mm'|kq\rangle$ is a Clebsch–Gordan coefficient. (See p. 215.) This gives, for example, $\alpha_{11:00} = \sqrt{\frac{1}{3}}(\alpha_{xx} + \alpha_{yy} + \alpha_{zz})$ (a scalar, $k = q = 0$) and $\alpha_{11:20} = \sqrt{\frac{1}{6}}(2\alpha_{zz} - \alpha_{xx} - \alpha_{yy}) = \sqrt{\frac{2}{3}}\Delta\alpha$. For linear or axially symmetric molecules these are the only non-zero components (see Section 2.6). The components with $k = 1$ are always zero; for example, $\alpha_{11:10} = \sqrt{\frac{1}{2}} i(\alpha_{xy} - \alpha_{yx}) = 0$. It is usual to use the symbol $\bar{\alpha}$ for the *mean polarizability* $\frac{1}{3}(\alpha_{xx} + \alpha_{yy} + \alpha_{zz})$, and the symbol $\Delta\alpha$ for the *anisotropy* $\alpha_{zz} - \frac{1}{2}(\alpha_{xx} + \alpha_{yy})$, so $\alpha_{11:00} = \sqrt{3}\bar{\alpha}$ and $\alpha_{11:20} = \sqrt{2/3}\Delta\alpha$. This definition of the anisotropy is valid only in axial symmetry, where $\alpha_{xx} = \alpha_{yy}$ and the off-diagonal elements of α are zero. A more general definition, valid in any symmetry and invariant under change of axes (but ambiguous as to sign) is $(\Delta\alpha)^2 = \frac{3}{2}\sum_q |\alpha_{11:2q}|^2$.

2.3.1 *Units and magnitudes*

The atomic unit of dipole–dipole polarizability is obtained from the relationship $\Delta\mu = \alpha F$. Since the atomic units of dipole and electric field are ea_0 and $e/(4\pi\varepsilon_0 a_0^2)$, the unit of polarizability is $4\pi\varepsilon_0 a_0^3$, i.e., $4\pi\varepsilon_0$ times a volume. In the obsolete esu system of units, $4\pi\varepsilon_0$ is dimensionless and equal to 1, so the unit of polarizability in this system is just a volume, typically $\mathbf{r}A^3$. Many authors still quote numerical values in such units. In the SI, the unit is $\mathrm{Fm^2}$, and $4\pi\varepsilon_0 a_0^3 = 0.14818\,\mathbf{r}A^3 = 0.16488 \times 10^{-40}\,\mathrm{Fm^2}$.

Roughly speaking, the magnitude of the polarizability increases with the molecular volume. However, we can express $\bar{\alpha}$, using eqn (2.3.2), in the form

$$\bar{\alpha} = \frac{2}{3}\sum_n{}' \frac{|\langle 0|\mu|n\rangle|^2}{W_n - W_0},$$

in which the numerator is the intensity of the electric dipole transition between states 0 and n according to Fermi's golden rule, except for a numerical factor. Accordingly the polarizability is larger for molecules that have strong electric dipole transitions to low-lying excited states. Thus the He atom is small and its excited states are high in energy, so its polarizability is very small, at 1.39 a.u. The other inert gas atoms are progressively larger and their excitation energies smaller, so the polarizabilities increase through 2.67 for Ne, 11.08 for Ar, 16.76 for Kr to 27.2 for Xe.

For molecules, the polarizability is related to the freedom of movement of the electrons. In benzene, for example, the electrons can move more freely in the plane of the ring (the xy plane) so $\alpha_{xx} = \alpha_{yy} = 84.2$, but the polarizability perpendicular to the plane is smaller: $\alpha_{zz} = 46.2$. Another way to view this is to note that there are low-lying $\pi - \pi^*$ transitions that are polarized in the plane of the ring and contribute to α_{xx} and α_{yy}, while the $\sigma - \pi^*$ and $\pi - \sigma^*$ transitions that are polarized perpendicular to the ring and contribute to α_{zz} lie at higher energy.

2.4 Hyperpolarizabilities

The atomic unit of electric field, $e/(4\pi\varepsilon_0 a_0^2)$, is about $5 \times 10^{11}\,\mathrm{V\,m^{-1}}$. This is very much larger than can be achieved on a macroscopic scale in the laboratory, where the largest static field that can be achieved is in the region of $10^6\,\mathrm{V\,m^{-1}} = 2 \times 10^{-6}$ a.u. However the most powerful pulsed lasers in current use can produce intensities in the region of $10^{11}\,\mathrm{W\,cm^{-2}}$, corresponding to electric fields of the order of 10^{-3} a.u. (Eberly 1989); while the fields due to other molecules can easily reach magnitudes of 0.1 a.u. In both of these situations it becomes necessary to consider the possibility of non-linear polarizability effects, or hyperpolarizabilities.

If we continue the expansion of the electrostatic energy in powers of the electrostatic field and its derivatives, we obtain (in spherical tensor form)

$$\begin{aligned} W &= W^0 + \sum_{l\kappa} Q_{l\kappa} V_{l\kappa} - \frac{1}{2} \sum_{l\kappa l'\kappa'} \alpha_{l\kappa,l'\kappa'} V_{l\kappa} V_{l'\kappa'} \\ &\quad + \frac{1}{3!} \sum_{l\kappa l'\kappa' l''\kappa''} \beta_{l\kappa,l'\kappa',l''\kappa''} V_{l\kappa} V_{l'\kappa'} V_{l''\kappa''} \\ &\quad - \frac{1}{4!} \sum_{l\kappa l'\kappa' l''\kappa'' l'''\kappa'''} \gamma_{l\kappa,l'\kappa',l''\kappa'',l'''\kappa'''} V_{l\kappa} V_{l'\kappa'} V_{l''\kappa''} V_{l'''\kappa'''} + \cdots. \end{aligned} \tag{2.4.1}$$

The quantity β is a (first) *hyperpolarizability*, γ is a second hyperpolarizability, and so on. We can obtain expressions for them from third-order and fourth-order perturbation theory, respectively. For example

$$W''' = \sum_{np}{}' \frac{\langle 0|\mathcal{H}'|n\rangle\langle n|\mathcal{H}'|p\rangle\langle p|\mathcal{H}'|0\rangle}{(W_n - W_0)(W_p - W_0)} - \langle 0|\mathcal{H}'|0\rangle \sum_n{}' \frac{\langle 0|\mathcal{H}'|n\rangle\langle n|\mathcal{H}'|0\rangle}{(W_n - W_0)^2},$$

and by substituting the expression for $\mathcal{H}'$ in terms of the spherical-tensor moments, we obtain

$$\begin{aligned} W''' = V_{l\kappa} V_{l'\kappa'} V_{l''\kappa''} \Bigg\{ & \sum_{np}{}' \frac{\langle 0|\hat{Q}_{l\kappa}|n\rangle\langle n|\hat{Q}_{l'\kappa'}|p\rangle\langle p|\hat{Q}_{l''\kappa''}|0\rangle}{(W_n - W_0)(W_p - W_0)} \\ & - \langle 0|\hat{Q}_{l\kappa}|0\rangle \sum_n{}' \frac{\langle 0|\hat{Q}_{l'\kappa'}|n\rangle\langle n|\hat{Q}_{l''\kappa''}|0\rangle}{(W_n - W_0)^2} \Bigg\} \end{aligned}$$

Hence the first hyperpolarizability is

$$\begin{aligned} \beta_{l\kappa,l'\kappa',l''\kappa''} = 6\mathcal{S} \Bigg\{ & \sum_{np}{}' \frac{\langle 0|\hat{Q}_{l\kappa}|n\rangle\langle n|\hat{Q}_{l'\kappa'}|p\rangle\langle p|\hat{Q}_{l''\kappa''}|0\rangle}{(W_n - W_0)(W_p - W_0)} \\ & - \langle 0|\hat{Q}_{l\kappa}|0\rangle \sum_n{}' \frac{\langle 0|\hat{Q}_{l'\kappa'}|n\rangle\langle n|\hat{Q}_{l''\kappa''}|0\rangle}{(W_n - W_0)^2} \Bigg\}. \end{aligned} \tag{2.4.2}$$

Here the initial $\mathcal{S}$ on the right-hand side means 'symmetrize': that is, we average over the 3! permutations of the labels $l\kappa$, $l'\kappa'$ and $l''\kappa''$ in order to obtain an expression that is symmetric with respect to such permutations. The unsymmetrical part of the expression in braces in (2.4.2) that is removed by this procedure contributes nothing to the energy (2.4.1).

The most important hyperpolarizability is the dipole–dipole–dipole polarizability, for which $l = l' = l'' = 1$. In this case the operators are components of the dipole moment, and we can express β in the form

$$\beta_{\alpha\beta\gamma} = 6\mathcal{S}\left\{\sum_{np}{}' \frac{\langle 0|\hat{\mu}_\alpha|n\rangle\langle n|\hat{\mu}_\beta|p\rangle\langle p|\hat{\mu}_\gamma|0\rangle}{(W_n - W_0)(W_p - W_0)} - \langle 0|\hat{\mu}_\alpha|0\rangle \sum_n{}' \frac{\langle 0|\hat{\mu}_\beta|n\rangle\langle n|\hat{\mu}_\gamma|0\rangle}{(W_n - W_0)^2}\right\}. \quad (2.4.3)$$

When these higher polarizabilities are included, the dipole moment takes the form

$$\mu_\alpha = \mu^0_\alpha + \alpha_{\alpha\beta}F_\beta + \frac{1}{2}\beta_{\alpha\beta\gamma}F_\beta F_\gamma + \frac{1}{3!}\gamma_{\alpha\beta\gamma\delta}F_\beta F_\gamma F_\delta + \cdots$$

(ignoring terms in the derivatives $F_{\beta\gamma}$, etc, of the electric field). So $\beta_{\alpha\beta\gamma}$ and $\gamma_{\alpha\beta\gamma\delta}$ describe the non-linearity in the response to electric fields.

2.5 The response to oscillating electric fields

If the applied electric field is not static, but oscillates in time, we have to use time-dependent perturbation theory to determine the response of the molecule. The Hamiltonian $\mathcal{H} = \mathcal{H}^0 + \mathcal{H}'$ consists of a time-independent part $\mathcal{H}^0$ and a time-dependent perturbation $\mathcal{H}' = \hat{V} f(t)$ that is a product of a time-independent operator $\hat{V}$ and a time factor $f(t)$.

According to standard time-dependent perturbation theory (see Appendix C), the wavefunction is written in the form

$$\Psi = \sum_k a_k(t)\Psi_k(t) = \sum_k a_k(t)\psi_k \exp(-\mathrm{i}\omega_k t). \quad (2.5.1)$$

On the assumptions that the perturbation is small and that the system is initially in state $|n\rangle$, the coefficient $a_k(t)$ satisfies the equation

$$\frac{\partial}{\partial t}a_k(t) = -\frac{\mathrm{i}}{\hbar}V_{kn}f(t)\exp(\mathrm{i}\omega_{kn}t), \quad k \neq n, \quad (2.5.2)$$

and integration gives

$$a_k(t) = -\frac{\mathrm{i}}{\hbar}V_{kn}\int_0^t f(\tau)\exp(\mathrm{i}\omega_{kn}\tau)\,\mathrm{d}\tau. \quad (2.5.3)$$

For electric fields at optical frequencies, the wavelength is so long compared with the molecular size that the field gradient and the higher derivatives can be neglected. Also we suppose that the field is gradually turned on in the distant past, so we take the perturbation to be

$$\mathcal{H}'(t) = 2\hat{V}\exp\varepsilon t\cos(\omega t) = \hat{V}\left(\exp[(\varepsilon + \mathrm{i}\omega)t] + \exp[(\varepsilon - \mathrm{i}\omega)t]\right).$$

Here ε is small, and will eventually be allowed to tend to zero, so that we obtain the steady-state response when the oscillating field has been on for a long time.

Now we can integrate (2.5.3) directly to obtain

$$
\begin{aligned}
a_k(t) &= -\frac{\mathrm{i}}{\hbar}V_{kn}\int_{-\infty}^{t}\Big(\exp[(\varepsilon+\mathrm{i}\omega_{kn}+\mathrm{i}\omega)\tau]+\exp[(\varepsilon+\mathrm{i}\omega_{kn}-\mathrm{i}\omega)\tau]\Big)\,\mathrm{d}\tau \\
&= -\frac{V_{kn}}{\hbar}\left[\frac{\exp[(\varepsilon+\mathrm{i}\omega_{kn}+\mathrm{i}\omega)\tau]}{\omega_{kn}+\omega-\mathrm{i}\varepsilon}+\frac{\exp[(\varepsilon+\mathrm{i}\omega_{kn}-\mathrm{i}\omega)\tau]}{\omega_{kn}-\omega-\mathrm{i}\varepsilon}\right]_{-\infty}^{t} \\
&= -\frac{V_{kn}}{\hbar}\left(\frac{\exp[(\varepsilon+\mathrm{i}\omega_{kn}+\mathrm{i}\omega)t]}{\omega_{kn}+\omega-\mathrm{i}\varepsilon}+\frac{\exp[(\varepsilon+\mathrm{i}\omega_{kn}-\mathrm{i}\omega)t]}{\omega_{kn}-\omega-\mathrm{i}\varepsilon}\right) \\
&= -\frac{V_{kn}}{\hbar}\left(\frac{\exp[\mathrm{i}(\omega_{kn}+\omega)t]}{\omega_{kn}+\omega}+\frac{\exp[\mathrm{i}(\omega_{kn}-\omega)t]}{\omega_{kn}-\omega}\right),
\end{aligned}
\tag{2.5.4}
$$

where the last line results from allowing ε to tend to zero. We see that the coefficient $\mathbf{a}_k(t)$ remains small at all times provided that V_{kn} is small compared with $\hbar(\omega_{kn}\pm\omega)$.

Now we evaluate the component μ_α of the dipole moment for the molecule in its ground state, in the presence of the perturbation $\mathcal{H}'(t) = 2\hat{V}\cos(\omega t) = -\hat{\mu}_\beta F_\beta\cos(\omega t)$ due to an electromagnetic field polarized in the β direction. To first order, it is

$$
\begin{aligned}
\mu_\alpha(t) &= \Big\langle\Psi_0+\sum_k{}' a_k(t)\Psi_k\Big|\hat{\mu}_\alpha\Big|\Psi_0+\sum_k{}' a_k(t)\Psi_k\Big\rangle \\
&= \langle 0|\hat{\mu}_\alpha|0\rangle+\Big\{\sum_k{}' a_k(t)\langle 0|\hat{\mu}_\alpha|k\rangle\exp(\mathrm{i}\omega_{0k}t)+\text{c.c.}\Big\}.
\end{aligned}
\tag{2.5.5}
$$

(Here 'c.c.' means 'complex conjugate'.) We substitute the expression (2.5.4) for $a_k(t)$, to obtain

$$
\begin{aligned}
\mu_\alpha(t) &= \langle 0|\hat{\mu}_\alpha|0\rangle+\sum_k{}'\left\{\frac{V_{k0}}{\hbar}\langle 0|\hat{\mu}_\alpha|k\rangle\left(\frac{\exp(\mathrm{i}\omega t)}{\omega_{k0}+\omega}+\frac{\exp(-\mathrm{i}\omega t)}{\omega_{k0}-\omega}\right)+\text{c.c.}\right\} \\
&= \langle 0|\hat{\mu}_\alpha|0\rangle+F_\beta\cos(\omega t)\sum_k{}'\frac{\omega_{k0}(\langle 0|\hat{\mu}_\alpha|k\rangle\langle k|\hat{\mu}_\beta|0\rangle+\langle 0|\hat{\mu}_\beta|k\rangle\langle k|\hat{\mu}_\alpha|0\rangle)}{\hbar(\omega_{k0}^2-\omega^2)} \\
&\quad -\mathrm{i}F_\beta\sin(\omega t)\sum_k{}'\frac{\omega(\langle 0|\hat{\mu}_\alpha|k\rangle\langle k|\hat{\mu}_\beta|0\rangle-\langle 0|\hat{\mu}_\beta|k\rangle\langle k|\hat{\mu}_\alpha|0\rangle)}{\hbar(\omega_{k0}^2-\omega^2)}.
\end{aligned}
\tag{2.5.6}
$$

If the molecule is in a non-degenerate (and therefore real) state, then $\langle 0|\hat{\mu}_\alpha|k\rangle\langle k|\hat{\mu}_\beta|0\rangle$ is real, and the out-of-phase final term in eqn (2.5.6) vanishes. We see that the expectation value of the dipole moment operator is time-dependent:

$$\mu_\alpha(t) = \langle 0|\mu_\alpha|0\rangle+\alpha_{\alpha\beta}(\omega)F_\beta\cos(\omega t),$$

where the polarizability $\alpha_{\alpha\beta}(\omega)$ at frequency ω is

$$\alpha_{\alpha\beta}(\omega) = \sum_n{}'\frac{\omega_{n0}(\langle 0|\hat{\mu}_\alpha|n\rangle\langle n|\hat{\mu}_\beta|0\rangle+\langle 0|\hat{\mu}_\beta|n\rangle\langle n|\hat{\mu}_\alpha|0\rangle)}{\hbar(\omega_{n0}^2-\omega^2)}. \tag{2.5.7}$$

Here ω_{n0} is the angular frequency of the transition between state 0 and state n: $\hbar\omega_{n0} = W_n^0 - W_0^0$.

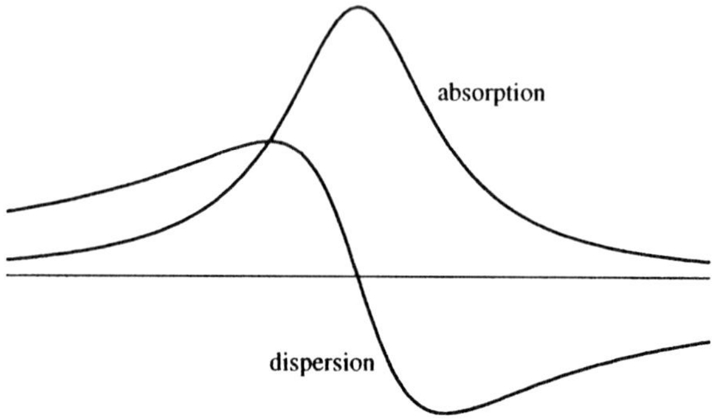

FIG. 2.4. Real and imaginary parts of the polarizability near an absorption frequency.

This describes the response of the molecule to the oscillating field, but it leaves out the important effect of *spontaneous emission*. To describe this properly we should use a quantum description of the electromagnetic field. Alternatively, the effect can be included phenomenologically by noting that a system in state k may emit a photon spontaneously and undergo a transition to another state. This is described by adding another term to eqn (2.5.2):

$$\frac{\partial}{\partial t}a_k(t) = -\frac{\mathrm{i}}{\hbar}V_{kn}f(t)\mathrm{e}^{\mathrm{i}\omega_{kn}t} - \tfrac{1}{2}\Gamma_k a_k(t), \tag{2.5.8}$$

Loudon (1973) describes both approaches. In either case, the effect is that the polarizability is slightly modified, and takes the form

$$\alpha_{\alpha\beta}(\omega) = 2{\sum_k}' \frac{\omega_{k0}\langle 0|\hat{\mu}_\alpha|k\rangle\langle k|\hat{\mu}_\beta|0\rangle}{\hbar[\omega_{k0}^2 - (\omega + \frac{1}{2}\mathrm{i}\Gamma_k)^2]}, \tag{2.5.9}$$

where Γ_k is a constant describing the probability of transitions from state k. The singularities in (2.5.7) at the transition frequencies have been removed, because the denominators are now non-zero for all real ω, but they are complex. The frequency dependence $\omega_{k0}/[\omega_{k0}^2 - (\omega + \frac{1}{2}\mathrm{i}\Gamma_k)^2]$ has a real part that goes through zero near ω_{k0} and an imaginary part that peaks near ω_{k0} (see Fig. 2.4). The real part describes anomalous dispersion (variation of refractive index with frequency), and the imaginary part describes absorption. If the matrix elements $\langle 0|\hat{\mu}_\alpha|k\rangle$ are complex, further complications arise (Buckingham 1978), but we do not need to consider them here.

2.6 Symmetry properties of the multipole moments and polarizabilities

If we assume that the molecule is in a non-degenerate state, so that its charge distribution is fully symmetric (symmetry species Γ_1 in the molecular point group), the symmetry species of the multipole moment integral $\langle 0|\hat{Q}_{l\kappa}|0\rangle = \int \rho(\mathbf{r})R_{l\kappa}(\mathbf{r})\,\mathrm{d}^3r$ is just the symmetry of the multipole operator itself, and it must vanish unless totally symmetric. We can use standard group-theoretical methods to discover the number of non-zero components of a multipole moment of a given rank, or to determine whether a particular moment may be non-zero. For an account of these methods, see for example Tinkham (1964) or Elliott and Dawber (1979).

For individual moments it is usually easier to use the cartesian forms. From the C_{2v} character table in its usual form we can see immediately that the x, y and z components of the dipole moment transform according to B_2, B_1 and A_1 respectively. Consequently a C_{2v} molecule such as water may have a non-zero μ_z, but μ_x and μ_y must be zero. Similarly the xy, xz and yz components of the quadrupole moment must vanish, since they transform as $B_2 \otimes B_1 = A_2, B_2 \otimes A_1 = B_2$ and $B_1 \otimes A_1 = B_1$ respectively, but the xx, yy and zz components may be non-zero. Remember, though, that the xx, yy and zz components are not independent; their sum is zero, so we have only two independent components in this case.

To examine the properties of a whole set of multipole moments of a given rank, we can use the fact that the character of a set of spherical harmonics under rotation through an angle α is independent of the axis of rotation and is given by the formula

$$\chi^{(k)}(\alpha) = \frac{\sin(k+\frac{1}{2})\alpha}{\sin\frac{1}{2}\alpha},$$

where k is the rank of the harmonics. This formula applies to proper rotations; we write an improper rotation as the product of the inversion with a suitable proper rotation, noting that inversion changes the sign of a spherical harmonic if its rank is odd: $\hat{\imath}R_{km} = (-1)^k R_{km}$. In this way we arrive at the following characters for H_2O:

C_{2v}	E	C_2^z	$\sigma^{xz} = \hat{\imath}C_2^y$	$\sigma^{yz} = \hat{\imath}C_2^x$	
Q_{1m}	3	−1	1	1	$A_1 + B_1 + B_2$
Q_{2m}	5	1	1	1	$2A_1 + A_2 + B_1 + B_2$

So we see as before that the dipole moment has one non-zero component and the quadrupole has two.

As another example, we consider a tetrahedral molecule, where the answers are less obvious. The results are

T_d	E	$8C_3$	$3C_2$	$6\sigma_d$	$6S_4 = \hat{\imath}C_4^{-1}$	
Q_{1m}	3	0	−1	1	−1	T_2
Q_{2m}	5	−1	1	1	−1	$E + T_2$
Q_{3m}	7	1	−1	1	1	$A_1 + T_1 + T_2$
Q_{4m}	9	0	1	1	1	$A_1 + E + T_1 + T_2$

So a tetrahedral molecule like methane has zero dipole and quadrupole moments, but there is one non-zero component of the octopole and one of the hexadecapole (rank 4). To find the non-zero components, we use the usual projection formula. It is quite easy to show in this way (using the cartesian form) that the non-zero component of the octopole is Ω_{xyz} or Ω_{32s}. A similar calculation for the hexadecapole gives either $\Phi_{xxxx} + \Phi_{yyyy} + \Phi_{zzzz}$ or $\Phi_{xxyy} + \Phi_{xxzz} + \Phi_{yyzz}$, depending which component one starts from. However this does not mean that there are two *independent* non-zero components; because of the tracelessness property these are equal to within a numerical factor: $\Phi_{xxxx}+\Phi_{yyyy}+\Phi_{zzzz}=-2(\Phi_{xxyy}+\Phi_{xxzz}+\Phi_{yyzz})$. In spherical-tensor form the equivalent expression is $Q_{40}+\sqrt{\frac{5}{14}}(Q_{44}+Q_{4,-4})=Q_{40}+\sqrt{\frac{5}{7}}Q_{44c}$. In work on tetrahedral molecules it is usual to use the abbreviation Ω for Ω_{xyz} and Φ for Φ_{zzzz}.

General symmetry properties include the results that for all linear molecules $Q_{km}=0$ unless $m=0$, and $Q_{k0}=0$ for odd k when there is a centre of inversion symmetry. If a molecule has axial symmetry of order p, then $Q_{km}=0$ unless $m=0 \bmod p$. If in addition there are symmetry planes containing the rotation axis, then only one of Q_{kmc} and Q_{kms} can be non-zero. Which one this is will depend on the choice of axes; if the x axis lies in a symmetry plane it will be Q_{kmc}. If there is a symmetry plane perpendicular to the principal axis, then Q_{kmc} and Q_{kms} both vanish when $k-m$ is odd. The first non-vanishing moment in tetrahedral symmetry is the octopole Ω_{xyz}, as we have seen; in octahedral symmetry it is the hexadecapole $Q_{40}+\sqrt{\frac{5}{7}}Q_{44c}$, and in icosahedral symmetry it is a single combination of the Q_{6m}.

The symmetries of the polarizabilities can be determined from those of the multipole moments. The equation defining $\alpha_{l\kappa,l'\kappa'}$ shows that if the set $\{Q_{l\kappa},\ \kappa=0,\ 1c,\ 1s,\ \ldots,\ ls\}$, for a given l, transforms according to $\Gamma^{(l)}$, then the set $\alpha_{l\kappa,l'\kappa'}$ for given l and l' transforms according to the direct product $\Gamma^{(l)}\otimes\Gamma^{(l')}$. Thus for H_2O, the dipole–quadrupole polarizability transforms as

$$\begin{aligned}\Gamma^{(1)}\otimes\Gamma^{(2)} &= (A_1+B_1+B_2)\otimes(2A_1+A_2+B_1+B_2)\\ &= 4A_1+3A_2+4B_1+4B_2,\end{aligned}$$

so that four of the components have A_1 symmetry and can be non-zero.

This is valid when $l\neq l'$. When $l=l'$, the calculation is slightly less straightforward. In this case the polarizability is symmetric with respect to the interchange of κ and κ': $\alpha_{l\kappa,l\kappa'}=\alpha_{l\kappa',l\kappa}$. Thus the antisymmetric combinations $\alpha_{l\kappa,l\kappa'}-\alpha_{l\kappa',l\kappa}$ are all zero, and we have to consider the symmetric combinations $\alpha_{l\kappa,l\kappa'}+\alpha_{l\kappa',l\kappa}$, which transform according to the symmetrized square, $(\Gamma^{(l)}\otimes\Gamma^{(l')})_+$. The standard formula for the character of the symmetrized square is

$$\chi^{(\Gamma\otimes\Gamma)_+}(R)=\tfrac{1}{2}\left[(\chi^{\Gamma}(R))^2+\chi^{\Gamma}(R^2)\right].$$

For the dipole–dipole polarizability of water this gives

C_{2v}	E	C_2^z	$\sigma^{xz}=iC_2^y$	$\sigma^{yz}=iC_2^x$	
$\Gamma^{(1)}$	3	-1	1	1	
$(\Gamma^{(1)}\otimes\Gamma^{(1)})_+$	6	2	2	2	$3A_1+A_2+B_1+B_2$

so there are three non-zero components. In the case of methane, the dipole operator transforms according to the T_2 representation, and we find

$\mathcal{T}_d$	E	$8C_3$	$3C_2$	$6\sigma_d$	$6S_4 = iC_4^{-1}$	
T_2	3	0	-1	1	-1	
$(T_2 \otimes T_2)_+$	6	0	2	2	0	$A_1 + E + T_2$

So the symmetrized square contains the symmetric representation A_1 once, and there is one independent non-zero component of the dipole polarizability.

An alternative approach, which avoids the need for the symmetrized square, is based on the fact that the direct product of two real irreducible representations Γ_1 and Γ_2 contains the symmetric representation exactly once if Γ_1 and Γ_2 are equivalent ($\Gamma_1 \sim \Gamma_2$) and not at all otherwise. Given a set of multipoles Q_i that transform according to Γ_1, and a set Q'_j that transform according to Γ_2, we can construct a set of products $Q_iQ'_j$, and these transform according to $\Gamma_1 \otimes \Gamma_2$. If $\Gamma_1 \sim \Gamma_2$ we can construct just one symmetric combination of these products; and correspondingly there is just one symmetric (and therefore non-zero) combination of the α_{ij}. If Γ_1 and Γ_2 are identical rather than just equivalent, the symmetric combination of the products $Q_iQ'_j$ is $\sum_i Q_iQ'_i$. Taking the dipole–quadrupole polarizability of the water molecule once again, we see that the dipole has components of symmetries A_1, B_1 and B_2 (μ_z, μ_x and μ_y respectively) while the quadrupole has two components of symmetry A_1 (Θ_{zz} and $\Theta_{xx} - \Theta_{yy}$) and one each of A_2, B_1 and B_2 (Θ_{xy}, Θ_{xz} and Θ_{yz} respectively). Thus we can take A_1 with A_1 to get the symmetric combinations $\mu_z\Theta_{zz}$ and $\mu_z(\Theta_{xx} - \Theta_{yy})$, or B_1 with B_1 to get $\mu_x\Theta_{xz}$, or B_2 with B_2 to get $\mu_y\Theta_{yz}$. Corresponding to each of these symmetric combinations there is a non-zero component of the dipole–quadrupole polarizability, i.e., $A_{z,zz}$, $A_{z,xx} - A_{z,yy}$, $A_{x,xz}$ and $A_{y,yz}$. The same approach for the dipole–dipole case gives the symmetric combinations $\mu_z\mu_z$, $\mu_x\mu_x$ and $\mu_y\mu_y$, corresponding to non-zero polarizabilities α_{zz}, α_{xx} and α_{yy} respectively.

For a higher-symmetry molecule such as methane, there are fewer non-zero components. In this case, the set of μ_i transform as T_2. For the dipole–dipole polarizability, therefore, we can match T_2 with T_2; the set of nine dipole–dipole products $\mu_i\mu_j$ contains just one symmetric combination, which is $\mu_x^2 + \mu_y^2 + \mu_z^2$. From this we deduce that there is just one non-zero component of the dipole–dipole polarizability, namely $\alpha_{xx} + \alpha_{yy} + \alpha_{zz}$. The fact that $\mu_x^2 - \mu_y^2$ and $2\mu_z^2 - \mu_x^2 - \mu_y^2$ are not symmetric (they are the components of the E representation that arises from the direct product $T_2 \otimes T_2$) tells us that $\alpha_{xx} - \alpha_{yy} = 0$ and $2\alpha_{zz} - \alpha_{xx} - \alpha_{yy} = 0$, and hence that $\alpha_{xx} = \alpha_{yy} = \alpha_{zz}$.

As a final example of this approach, we note that the quadrupole moment transforms according to $E + T_2$ in tetrahedral symmetry. For the dipole–quadrupole polarizability, we again match T_2 with T_2; the quadrupole components transforming like μ_x, μ_y and μ_z are Θ_{yz}, Θ_{xz} and Θ_{xy} respectively, and the symmetric combination is $\mu_x\Theta_{yz} + \mu_y\Theta_{xz} + \mu_z\Theta_{xy}$. Consequently the single independent non-zero component of the dipole–quadrupole polarizability is $A_{x,yz} + A_{y,xz} + A_{z,xy}$.

Yet another approach uses the recoupled polarizabilities $\alpha_{ll':kq}$ defined on p. 24. They transform under rotations like spherical harmonics of rank k, so we can use the same methods as for the multipole moments. Thus the static dipole polarizability has six components, $\alpha_{11:00}$ and the five $\alpha_{11:2q}$; the components of $\alpha_{11:1q}$ are combinations of the antisymmetric part $\alpha_{\alpha\beta} - \alpha_{\beta\alpha}$ of the polarizability, and are zero. The $k = 0$ component is

$$\alpha_{11:00} = \sqrt{3}\bar{\alpha} = \sqrt{\tfrac{1}{3}}(\alpha_{xx} + \alpha_{yy} + \alpha_{zz});$$

it is always symmetric and so is always non-zero. For molecules with axial symmetry, the only other non-zero component is

$$\alpha_{11:20} = \sqrt{\tfrac{1}{6}}\left(2\alpha_{zz} - \tfrac{1}{2}(\alpha_{xx} + \alpha_{yy})\right) = \sqrt{\tfrac{2}{3}}\Delta\alpha.$$

It is customary in this case to use the notation $\alpha_\parallel$ for α_{zz} and $\alpha_\perp$ for α_{xx} and α_{yy}, which are equal.

Care must be taken to account correctly for the effects of inversion. For the dipole–quadrupole polarizabilities, for example, the recoupled polarizabilities are $\alpha_{12:kq}$ with $k = 1, 2$ or 3. These behave under rotations like spherical harmonics of rank k, *but they all change sign under inversion, even when* $k = 2$, because the operators $\hat{Q}_{1m}$ in the polarizability expression change sign but the $\hat{Q}_{2m}$ do not. In spherical symmetry, we can classify these polarizability components as P_u, D_u and F_u. It follows that all of them vanish for any centrosymmetric molecule, such as CO_2 or benzene. For a $C_{\infty v}$ molecule, we find by descent in symmetry that P_u reduces to $\Sigma^+ \oplus \Pi$ (one non-zero component), D_u to $\Sigma^- \oplus \Pi \oplus \Delta$ (all zero) and F_u to $\Sigma^+ \oplus \Pi \oplus \Delta \oplus \Phi$ (one non-zero component) so that there are two independent non-zero components in this symmetry.

Where there are only a few non-zero components it is conventional to use a special notation for them (Buckingham 1967). For axially symmetric molecules (linear, or with a unique rotation axis of order 3 or more) we have already met the notations $\alpha_\parallel = \alpha_{zz}$ and $\alpha_\perp = \alpha_{xx} = \alpha_{yy}$ for dipole polarizabilities. In addition, the notation $A_\parallel$ is used for $A_{z,zz}$ and $A_\perp$ for $A_{x,xz} = A_{y,yz}$. Similarly, $\beta_\parallel = \beta_{zzz}$ and $\beta_\perp = \beta_{xxz} = \beta_{yyz}$. In all of these cases, the z direction is parallel to the unique axis. In tetrahedral molecules, where there is only one non-zero dipole–quadrupole polarizability and one non-zero hyperpolarizability, they are usually denoted simply $A = A_{x,yz}$ and $\beta = \beta_{xyz}$ respectively. In the case of the quadrupole–quadrupole polarizability, the symbol C denotes $\frac{1}{5}C_{\alpha\beta,\alpha\beta} = \frac{3}{10}\sum_\kappa \alpha_{2\kappa,2\kappa}$.

2.7 Change of origin

Consider the OH^- ion. It has a negative charge; can we say anything about where the charge resides? It has higher moments too; how do their values depend on our choice of origin? Is there an optimum choice of origin? In many applications it is convenient to take the origin at the centre of mass, but this is not necessarily the optimum choice for describing the electrostatic properties. Indeed, the centre of mass is different for OH^- and OD^-, whereas the charge distribution is the same for both, except for some very small differences due to breakdown of the Born–Oppenheimer approximation.

We take the origin initially at the oxygen nucleus and the z axis from O to H. The total charge $q = -e$. The dipole moment is $\mu_z^O = \langle 0|\sum e_a a_z|0\rangle$ relative to this origin. Now consider a new origin C at $(0, 0, c)$. The dipole moment with respect to this origin is

$$\mu_z^C = \langle 0|\sum_a (a_z - c)|0\rangle$$

$$= \mu_z^O - cq. \tag{2.7.1}$$

So if $q \neq 0$, as in this case, the dipole moment is not invariant under change of origin.

It is possible if $q \neq 0$ to find an origin for which $\mu_z^C = 0$. This happens when $\mu_z^O = cq$ or $c = \mu_z^O/q$. This origin is called the *centre of charge*. If we express the electrostatic potential as a multipole series (as described in the next chapter), the leading term is the usual point-charge term $q/4\pi\varepsilon_0 R$, the next depends on the dipole and is proportional to μ/R^2, the next to Θ/R^3, and so on. Thus the use of the centre of charge as the origin eliminates the dipole contribution and will often be a good choice for an ion. However it is undefined for a neutral molecule; in this case the dipole moment is independent of origin, as we can see from eqn (2.7.1).

The quadrupole moment too may be affected by change of origin:

$$\begin{aligned}\hat{\Theta}_{zz}^C &= \sum_a e_a\left[\tfrac{3}{2}(a_z - c)^2 - \tfrac{1}{2}\left(a_x^2 + a_y^2 + (a_z - c)^2\right)\right] \\ &= \sum_a e_a\left[\left(\tfrac{3}{2}a_z^2 - \tfrac{1}{2}a^2\right) - 2a_z c + c^2\right] \\ &= \hat{\Theta}_{zz}^O - 2c\hat{\mu}_z^O + qc^2. \end{aligned} \tag{2.7.2}$$

This is a special case ($c_x = c_y = 0$). A more general expression can be constructed for an arbitrary displacement and for the other components of the quadrupole moment, but the algebra rapidly becomes heavy for higher ranks. This is a case where the spherical-tensor form gives a much more compact expression. There is a standard addition theorem for regular spherical harmonics (Brink and Satchler 1993):

$$\begin{aligned} R_{LM}(\mathbf{a}+\mathbf{b}) &= \sum_{l_1 l_2}\sum_{m_1 m_2} \delta_{l_1+l_2,L}(-1)^{L+M}\left[\frac{(2L+1)!}{(2l_1)!(2l_2)!}\right]^{1/2} \\ &\quad \times R_{l_1 m_1}(\mathbf{a}) R_{l_2 m_2}(\mathbf{b}) \begin{pmatrix} l_1 & l_2 & L \\ m_1 & m_2 & -M \end{pmatrix}, \end{aligned} \tag{2.7.3}$$

where $\begin{pmatrix} l_1 & l_2 & L \\ m_1 & m_2 & M \end{pmatrix}$ is a Wigner $3j$ coefficient (see Appendix B). When $L = l_1 + l_2$, as here, this $3j$ coefficient can be written in the explicit form

$$\begin{aligned} \begin{pmatrix} l_1 & l_2 & l_1+l_2 \\ m_1 & m_2 & -M \end{pmatrix} &= (-1)^{l_1-l_2+M}\delta_{m_1+m_2,M}\left[\frac{(2l_1)!(2l_2)!}{(2l_1+2l_2+1)!}\right]^{1/2} \\ &\quad \times \left[\binom{l_1+l_2+m_1+m_2}{l_1+m_1}\binom{l_1+l_2-m_1-m_2}{l_1-m_1}\right]^{1/2}, \end{aligned} \tag{2.7.4}$$

where $\binom{n}{m}$ is the binomial coefficient $n!/(m!(n-m)!)$, so that we can write

$$R_{lk}(\mathbf{r}-\mathbf{c}) = \sum_{l'=0}^{l}\sum_{k'=-l'}^{l'}\left[\binom{l+k}{l'+k'}\binom{l-k}{l'-k'}\right]^{1/2} R_{l'k'}(\mathbf{r})R_{l-l',k-k'}(-\mathbf{c}). \tag{2.7.5}$$

Using this result we immediately find

$$Q_{lk}^{C} = \sum_{l'=0}^{l} \sum_{k'=-l'}^{l'} \left[\binom{l+k}{l'+k'} \binom{l-k}{l'-k'} \right]^{1/2} Q_{l'k'}^{O} R_{l-l',k-k'}(-\mathbf{c}). \tag{2.7.6}$$

From this formula we see that Q_{lk} is invariant under arbitrary changes of origin if and only if $Q_{l'k'} = 0$ for all $l' < l$.

A similar formula can easily be derived for change of origin of the real components, by using eqns (B.1.8)–(B.1.10):

$$Q_{l\kappa}^{C} = \sum_{l'\kappa'} Q_{l'\kappa'}^{O} W_{l'\kappa',l\kappa}(-\mathbf{c}), \tag{2.7.7}$$

where

$$\begin{aligned} W_{l'k'c,lkc}(\mathbf{x}) &= W_{l'k's,lks}(\mathbf{x}) \\ &= \left[\binom{l+k}{l'+k'} \binom{l-k}{l'-k'} \right] \frac{b_k}{2b_{k'}b_{k-k'}} R_{l-l',k-k',c}(\mathbf{x}), \\ W_{l'k'c,lks}(\mathbf{x}) &= -W_{l'k's,lkc}(\mathbf{x}) \\ &= \left[\binom{l+k}{l'+k'} \binom{l-k}{l'-k'} \right] \frac{b_k}{2b_{k'}b_{k-k'}} R_{l-l',k-k',s}(\mathbf{x}). \end{aligned} \tag{2.7.8}$$

Even if a charge distribution can be described exactly in terms of a finite number of multipole moments at one origin, there will be multipole moments of ranks up to infinity if a different origin is chosen. For example, the Na^+ ion has a charge of +1, and all higher moments vanish. If we move the origin from $(0,0,0)$ to $(0,0,c)$ we find, using either the formula for change of origin or (more simply in this case) the original definitions of the multipole moments,

$$\begin{aligned} q^C &= +1, \\ \mu_z^C &= -qc, \\ \Theta_{zz}^C &= +qc^2, \\ \Omega_{zzz}^C &= -qc^3, \end{aligned}$$

and so on. If the new origin is too far from the old one, these multipole moments rapidly become very large. If we wish to calculate the electrostatic potential using multipole moments defined with respect to the new origin, we must include these higher moments. If we were to choose a new origin (such as $(0,0,c)$ in the example of the Na^+ ion) and calculate the potential of a point charge as if it were at this origin, we would evidently get the wrong answer, even though the magnitude of the charge does not change when we move to this origin. To get the right answer, we have to take account of the higher moments at the new origin too.

This may seem obvious enough in the case of a point charge. It is not always appreciated, however, that the same applies in the case of higher moments such as dipole moments. The energy of a dipole in an electric field $\mathbf{F}$ is $-\boldsymbol{\mu}\cdot\mathbf{F}$; if $\mathbf{F}$ is non-uniform the energy depends on the position ascribed to the dipole moment. If we change the origin, the dipole moment may not change (it will not if the molecule is neutral) but the energy calculated for the new origin will be different. We have to take the higher moments into account if we are to get the same answer in both cases.

Change of origin affects polarizabilities as it does multipole moments. There is one simplification: the matrix elements of the total charge that might occur in the sum over states are all zero, because q is just a constant and the states are orthogonal. This means that the dipole–dipole polarizability is independent of origin, even for ions. However, if we refer the polarizability to a new origin, then the higher-rank polarizabilities will change. We have seen that the potential due to a charge will be wrong if we calculate it as if it were at a different origin, without taking account of the higher moments that arise at the new origin; it is equally true, though not quite so obvious, that the response to an external field will be calculated incorrectly if we assign the polarizability to a new origin without taking account of the changes to the higher polarizabilities.

The higher polarizabilities are always origin-dependent, because the dipole–dipole polarizability is always non-zero. Consider $A_{z,zz}$ for a linear molecule. It involves the operator $\hat{\Theta}_{zz}$, which is origin-dependent:

$$\hat{\Theta}^C_{zz} = \hat{\Theta}^O_{zz} - 2c\hat{\mu}^O_z.$$

(Again, we are dealing here with the special case of moving the origin from $(0,0,0)$ to $(0,0,c)$.) So

$$\begin{aligned} A^C_{z,zz} &= 2\sum_n{}' \frac{\langle 0|\hat{\mu}^O_z|n\rangle\langle n|\hat{\Theta}^O_{zz} - 2c\hat{\mu}^O_z|0\rangle}{W_n - W_0}, \\ &= A^O_{z,zz} - 2c\alpha_{zz}. \end{aligned}$$

This means that the origin should always be specified when values of $A_{\alpha,\beta\gamma}$ are given. The general formula for change of origin in polarizabilities is obtained by applying (2.7.6) or (2.7.7) to the operators in the formula for the polarizability.

3

ELECTROSTATIC INTERACTIONS BETWEEN MOLECULES

3.1 The electric field of a molecule

Suppose that molecule A is located at position $\mathbf{A}$ in some global coordinate system. The particles of this molecule are at positions $\mathbf{a}$ relative to $\mathbf{A}$, i.e., at positions $\mathbf{A}+\mathbf{a}$. We want to evaluate the potential at a point $\mathbf{B}$ where we shall in due course put another molecule. In terms of the positions and charges of the particles of molecule A, the potential is

$$V^A(\mathbf{B}) = \sum_a \frac{e_a}{4\pi\varepsilon_0|\mathbf{B}-\mathbf{A}-\mathbf{a}|} = \sum_a \frac{e_a}{4\pi\varepsilon_0|\mathbf{R}-\mathbf{a}|}, \tag{3.1.1}$$

where $\mathbf{R} = \mathbf{B} - \mathbf{A}$. (See Fig. 3.1.) We expand this potential as a Taylor series. We have

$$\begin{aligned}
V^A(\mathbf{B}) &= \sum_a \frac{e_a}{4\pi\varepsilon_0|\mathbf{R}-\mathbf{a}|} \\
&= \sum_a \frac{e_a}{4\pi\varepsilon_0}\left\{\frac{1}{R} + a_\alpha\left(\frac{\partial}{\partial a_\alpha}\frac{1}{|\mathbf{R}-\mathbf{a}|}\right)_{\mathbf{a}=0} + \frac{1}{2}a_\alpha a_\beta\left(\frac{\partial^2}{\partial a_\alpha \partial a_\beta}\frac{1}{|\mathbf{R}-\mathbf{a}|}\right)_{\mathbf{a}=0} + \cdots\right\} \\
&= \sum_a \frac{e_a}{4\pi\varepsilon_0}\left\{\frac{1}{R} - a_\alpha\left(\frac{\partial}{\partial R_\alpha}\frac{1}{|\mathbf{R}-\mathbf{a}|}\right)_{\mathbf{a}=0} + \frac{1}{2}a_\alpha a_\beta\left(\frac{\partial^2}{\partial R_\alpha \partial R_\beta}\frac{1}{|\mathbf{R}-\mathbf{a}|}\right)_{\mathbf{a}=0} - \cdots\right\} \\
&= \sum_a \frac{e_a}{4\pi\varepsilon_0}\left\{\frac{1}{R} - a_\alpha\nabla_\alpha\frac{1}{R} + \frac{1}{2}a_\alpha a_\beta\nabla_\alpha\nabla_\beta\frac{1}{R} - \cdots\right\}.
\end{aligned}$$

As before, we can replace the second moment $M_{\alpha\beta} = \sum_a e_a a_\alpha a_\beta$ by the quadrupole moment $\Theta_{\alpha\beta}$, because $1/R$ satisfies Laplace's equation and there is no contribution to the potential from the trace $M_{\alpha\alpha}$. The higher moments are treated similarly. When we do this, we get

$$V^A(\mathbf{B}) = \frac{1}{4\pi\varepsilon_0}\left\{q\left(\frac{1}{R}\right) - \hat{\mu}_\alpha\nabla_\alpha\left(\frac{1}{R}\right) + \frac{1}{3}\hat{\Theta}_{\alpha\beta}\nabla_\alpha\nabla_\beta\left(\frac{1}{R}\right) - \cdots\right\}$$

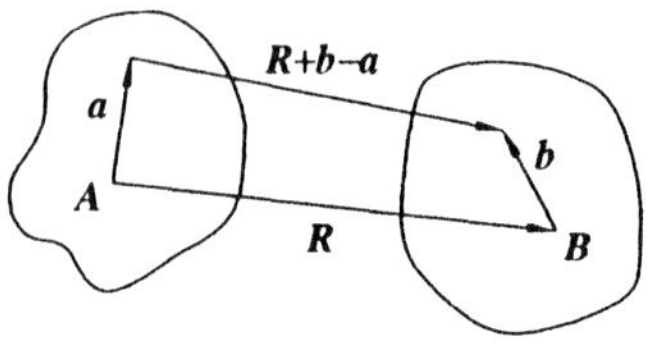

FIG. 3.1. Definition of position vectors in two interacting molecules

$$\equiv Tq - T_\alpha \hat{\mu}_\alpha + \frac{1}{3} T_{\alpha\beta} \hat{\Theta}_{\alpha\beta} - \cdots + \frac{(-1)^n}{(2n-1)!!} T^{(n)}_{\alpha\beta\ldots\nu} \hat{\xi}^{(n)}_{\alpha\beta\ldots\nu} + \cdots, \tag{3.1.2}$$

where

$$T = \frac{1}{4\pi\varepsilon_0 R}, \tag{3.1.3}$$

$$T_\alpha = \frac{1}{4\pi\varepsilon_0} \nabla_\alpha \frac{1}{R} = -\frac{R_\alpha}{4\pi\varepsilon_0 R^3}, \tag{3.1.4}$$

$$T_{\alpha\beta} = \frac{1}{4\pi\varepsilon_0} \nabla_\alpha \nabla_\beta \frac{1}{R} = \frac{3R_\alpha R_\beta - R^2 \delta_{\alpha\beta}}{4\pi\varepsilon_0 R^5}, \tag{3.1.5}$$

$$\begin{aligned} T_{\alpha\beta\gamma} &= \frac{1}{4\pi\varepsilon_0} \nabla_\alpha \nabla_\beta \nabla_\gamma \frac{1}{R} \\ &= -\frac{15 R_\alpha R_\beta R_\gamma - 3R^2 (R_\alpha \delta_{\beta\gamma} + R_\beta \delta_{\alpha\gamma} + R_\gamma \delta_{\alpha\beta})}{4\pi\varepsilon_0 R^7}, \end{aligned} \tag{3.1.6}$$

$$\begin{aligned} T_{\alpha\beta\gamma\delta} &= \frac{1}{4\pi\varepsilon_0} \nabla_\alpha \nabla_\beta \nabla_\gamma \nabla_\delta \frac{1}{R} \\ &= \frac{1}{4\pi\varepsilon_0 R^9} \big[105 R_\alpha R_\beta R_\gamma R_\delta \\ &\qquad - 15R^2 (R_\alpha R_\beta \delta_{\gamma\delta} + R_\alpha R_\gamma \delta_{\beta\delta} + R_\alpha R_\delta \delta_{\beta\gamma} \\ &\qquad\qquad + R_\beta R_\gamma \delta_{\alpha\delta} + R_\beta R_\delta \delta_{\alpha\gamma} + R_\gamma R_\delta \delta_{\alpha\beta}) \\ &\qquad + 3R^4 (\delta_{\alpha\beta}\delta_{\gamma\delta} + \delta_{\alpha\gamma}\delta_{\beta\delta} + \delta_{\alpha\delta}\delta_{\beta\gamma}) \big] \end{aligned} \tag{3.1.7}$$

and in general

$$T^{(n)}_{\alpha\beta\ldots\nu} = \frac{1}{4\pi\varepsilon_0} \nabla_\alpha \nabla_\beta \ldots \nabla_\nu \frac{1}{R}. \tag{3.1.8}$$

The superscript (n) specifies the number of subcripts, but is normally omitted when the number is obvious. If we wish to avoid ambiguity when dealing with a system of more than two molecules, we can label the T tensors with the molecular labels: T^{AB}, T^{AB}_α, etc. This tends to make the notation rather cumbersome, however, and we omit the labels in the two-molecule case. Notice though that it is important to establish whether we are dealing with T^{AB} or T^{BA}, i.e., whether $\mathbf{R} = \mathbf{B} - \mathbf{A}$, as above, or $\mathbf{R} = \mathbf{A} - \mathbf{B}$. The definitions above show that $T^{BA(n)}_{\alpha\beta\ldots\nu} = (-1)^n T^{AB(n)}_{\alpha\beta\ldots\nu}$.

Returning to eqn (3.1.2), we see that the potential due to a charge is $q/4\pi\varepsilon_0 R$; the potential due to a dipole $\boldsymbol{\mu}$ is $-\mu_\alpha T_\alpha = +\mu_\alpha R_\alpha / 4\pi\varepsilon_0 R^3 = \boldsymbol{\mu} \cdot \mathbf{R}/4\pi\varepsilon_0 R^3$, and so on.

Having found the potential as a function of position $\mathbf{R}$, it is now very easy to determine the electric field, the field gradient and the higher derivatives. Thus from the potential qT arising from the charge q, we obtain the electric field $F^A_\alpha(\mathbf{B}) = -\nabla_\alpha qT = -qT_\alpha$ and the field gradient $F^A_{\alpha\beta}(\mathbf{B}) = -\nabla_\alpha \nabla_\beta qT = -qT_{\alpha\beta}$. For the dipole potential we need to be a little more careful with the suffixes, to avoid clashes; so we write the potential as $-\mu_\gamma T_\gamma$ and then the electric field is $F^A_\alpha(\mathbf{B}) = -\nabla_\alpha(-\mu_\gamma T_\gamma) = +\mu_\gamma T_{\alpha\gamma}$. Similarly the field gradient is $F^A_{\alpha\beta}(\mathbf{B}) = -\nabla_\alpha\nabla_\beta(-\mu_\gamma T_\gamma) = +\mu_\gamma T_{\alpha\beta\gamma}$. In this way we find, for the complete field,

$$\begin{aligned} F^A_\alpha(\mathbf{B}) &= -\nabla_\alpha V^A(\mathbf{B}) \\ &= -T_\alpha q + T_{\alpha\beta}\hat{\mu}_\beta - \frac{1}{3}T_{\alpha\beta\gamma}\hat{\Theta}_{\beta\gamma} + \cdots \\ &\quad \cdots - \frac{(-1)^n}{(2n-1)!!}T^{(n+1)}_{\alpha\beta\ldots\nu\sigma}\hat{\xi}^{(n)}_{\beta\gamma\ldots\nu\sigma} - \cdots, \end{aligned} \tag{3.1.9}$$

and for the field gradient,

$$\begin{aligned} F^A_{\alpha\beta}(\mathbf{B}) &= -\nabla_\alpha\nabla_\beta V^A(\mathbf{B}) \\ &= -T_{\alpha\beta}q + T_{\alpha\beta\gamma}\hat{\mu}_\gamma - \frac{1}{3}T_{\alpha\beta\gamma\delta}\hat{\Theta}_{\gamma\delta} + \cdots \\ &\quad \cdots - \frac{(-1)^n}{(2n-1)!!}T^{(n+2)}_{\alpha\beta\ldots\nu\sigma\tau}\hat{\xi}^{(n)}_{\gamma\delta\ldots\nu\sigma\tau} - \cdots. \end{aligned} \tag{3.1.10}$$

This representation is quite compact and economical, but it is rather terse on first acquaintance. We will look at some examples shortly. Notice that T_α describes both the electric field due to a point charge (regarding T_α as a vector function of position) and also the potential due to a point dipole (where we take the scalar product $-\mu_\alpha T_\alpha$ to obtain a scalar function of position, as required for a potential).

Two important general properties of the T tensors with two or more suffixes are (i) invariance with respect to interchange of suffixes, so that for example $T_{xy} = T_{yx}$; and (ii) tracelessness: $T_{\alpha\alpha\gamma\ldots\nu} = 0$, for any $\gamma\ldots\nu$. These results follow from the fact that the differential operators commute, and from the fact that $\nabla^2(1/R) = 0$ (provided that $R \neq 0$). It follows from these properties that $T^{(n)}_{\alpha\beta\ldots\nu}$, like $\xi^{(n)}_{\alpha\beta\ldots\nu}$, has $2n+1$ components. (See p. 16 for the proof.) It also satisfies Laplace's equation, $\nabla^2 T^{(n)}_{\alpha\beta\ldots\nu} = 0$, and is proportional to R^{-n-1} (because it is obtained by differentiating R^{-1} n times), so its components must be linear combinations of the irregular spherical harmonics of rank n, $I_{nm} = r^{-l-1}C_{lm}$. (See Appendix B). A detailed discussion of the relationship may be found in Tough and Stone (1977). From this and the orthogonality property of the spherical harmonics a further important property follows: if $T^{(n)}_{\alpha\beta\ldots\nu}$ is averaged over all directions of the intermolecular vector $\mathbf{R}$, the result is zero, except for $n = 0$, i.e., for $T = 1/R$.

3.2 Electrostatic interaction between molecules

We are now in a position to calculate the interaction between a pair of molecules. Molecule A has its local origin at position $\mathbf{A}$ in the global coordinate system, and molecule B has its origin at $\mathbf{B}$. We know the potential V^A at $\mathbf{B}$ due to molecule A from eqn (3.1.2), and we can write down the energy of a molecule in a given potential from the results of Chapter 2. Combining these formulae gives the interaction operator:

$$\mathcal{H}' = q^B V^A + \hat{\mu}^B_\alpha V^A_\alpha + \frac{1}{3}\hat{\Theta}^B_{\alpha\beta}V^A_{\alpha\beta} + \cdots$$

$$
\begin{aligned}
&= q^B[Tq^A - T_\alpha\hat{\mu}^A_\alpha + \tfrac{1}{3}T_{\alpha\beta}\hat{\Theta}^A_{\alpha\beta} - \cdots] \\
&\quad + \hat{\mu}^B_\alpha[T_\alpha q^A - T_{\alpha\beta}\hat{\mu}^A_\beta + \tfrac{1}{3}T_{\alpha\beta\gamma}\hat{\Theta}^A_{\beta\gamma} - \cdots] \\
&\quad + \tfrac{1}{3}\hat{\Theta}^B_{\alpha\beta}[T_{\alpha\beta}q^A - T_{\alpha\beta\gamma}\hat{\mu}^A_\gamma + \tfrac{1}{3}T_{\alpha\beta\gamma\delta}\hat{\Theta}^A_{\gamma\delta} - \cdots], \\
&= Tq^Aq^B + T_\alpha(q^A\hat{\mu}^B_\alpha - \hat{\mu}^A_\alpha q^B) \\
&\quad + T_{\alpha\beta}(\tfrac{1}{3}q^A\hat{\Theta}^B_{\alpha\beta} - \hat{\mu}^A_\alpha\hat{\mu}^B_\beta + \tfrac{1}{3}\hat{\Theta}^A_{\alpha\beta}q^B) + \cdots.
\end{aligned}
\tag{3.2.1}
$$

Notice that some relabelling of subscripts has been necessary to avoid clashes. For neutral species, the charges are zero, and the leading term is the dipole–dipole interaction:

$$
\begin{aligned}
\mathcal{H}' &= -T_{\alpha\beta}\hat{\mu}^A_\alpha\hat{\mu}^B_\beta - \tfrac{1}{3}T_{\alpha\beta\gamma}(\hat{\mu}^A_\alpha\hat{\Theta}^B_{\beta\gamma} - \hat{\Theta}^A_{\alpha\beta}\hat{\mu}^B_\gamma) \\
&\quad - T_{\alpha\beta\gamma\delta}(\tfrac{1}{15}\hat{\mu}^A_\alpha\hat{\Omega}^B_{\beta\gamma\delta} - \tfrac{1}{9}\hat{\Theta}^A_{\alpha\beta}\hat{\Theta}^B_{\gamma\delta} + \tfrac{1}{15}\hat{\Omega}^A_{\alpha\beta\gamma}\hat{\mu}^B_\delta) + \cdots.
\end{aligned}
\tag{3.2.2}
$$

This expression, like the preceding one, is an operator. If we require the electrostatic interaction U_{es} between two molecules in non-degenerate states, then we need the expectation value of this operator, which we obtain by replacing each multipole operator by its expectation value. Thus for two neutral molecules the result is

$$
\begin{aligned}
U_{\text{es}} &= -T_{\alpha\beta}\mu^A_\alpha\mu^B_\beta - \tfrac{1}{3}T_{\alpha\beta\gamma}(\mu^A_\alpha\Theta^B_{\beta\gamma} - \Theta^A_{\alpha\beta}\mu^B_\gamma) \\
&\quad - T_{\alpha\beta\gamma\delta}(\tfrac{1}{15}\mu^A_\alpha\Omega^B_{\beta\gamma\delta} - \tfrac{1}{9}\Theta^A_{\alpha\beta}\Theta^B_{\gamma\delta} + \tfrac{1}{15}\Omega^A_{\alpha\beta\gamma}\mu^B_\delta) + \cdots.
\end{aligned}
\tag{3.2.3}
$$

Eqns (3.2.2) and (3.2.3) have been derived for a pair of molecules, isolated from any others. However they are based on the Coulomb interactions between nuclear and electronic charges, which are strictly additive, so we can generalize them to an assembly of molecules by summing over the distinct pairs.

Similar expressions have been derived by a number of authors. See, for example, Hirschfelder *et al.* (1954), Jansen (1957, 1958), Buckingham (1967) and Leavitt (1980).

3.2.1 *Explicit formulae*

The formulae just derived are general but somewhat opaque, and for practical application we need more transparent forms. We substitute the explicit expression for $T_{\alpha\beta}$, eqn (3.1.8), to obtain the dipole–dipole interaction in the form

$$
U_{\mu\mu} = -\mu^A_\alpha\mu^B_\beta\frac{3R_\alpha R_\beta - R^2\delta_{\alpha\beta}}{4\pi\varepsilon_0R^5} \tag{3.2.4}
$$

$$
= \frac{R^2\boldsymbol{\mu}^A\cdot\boldsymbol{\mu}^B - 3(\boldsymbol{\mu}^A\cdot\mathbf{R})(\boldsymbol{\mu}^B\cdot\mathbf{R})}{4\pi\varepsilon_0R^5}. \tag{3.2.5}
$$

It is often convenient to choose coordinates with the z axis along $\mathbf{R}$, with the origin at $\mathbf{A}$. The direction of $\boldsymbol{\mu}^A$ is specified by polar angles θ_A and φ_A and the direction of $\boldsymbol{\mu}^B$ by θ_B and φ_B. (See Fig. 1.4.) Then

$$\begin{aligned}
\boldsymbol{\mu}^A \cdot \mathbf{R} &= \mu^A R \cos\theta_A, \\
\boldsymbol{\mu}^B \cdot \mathbf{R} &= \mu^B R \cos\theta_B, \\
\boldsymbol{\mu}^A \cdot \boldsymbol{\mu}^B &= \mu^A \mu^B (\sin\theta_A \cos\varphi_A \sin\theta_B \cos\varphi_A \\
&\qquad + \sin\theta_A \sin\varphi_A \sin\theta_B \sin\varphi_B + \cos\theta_A \cos\theta_B) \\
&= \mu^A \mu^B \big(\cos\theta_A \cos\theta_B + \sin\theta_A \sin\theta_B \cos(\varphi_B - \varphi_A)\big),
\end{aligned}$$

so that the dipole–dipole interaction becomes

$$U_{\mu\mu} = -\frac{\mu^A \mu^B}{4\pi\varepsilon_0 R^3}(2\cos\theta_A \cos\theta_B - \sin\theta_A \sin\theta_B \cos\varphi), \tag{3.2.6}$$

where $\varphi = \varphi_B - \varphi_A$.

Another important interaction is the one between two quadrupolar but non-polar molecules ($\Theta \neq 0$ but $\boldsymbol{\mu} = 0$). A general dipole, like any vector, can be described in terms of a magnitude and two polar angles, but the description of a general quadrupole is rather more complicated and requires five numbers. However, for linear molecules (and symmetric tops) there is only one independent non-zero component. In a coordinate system with the z axis along the molecular axis such a molecule has $\Theta_{zz} = \Theta$ and $\Theta_{xx} = \Theta_{yy} = -\frac{1}{2}\Theta$. In a general coordinate system this can be expressed as

$$\Theta_{\alpha\beta} = \Theta(\tfrac{3}{2} n_\alpha n_\beta - \tfrac{1}{2}\delta_{\alpha\beta}),$$

where $\mathbf{n}$ is a unit vector in the direction of the molecular axis (see Appendix A). So the quadrupole–quadrupole interaction is

$$U_{\Theta\Theta} = \tfrac{1}{9}\Theta^A \Theta^B (\tfrac{3}{2} n_\alpha^A n_\beta^A - \tfrac{1}{2}\delta_{\alpha\beta}) T_{\alpha\beta\gamma\delta} (\tfrac{3}{2} n_\gamma^B n_\delta^B - \tfrac{1}{2}\delta_{\gamma\delta}). \tag{3.2.7}$$

The evaluation of this expression is straightforward, if laborious, for the case when both molecules lie on the z axis, and gives the result

$$\begin{aligned}
U_{\Theta\Theta} = \frac{\Theta^A \Theta^B}{4\pi\varepsilon_0 R^5} \cdot \frac{3}{4} \big[&1 - 5\cos^2\theta_A - 5\cos^2\theta_B - 15\cos^2\theta_A \cos^2\theta_B \\
&+ 2(4\cos\theta_A \cos\theta_B - \sin\theta_A \sin\theta_B \cos\varphi)^2\big].
\end{aligned} \tag{3.2.8}$$

The interaction between a dipole and a linear quadrupole is calculated in a similar way, and is

$$U_{\mu\Theta} = \frac{\mu^A \Theta^B}{4\pi\varepsilon_0 R^4} \cdot \frac{3}{2}[\cos\theta_A (3\cos^2\theta_B - 1) - \sin\theta_A \sin 2\theta_B \cos\varphi]. \tag{3.2.9}$$

Note the R dependence in these formulae. The dipole–dipole interaction is proportional to R^{-3}, the dipole–quadrupole to R^{-4} and the quadrupole–quadrupole to R^{-5}. In general, the interaction between multipoles of ranks l and l' is proportional to $R^{-l-l'-1}$. At large R, therefore, only the lowest-rank non-vanishing moment is important, but the others become increasingly significant as R decreases.

3.3 Spherical tensor formulation

For many purposes, a spherical tensor formulation of the interaction is more convenient. It is best obtained by a somewhat different route. The derivation starts with an expansion of $1/r_{ab}$, just as the cartesian formulation does, but this time we use the expansion in terms of spherical harmonics (the *spherical harmonic addition theorem*), which takes the form

$$\frac{1}{|\mathbf{r}_1 - \mathbf{r}_2|} = \sum_{lm} \frac{r_<^l}{r_>^{l+1}} (-1)^m C_{l,-m}(\theta_1, \varphi_1) C_{lm}(\theta_2, \varphi_2). \tag{3.3.1}$$

In this expression, which is proved in many textbooks of quantum mechanics and mathematical physics (Arfken 1970, Zare 1988), $r_<$ is the smaller and $r_>$ the larger of r_1 and r_2. For our purposes, we want $1/r_{ab} = 1/|\mathbf{B}+\mathbf{b}-\mathbf{A}-\mathbf{a}|$, and we take $\mathbf{r}_1 = \mathbf{B}-\mathbf{A} = \mathbf{R}$ and $\mathbf{r}_2 = \mathbf{a}-\mathbf{b}$, and assuming that $|\mathbf{a}-\mathbf{b}| < R$ we obtain

$$\frac{1}{|\mathbf{R}+\mathbf{b}-\mathbf{a}|} = \sum_{l=0}^{\infty} \sum_{m=-l}^{l} (-1)^m R_{l,-m}(\mathbf{a}-\mathbf{b}) I_{lm}(\mathbf{R}), \tag{3.3.2}$$

where $R_{km}(\mathbf{r})$ and $I_{km}(\mathbf{r})$ are the regular and irregular spherical harmonics. (See Appendix B). We note that eqn (3.3.2) is valid only if $|\mathbf{a}-\mathbf{b}| < R$. Now we use the addition theorem already quoted in eqn (2.7.3) on p. 33, and, remembering that $R_{lm}(-\mathbf{r}) = (-1)^l R_{lm}(\mathbf{r})$, we find

$$\begin{aligned}
\mathcal{H}' &= \frac{1}{4\pi\varepsilon_0} \sum_{a\in A} \sum_{b\in B} \frac{e_a e_b}{|\mathbf{R}+\mathbf{b}-\mathbf{a}|} \\
&= \frac{1}{4\pi\varepsilon_0} \sum_{l_1,l_2} \sum_{m_1 m_2 m} (-1)^{l_1} \left(\frac{(2l_1+2l_2+1)!}{(2l_1)!(2l_2)!} \right)^{1/2} \\
&\qquad \times \sum_{a\in A} e_a R_{l_1 m_1}(\mathbf{a}) \sum_{b\in B} e_b R_{l_2 m_2}(\mathbf{b}) I_{l_1+l_2,m}(\mathbf{R}) \begin{pmatrix} l_1 & l_2 & l_1+l_2 \\ m_1 & m_2 & m \end{pmatrix} \\
&= \frac{1}{4\pi\varepsilon_0} \sum_{l_1,l_2} \sum_{m_1 m_2 m} (-1)^{l_1} \left(\frac{(2l_1+2l_2+1)!}{(2l_1)!(2l_2)!} \right)^{1/2} \\
&\qquad \times \hat{Q}^{A(G)}_{l_1 m_1} \hat{Q}^{B(G)}_{l_2 m_2} I_{l_1+l_2,m}(\mathbf{R}) \begin{pmatrix} l_1 & l_2 & l_1+l_2 \\ m_1 & m_2 & m \end{pmatrix}.
\end{aligned} \tag{3.3.3}$$

where in the last line we have introduced the multipole moment operators:

$$\hat{Q}^{A(G)}_{lm} = \sum_{a\in A} e_a R_{lm}(\mathbf{a}).$$

Now the expression (3.3.3) is expressed in the global coordinate system, which we have been using throughout; the superscript (G) is to remind us of this fact. As we have seen, however, it is much more convenient to express the interaction in terms of multipole moments defined in the local coordinate system of each molecule. The components in the global system are related to those in the local system by

$$Q_{lk}^{(L)} = \sum_m Q_{lm}^{(G)} D_{mk}^l(\Omega) \tag{3.3.4}$$

where $\Omega = (\alpha, \beta, \gamma)$ is the rotation that takes the global axes to the local axes, and $D_{mk}^l(\Omega)$ is the Wigner rotation matrix element for this rotation. (See eqn (B.2.1) on p. 215. We take the component $Q_{lk}^{(L)}$ initially in an orientation where the local axes coincide with the global axes, and rotate it and the local axes to the required orientation.) Equivalently, we can write the global components in terms of the local ones:

$$Q_{lm}^{(G)} = \sum_k Q_{lk}^{(L)} D_{km}^l(\Omega^{-1}) = \sum_k Q_{lk}^{(L)} \left[D_{mk}^l(\Omega)\right]^*, \tag{3.3.5}$$

where the last step follows because the Wigner matrices are hermitian. Making this substitution for the multipole moment operators in global axes that occur in (3.3.3), we get

$$\begin{aligned}\mathcal{H}' &= \frac{1}{4\pi\varepsilon_0} \sum_{l_1, l_2} \sum_{k_1 k_2} (-1)^{l_1} \left(\frac{(2l_1+2l_2+1)!}{(2l_1)!(2l_2)!}\right)^{1/2} \hat{Q}_{l_1 k_1}^{A(L)} \hat{Q}_{l_2 k_2}^{B(L)} \\ &\quad \times \sum_{m_1 m_2 m} [D_{m_1 k_1}^{l_1}(\Omega_1)]^* [D_{m_2 k_2}^{l_2}(\Omega_2)]^* I_{l_1+l_2, m}(\mathbf{R}) \begin{pmatrix} l_1 & l_2 & l_1+l_2 \\ m_1 & m_2 & m \end{pmatrix}.\end{aligned} \tag{3.3.6}$$

The multipole moment operators are now referred to local molecular axes (different for the two molecules) and the orientational and distance dependence is all contained in the sum over Wigner functions and irregular spherical harmonics. We define new functions of the orientations by

$$\begin{aligned}\bar{S}_{l_1 l_2 j}^{k_1 k_2} &= \mathrm{i}^{l_1-l_2-j} \left[\begin{pmatrix} l_1 & l_2 & j \\ 0 & 0 & 0 \end{pmatrix}\right]^{-1} \\ &\quad \times \sum_{m_1 m_2 m} [D_{m_1 k_1}^{l_1}(\Omega_1)]^* [D_{m_2 k_2}^{l_2}(\Omega_2)]^* C_{jm}(\theta, \varphi) \begin{pmatrix} l_1 & l_2 & j \\ m_1 & m_2 & m \end{pmatrix}.\end{aligned} \tag{3.3.7}$$

Here θ and φ are the polar angles describing the direction of the intermolecular vector $\mathbf{R}$. In terms of these functions, (3.3.6) becomes

$$\begin{aligned}\mathcal{H}' &= \frac{1}{4\pi\varepsilon_0} \sum_{l_1 l_2} \sum_{k_1 k_2} (-1)^{l_1+l_2} \left(\frac{(2l_1+2l_2+1)!}{(2l_1)!(2l_2)!}\right)^{1/2} \\ &\quad \times \hat{Q}_{l_1 k_1}^{A} \hat{Q}_{l_2 k_2}^{B} R^{-l_1-l_2-1} \begin{pmatrix} l_1 & l_2 & l_1+l_2 \\ 0 & 0 & 0 \end{pmatrix} \bar{S}_{l_1 l_2 l_1+l_2}^{k_1 k_2},\end{aligned}$$

(dropping the superscript (L), since we shall be using local axes for the multipoles from now on) and if we insert the explicit formula for the $3j$ symbol (eqn (2.7.4) on p. 33), this reduces to

$$\mathcal{H}' = \frac{1}{4\pi\varepsilon_0} \sum_{l_1,l_2} \sum_{k_1 k_2} \binom{l_1+l_2}{l_1} \hat{Q}^A_{l_1 k_1} \hat{Q}^B_{l_2 k_2} \bar{S}^{k_1 k_2}_{l_1\, l_2\, l_1+l_2} R^{-l_1-l_2-1}. \tag{3.3.8}$$

This formulation explicitly separates each term in the interaction into an operator part, involving multipole moment operators in local molecular axes, a factor $\bar{S}^{k_1 k_2}_{l_1\, l_2\, l_1+l_2}$ that describes the orientation dependence, and a distance dependence $R^{-l_1-l_2-1}$. Notice that the orientational part (eqn (3.3.7)) involves a linear combination of products of the Wigner functions and a spherical harmonic, with coefficients that are Wigner $3j$ symbols. This ensures that the result is a scalar, invariant under rotations of the entire system, and takes account of the fact that only five of the eight angular coordinates are independent.

This is a general and powerful formulation, but a little cumbersome for routine use. Also it is at present expressed in terms of the complex components of the multipole moments. Accordingly we first transform to the real components, using eqn (B.1.6):

$$\hat{Q}_{l,k} = \sum_{\kappa} X_{\kappa,k} \hat{Q}_{l,\kappa}, \tag{3.3.9}$$

where the $X_{\kappa,k}$ are defined in eqn (B.1.7). This leads to an expression equivalent to (3.3.8) but in terms of the real components:

$$\mathcal{H}' = \frac{1}{4\pi\varepsilon_0} \sum_{l_1,l_2} \sum_{\kappa_1 \kappa_2} \binom{l_1+l_2}{l_1} \hat{Q}^A_{l_1 \kappa_1} \hat{Q}^B_{l_2 \kappa_2} \bar{S}^{\kappa_1 \kappa_2}_{l_1\, l_2\, l_1+l_2} R^{-l_1-l_2-1}, \tag{3.3.10}$$

where

$$\bar{S}^{\kappa_1 \kappa_2}_{l_1\, l_2\, j} = \sum_{k_1 k_2} X_{\kappa_1,k_1} X_{\kappa_2,k_2} \bar{S}^{k_1 k_2}_{l_1\, l_2\, j}. \tag{3.3.11}$$

We can now obtain a more compact representation by defining analogues to the T tensors of the cartesian formulation:

$$T_{l_1\kappa_1,l_2\kappa_2} = \frac{1}{4\pi\varepsilon_0} \binom{l_1+l_2}{l_1} \bar{S}^{\kappa_1 \kappa_2}_{l_1\, l_2\, l_1+l_2} R^{-l_1-l_2-1}, \tag{3.3.12}$$

and then the interaction is just

$$\mathcal{H}' = \sum_{l_1,l_2} \sum_{\kappa_1 \kappa_2} \hat{Q}^A_{l_1 \kappa_1} \hat{Q}^B_{l_2 \kappa_2} T_{l_1\kappa_1,l_2\kappa_2}. \tag{3.3.13}$$

We can abbreviate this even further by adopting a notation in which the successive multipole components are arranged in a vector, indexed by angular momentum labels 00, 10, 11c, 11s, 20, 21c, 21s, 22c, 22s, 30, ... If we agree always to order the components in this way, then we can use a single index, say t or u, to label them, and write the interaction as

$$\mathcal{H}' = \hat{Q}^A_t T^{AB}_{t\,u} \hat{Q}^B_u. \tag{3.3.14}$$

As a final abbreviation we have adopted the repeated-suffix summation convention in this last form. While this may appear excessively terse to the beginner, it is a very compact and convenient form for the manipulation of general formulae.

The interaction functions $T^{AB}_{t\,u}$ depend only on the relative positions of the two molecular axis systems, so they can be evaluated once and for all, and tabulated for further use. The functions for $l_1 + l_2 \leq 5$, which include all terms in the interaction up to R^{-6}, are tabulated in Appendix F. These formulae were first given by Price *et al.* (1984), in a different form, and by Stone (1991) in the form given here, for $l_1 + l_2 \leq 4$. Recently Hättig and Hess (1993) have described a much simpler method for deriving these formulae, and have used it to obtain the formulae for $l_1 + l_2 = 5$ and to check the earlier formulae.

One of the important features of these formulae, from the practical point of view, is that they do not involve any trigonometric functions. Although the Wigner functions are usually expressed in terms of sines and cosines, the interaction functions can be expressed entirely in terms of scalar products involving the unit vectors $\mathbf{e}_i^A$ and $\mathbf{e}_j^B$ along the local axes of each molecule, and a unit vector in the direction of $\mathbf{R}$. These vectors are readily available in circumstances such as a molecular dynamics calculation, so the evaluation of the interaction functions is not unduly time-consuming.

A study of the dipole–dipole and quadrupole–quadrupole interactions for linear molecules will perhaps illuminate the difference between this formulation and the cartesian one. The cartesian version of the dipole–dipole interaction is given by eqn (3.2.5), or by (3.2.4), which emphasizes that the expression involves the multipole components μ_α^A and μ_β^B *in the global coordinate system.* In order to evaluate the expression we need to transform the dipole moments into this coordinate system. The resulting expression involves (in effect) the scalar products $e_z^A \cdot \hat{\mathbf{R}}$, $e_z^B \cdot \hat{\mathbf{R}}$ and $e_z^A \cdot e_z^B$, though we previously used the notation $\mathbf{n}^A$ and $\mathbf{n}^B$ for e_z^A and e_z^B. In the spherical-tensor approach, we use eqn (3.3.13), and the interaction takes the form $\mu_{10}^A \mu_{10}^B T_{10,10}$. The dipole moments are the z components in local axes, while $T_{10,10}$ is obtained from Table F.1. In this case there is little difference between the cartesian and spherical-tensor expressions.

In the case of the cartesian quadrupole–quadrupole interaction the same procedure is required. However, even for Θ_{zz} the transformation to global coordinates is more cumbersome, and the expression involves the interaction tensor $T_{\alpha\beta\gamma\delta}$. In contrast, the spherical tensor expression involves the same scalar products as before:

$$U_{\Theta\Theta} = \frac{\Theta^A \Theta^B}{4\pi\varepsilon_0 R^5} \times \frac{3}{4}\left[35(\mathbf{n}^A \cdot \hat{R})^2 (\mathbf{n}^B \cdot \hat{R})^2 - 5(\mathbf{n}^A \cdot \hat{R})^2 - 5(\mathbf{n}^B \cdot \hat{R})^2 - 20(\mathbf{n}^A \cdot \hat{R})(\mathbf{n}^B \cdot \hat{R})(\mathbf{n}^A \cdot \mathbf{n}^B) + 2(\mathbf{n}^A \cdot \mathbf{n}^B)^2 + 1\right]. \tag{3.3.15}$$

As for the dipole–dipole case, when both molecules lie on the z axis the scalar products are

$$\begin{aligned} \mathbf{n}_A \cdot \hat{R} &= \cos\theta_A, \\ \mathbf{n}_B \cdot \hat{R} &= \cos\theta_B, \\ \mathbf{n}_A \cdot \mathbf{n}_B &= \cos\theta_A \cos\theta_B + \sin\theta_A \sin\theta_B \cos(\varphi_B - \varphi_A) \end{aligned}$$

and (3.3.15) is easily evaluated to give (3.2.8).

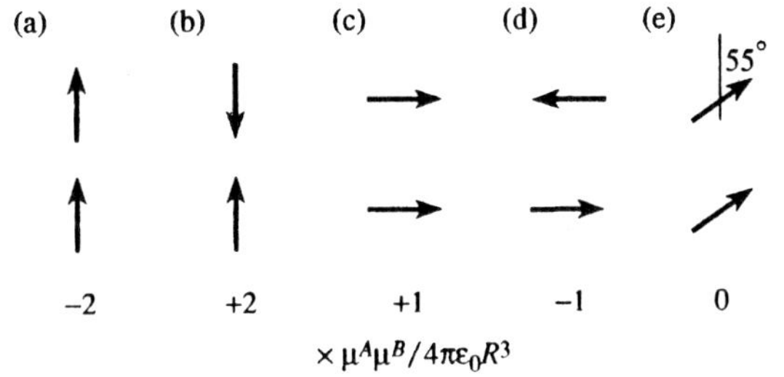

FIG. 3.2. Favourable and unfavourable geometries for dipole–dipole interactions.

3.4 Examples

We noted above that the T tensors are irregular spherical harmonics, and so have the important property that all of them, except T itself, average to zero over all directions of **R**. Thus the angular variation of the electrostatic interaction is much greater than that of other contributions to the interaction energy, such as the dispersion energy, which is always negative. This means that the electrostatic term often has a dominant effect in determining geometry, even when it is not the dominant contribution to the binding energy.

3.4.1 *The dipole–dipole interaction*

The dipole–dipole interaction has the simple angle-dependence given in eqn (3.2.6). At a given value of R, the most favourable orientation has $\theta_A = \theta_B = 0$. (See Fig. 3.2a.) In this case, the angular factor that multiplies the magnitude $\mu^A\mu^B/4\pi\varepsilon_0 R^3$ of the interaction is -2. Conversely the the most unfavourable orientation has $\theta_A = \pi$, $\theta_B = 0$ or *vice versa*, with an angular factor of $+2$ (Fig. 3.2b). If the molecules are linear, however, the orientation with $\theta_A = \theta_B = \pi/2$ and $\varphi = \pi$ may be preferable (Fig. 3.2d); in this case the angular factor is only -1, but it may be possible for the molecules to pack more closely, increasing the R^{-3} factor.

The HCN molecule provides a good illustration of these features. The HCN dimer adopts the linear head-to-tail structure (Fig. 3.2a) with $\theta_A = \theta_B = 0$ (Legon *et al.* 1977), while the crystal structure has chains of HCN molecules arranged head to tail (Dulmage and Lipscomb 1951).

3.4.2 *The quadrupole–quadrupole interaction*

The quadrupole–quadrupole interaction alone is important in determining the structure of many non-polar molecules, i.e., those with $\mu = 0$. For given R, the energy as a multiple of $\Theta^A\Theta^B/4\pi\varepsilon_0 R^5$ is largest when $\theta_A = \theta_B = 0$, when the angular factor is $+6$. (See Fig. 3.3a.) The configurations with $\theta_A = \theta_B = \pi/2$ (Figs 3.3c and 3.3d) are also repulsive, with angular factor $+2.25$ when $\varphi = 0$ and 0.75 when $\varphi = \pi/2$. There are two common attractive geometries, both planar ($\varphi = 0$): the 'T' structure, with $\theta_A = 0$ and $\theta_B = \pi/2$, or *vice versa* (Fig. 3.3b); and the 'slipped parallel' structure, with $\theta_A = \theta_B \approx \pi/4$ (Fig. 3.3e). These structures have angular factors of -3 and $-2\frac{7}{16}$ respectively, so they are quite similar in energy. In fact there is a range of attractive geometries with $\theta_B = \pi/2 - \theta_A$, and which of them is preferred in any given case usually depends on the

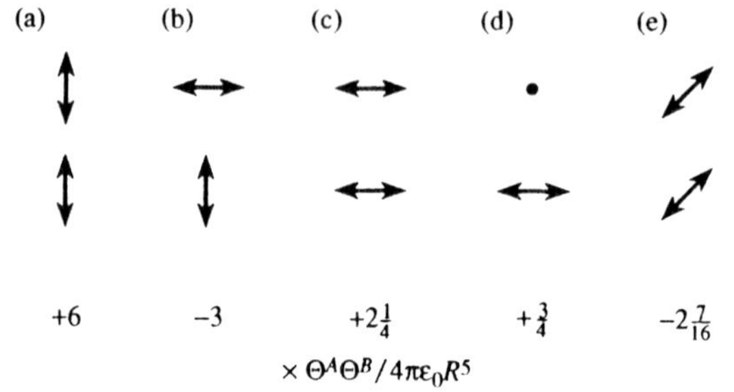

FIG. 3.3. Favourable and unfavourable geometries for quadrupole–quadrupole interactions. The double arrow represents a quadrupole moment Q_{20} with magnitude $\Theta > 0$.

molecular shape. For long narrow molecules, or planar ones, a structure of the slipped parallel type usually gives the smallest intermolecular distance and the lowest energy; for molecules that are more nearly spherical a T structure is usually favoured.

An interesting example is acetylene dimer. *Ab initio* calculations give the total binding energy to be about $400\,\mathrm{cm}^{-1}$ (Bone and Handy 1990). In the T structure the separation of the centres of mass is 4.344 Å, and the quadrupole–quadrupole interaction contributes about $-550\,\mathrm{cm}^{-1}$ to the interaction energy, with dispersion, repulsion and higher-rank electrostatic terms making up the difference. In the slipped parallel structure, the centres of mass are slightly closer, at 4.243 Å, which partly offsets the less favourable angular factor. Here the quadrupole–quadrupole term is $-500\,\mathrm{cm}^{-1}$. Consequently the energy difference between the two structures is very small, and there is an interconversion between different T structures via slipped-parallel transition states, the molecules rotating in opposite directions in a geared fashion. (See Fig. 3.4.) The consequences of this isomerization can be seen in the infrared spectrum as a tunnelling splitting (Fraser *et al.* 1988, Ohshima *et al.* 1988), from which the barrier to interconversion is estimated to be only $33\,\mathrm{cm}^{-1}$.

Acetylene trimer is an even more interesting case. Here too the electrostatic interaction dominates the binding, and the equilibrium geometry is a planar C_{3h} structure in which each pair of acetylene molecules is bound in an approximately T-shaped manner, with one H atom of each acetylene directed towards the π orbitals of the next. Here too there are isomerization possibilities, and there is a great variety of rearrangement pathways whose energetics can be well understood in terms of quadrupole–quadrupole interactions (Bone *et al.* 1991).

There are many non-polar linear or planar molecules that show the effects of quadrupole–quadrupole interactions in their crystal structures. Planar aromatic molecules such as benzene, naphthalene and anthracene are examples. Although the molecular packing is important in determining the crystal structure, the influence of quadrupole–quadrupole interactions is often apparent. Structures commonly consist of stacks of tilted parallel molecules, with neighbouring molecules in adjacent stacks often having a T geometry relative to each other.

If the two molecules are different, it is possible for Θ^A and Θ^B to have opposite signs. In the case of dipoles, a change of sign is merely a change of direction, but when Θ^A

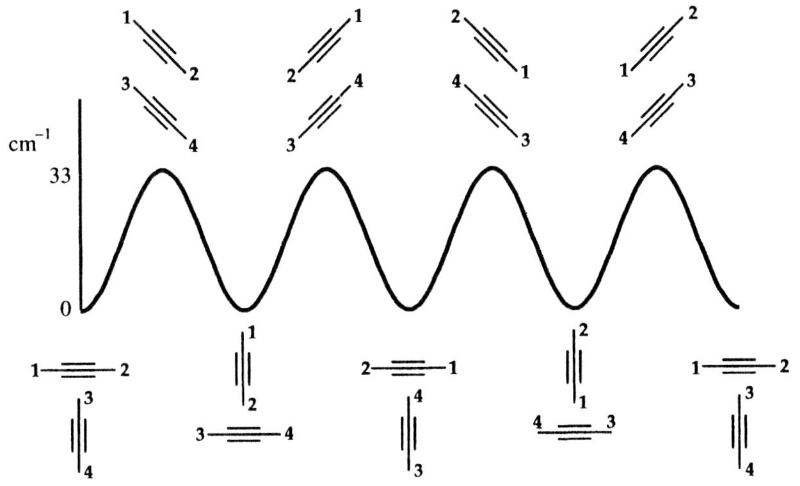

FIG. 3.4. Acetylene dimer: interconversion between different isomers of the T-shaped equilibrium structure via a slipped-parallel structure.

and Θ^B have opposite signs the preferred geometries are qualitatively different. In this case the preferred geometry for given R is $\theta_A = \theta_B = 0$, with its angular factor of $+6$ multiplying the negative value of $\Theta^A\Theta^B/4\pi\varepsilon_0 R^5$. An example is the complex between benzene ($\Theta = -6.7\, ea_0^2$) and hexafluorobenzene ($\Theta = +7.1\, ea_0^2$). Here the two molecules pack with their planes parallel in a C_{6v} structure (Williams 1993).

For linear molecules, the geometry with $\theta_A = \theta_B = \pi/2$ may be preferred to permit a smaller intermolecular distance. An example of this is the complex between CO_2 and acetylene, which have quadrupole moments of $-3.3\, ea_0^2$ and $5.6\, ea_0^2$ respectively. Here the experimental structure has $\theta_A = \theta_B = \pi/2$, so that the molecules are parallel to each other in a C_{2v} structure, with centre-of-mass separation of $3.29\,\text{Å} = 6.22a_0$ (Prichard *et al.* 1988, Muenter 1989, Huang and Miller 1989). In this geometry the quadrupole–quadrupole interaction is $(-3.3 \times 5.6 \times 9)/(4 \times 6.22^5) = -0.0045\,\text{a.u.} = -980\,\text{cm}^{-1}$. The structure with $\theta_A = \theta_B = 0$—the linear structure—is not observed spectroscopically; *ab initio* calculations (Bone and Handy 1990) predict that it is a local minimum on the potential energy surface, with a centre-of-mass separation of $5.233\,\text{Å} = 9.89a_0$. Here the quadrupole–quadrupole energy is much smaller: $(-3.3 \times 5.6 \times 6)/(9.89^5) = -0.0012\,\text{a.u.} = -257\,\text{cm}^{-1}$, the more favourable angular factor being more than outweighed by the increased separation. We should be aware however that such calculations are greatly over-simplified; not only do they ignore other terms in the interaction between the molecules, such as repulsion and dispersion, but even the electrostatic interaction itself is not reliably described by the multipole expansion when the molecules are so close together. We shall return to these matters in later chapters; for the moment we just note that the *ab initio* calculation of Bone and Handy gives binding energies of $568\,\text{cm}^{-1}$ and $358\,\text{cm}^{-1}$ respectively for the C_{2v} and linear structures.

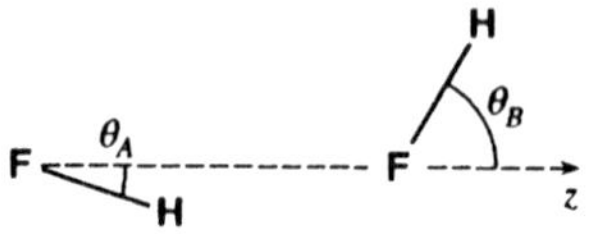

FIG. 3.5. Structure of the HF dimer.

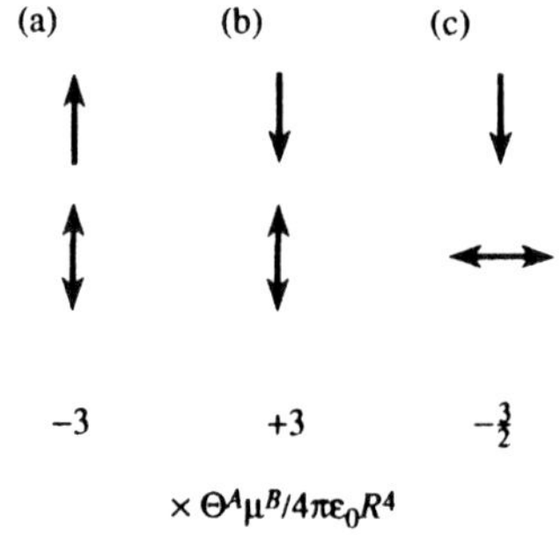

FIG. 3.6. Favourable and unfavourable geometries for dipole–quadrupole interactions.

3.4.3 *Competition between electrostatic terms*

It might be thought that the HF dimer would be another example where the electrostatic interaction favours a linear structure, because of the dipole–dipole interaction, but the experimental evidence (Dyke *et al.* 1972) leads to a structure (Fig. 3.5) in which the F–H···F hydrogen bond is indeed nearly linear ($\theta_A \approx 0$), but the other hydrogen atom is tilted away from the linear geometry by about 60°. At one time this was adduced as evidence for chemical bonding in the hydrogen bond. However HF has a significant quadrupole as well as a dipole moment, so we have to consider the effects of dipole–quadrupole and quadrupole–quadrupole interactions as well as dipole–dipole terms. The favourable orientations for dipole–quadrupole interactions are shown in Fig. 3.6. If we postulate a structure in which one of the HF molecules is at the origin and the other on the positive z axis, with $\theta_A = 0$ to give a linear hydrogen bond but with θ_B allowed to be non-zero, it is possible to see from the formulae given above that the linear structure is favoured by the dipole–dipole and quadrupole–dipole ($\Theta^A\mu^B$) terms, but that the dipole–quadrupole ($\mu^A\Theta^B$) and quadrupole–quadrupole terms favour a structure with $\theta_B = \pi/2$. In fact, there is a competition between these two effects, and the minimum energy, for this postulated structure, occurs at a geometry given by $\cos\theta_B = (2+3\lambda)/\big(9\lambda(1+2\lambda)\big)$, where $\lambda = \Theta/\mu R$. For HF, where $\mu = 0.72\,ea_0$, $\Theta = 1.875\,ea_0^2$ and $R = 5.1\,a_0$, this gives $\theta_B = 68°$, in reasonable agreement with the observed value. For HCl, the dipole moment is smaller, $0.433\,ea_0$, the quadrupole moment larger, $2.8\,ea_0^2$, and the separation also somewhat larger at $6.8\,a_0$. In this case the value of θ_B that emerges is also bigger, at 79°. This result too is in agreement with the experimental evidence. We conclude that there is no need to invoke chemical bonding effects in the description of the hydrogen bond, and that the structure can be understood in a purely classical fashion, provided that the description of the electrostatics is sufficiently complete and accurate. One of the limitations of conventional treatments is that the multipole expansion in the form that we have used here converges only slowly, or even diverges, for molecules of quite modest size. We return to this theme in Chapter 7.

4

PERTURBATION THEORY OF INTERMOLECULAR FORCES AT LONG RANGE

4.1 Introduction

Because intermolecular forces are relatively weak, it is natural to describe them using perturbation theory. If the molecules are far enough apart that the overlap between their wavefunctions can be ignored, the theory becomes relatively simple, so we will study this case first. It was first treated by London (1930*a*, 1937), and has been re-formulated by others since then (Margenau 1939, Longuet-Higgins 1956, Buckingham 1967).

The reason for the simplifications has to do with electron exchange. Suppose that we have a wavefunction $\Psi^A(1,2,\ldots,n_A)$ that describes molecule A (a function of the coordinates of its n_A electrons), and a wavefunction $\Psi^B(1',2',\ldots,n'_B)$ describing the n_B electrons of molecule B. The primes are used to distinguish the labels of the electrons of molecule B from those of molecule A. By hypothesis there is a region of space associated with Ψ^A, such that Ψ^A is non-zero only when all its electrons are in this region. Likewise there is another region associated with Ψ^B, and the two regions do not overlap. The wavefunction for the combined system should be written as an antisymmetrized product $\mathcal{A}\Psi^A\Psi^B$. But this antisymmetrized product contains terms like $\Psi^A(1',2,\ldots,n_A) \times \Psi^B(1,2',\ldots,n'_B)$, in which electron 1 of molecule A has been exchanged with electron 1′ of molecule B. The overlap between this and the original product function is

$$\int \Psi^A(1,2,\ldots,n_A)^*\Psi^B(1',2',\ldots,n'_B)^* \times \Psi^A(1',2,\ldots,n_A)\Psi^B(1,2',\ldots,n'_B)\,\mathrm{d}\tau$$

and when we integrate over the coordinates of electron 1 we get zero, because Ψ^A and Ψ^B are non-zero in different regions of space. So the terms in which the allocation of electrons between the two molecules is different do not mix with each other at all. The effect of this is that calculations may be done without antisymmetrization and the result will be the same.

In practice the overlap is never exactly zero, but the error made by ignoring it decreases exponentially as the distance between the molecules increases. Moreover the error is small enough to neglect at quite modest separations; indeed, if it were not, it would never be possible to discuss the electronic structure of an individual molecule in terms of the behaviour of 'its' electrons. However it becomes significant when molecules approach each other closely, and we shall have to examine this case later.

The consequence of all this is that we can identify a set of n_A electrons as belonging to molecule A, and we can define a Hamiltonian $\mathcal{H}^A$ for molecule A in terms of these electrons. Similarly, the Hamiltonian $\mathcal{H}^B$ for molecule B is defined in terms of its private set of electrons. The unperturbed Hamiltonian for the combined system is then $\mathcal{H}^0 = \mathcal{H}^A + \mathcal{H}^B$, and the perturbation consists of the electrostatic interaction between the particles of molecule A (electrons and nuclei) and those of B:

$$H' = \sum_{a\in A}\sum_{b\in B} \frac{e_a e_b}{4\pi\varepsilon_0 r_{ab}}. \tag{4.1.1}$$

Here e_a is the charge on particle a, one of the particles of molecule A, and r_{ab} is the distance between it and particle b in molecule B. This operator can be expressed in several other useful forms. Following Longuet-Higgins (1956) we define a charge density operator $\hat{\rho}^A(\mathbf{r})$ for molecule A:

$$\hat{\rho}^A(\mathbf{r}) = \sum_{a\in A} e_a \delta(\mathbf{r}-\mathbf{a}), \tag{4.1.2}$$

and $\hat{\rho}^B(\mathbf{r})$ similarly, using the Dirac delta function $\delta(\mathbf{r}-\mathbf{a})$ which as a function of $\mathbf{r}$ is an infinitely narrow spike of unit volume at $\mathbf{a}$.* In terms of these operators the perturbation becomes

$$\begin{aligned} H' &= \int \sum_{a\in A}\sum_{b\in B} \frac{e_a\delta(\mathbf{r}-\mathbf{a})e_b\delta(\mathbf{r}'-\mathbf{b})}{4\pi\varepsilon_0|\mathbf{r}-\mathbf{r}'|}\,\mathrm{d}^3\mathbf{r}\,\mathrm{d}^3\mathbf{r}' \\ &= \int \frac{\hat{\rho}^A(\mathbf{r})\hat{\rho}^B(\mathbf{r}')}{4\pi\varepsilon_0|\mathbf{r}-\mathbf{r}'|}\,\mathrm{d}^3\mathbf{r}\,\mathrm{d}^3\mathbf{r}'. \end{aligned} \tag{4.1.3}$$

Another useful expression results if we notice that the potential at $\mathbf{r}$ due to molecule B is

$$\hat{V}^B(\mathbf{r}) = \int \frac{\hat{\rho}^B(\mathbf{r}')}{4\pi\varepsilon_0|\mathbf{r}-\mathbf{r}'|}\,\mathrm{d}^3\mathbf{r}', \tag{4.1.4}$$

so that we can write

$$\mathcal{H}' = \int \hat{V}^B(\mathbf{r})\hat{\rho}^A(\mathbf{r})\,\mathrm{d}^3\mathbf{r}, \tag{4.1.5}$$

or equivalently

$$\mathcal{H}' = \int \hat{V}^A(\mathbf{r}')\hat{\rho}^B(\mathbf{r}')\,\mathrm{d}^3\mathbf{r}'. \tag{4.1.6}$$

(The variable that appears in these integrals is a dummy, so it does not matter whether it is $\mathbf{r}$ or $\mathbf{r}'$; the choice of variable is made simply to retain the connection with the earlier formulae.)

Now the unperturbed states are simple product functions $\Psi^A_m\Psi^B_n$, which we abbreviate to $|mn\rangle$, and they are eigenfunctions of $\mathcal{H}^0$:

$$\begin{aligned} \mathcal{H}^0|mn\rangle &= (\mathcal{H}^A + \mathcal{H}^B)|mn\rangle \\ &= (W^A_m + W^B_n)|mn\rangle \\ &= W^0_{mn}|mn\rangle \end{aligned}$$

*Properly, the 'delta function' is not a function at all but a 'distribution'. We should view it as a function of small but finite width, and let the width tend to zero, keeping the volume constant, only after we have integrated over it. See Lighthill (1958) for a full discussion.

For closed-shell molecules, ordinary non-degenerate Rayleigh–Schrödinger perturbation theory gives the energy to second order of the ground state† of the system, labelled by $m = n = 0$:

$$W_{00} = W_{00}^0 + W_{00}' + W_{00}'' + \cdots, \tag{4.1.7}$$

where

$$W_{00}^0 = W_0^A + W_0^B \tag{4.1.8}$$

$$W_{00}' = \langle 00|\mathcal{H}'|00\rangle, \tag{4.1.9}$$

$$W_{00}'' = -\sum_{mn}{}' \frac{\langle 00|\mathcal{H}'|mn\rangle\langle mn|\mathcal{H}'|00\rangle}{W_{mn}^0 - W_{00}^0}. \tag{4.1.10}$$

This is the long-range approximation to the interaction energy, sometimes called in this context the 'polarization approximation' (Hirschfelder 1967). The first-order energy, eqn (4.1.9), is just the expectation value of the electrostatic interaction $\mathcal{H}'$ for the ground state $|00\rangle$, which is the electrostatic interaction that we discussed in Chapter 3. Note, however, that if we write the perturbation in the form (4.1.3) we can express the interaction as

$$W_{00}' = \int \frac{\rho^A(\mathbf{r})\rho^B(\mathbf{r}')}{4\pi\varepsilon_0|\mathbf{r} - \mathbf{r}'|} \mathrm{d}^3\mathbf{r}\,\mathrm{d}^3\mathbf{r}', \tag{4.1.11}$$

since the integration over the coordinates of the particles in molecule A just replaces the operator $\hat{\rho}^A(\mathbf{r})$ by its expectation value $\rho^A(\mathbf{r})$. This is the exact classical interaction energy of the two molecular charge distributions in a form that does not depend on the multipole expansion.

The second-order energy (4.1.10) describes the induction and dispersion energies. To see this, we first separate it into three parts. Noting that the only term excluded in the sum is the one in which *both* molecules are in the ground state, we consider separately the terms in the sum for which molecule A is excited but B is in its ground state, the terms for which molecule B is excited but A is in its ground state, and the terms where both molecules are excited. This gives

$$W'' = U_{\text{ind}}^A + U_{\text{ind}}^B + U_{\text{disp}},$$

where

$$U_{\text{ind}}^A = -\sum_{m\neq 0} \frac{\langle 00|\mathcal{H}'|m0\rangle\langle m0|\mathcal{H}'|00\rangle}{W_m^A - W_0^A}, \tag{4.1.12}$$

$$U_{\text{ind}}^B = -\sum_{n\neq 0} \frac{\langle 00|\mathcal{H}'|0n\rangle\langle 0n|\mathcal{H}'|00\rangle}{W_n^B - W_0^B}, \tag{4.1.13}$$

$$U_{\text{disp}} = -\sum_{\substack{m\neq 0\\ n\neq 0}} \frac{\langle 00|\mathcal{H}'|mn\rangle\langle mn|\mathcal{H}'|00\rangle}{W_m^A + W_n^B - W_0^A - W_0^B}. \tag{4.1.14}$$

† In fact this need not be the ground state; it may be a state in which one molecule or both is excited. However it may not be a degenerate state of the combined system, which excludes all excited states in the case where the two molecules are identical. See the discussion of resonance interactions in Chapter 10.

These describe respectively the induction energy of molecule A, the induction energy of molecule B, and the dispersion energy.

4.2 The induction energy

As in the case of the first-order energy, we can write the perturbation in the form (4.1.3). This leads to 'non-expanded' expressions for the induction and dispersion energy that do not depend on the validity of the multipole expansion, and so can be used at short range where it does not converge (Longuet-Higgins 1956). However, if the molecular wavefunctions do not overlap there are better ways of dealing with the non-convergence of the multipole expansion (see Chapter 7), and if they do overlap the long-range approximation ceases to be valid, and different methods are needed (Chapter 6). Normally, therefore, we use the multipole expansion, so that the operator $\mathcal{H}'$ is given by eqn (3.2.2):

$$\mathcal{H}' = Tq^A q^B + T_\alpha(q^A \hat{\mu}_\alpha^B - \hat{\mu}_\alpha^A q^B) - T_{\alpha\beta}\hat{\mu}_\alpha^A \hat{\mu}_\beta^B + \cdots.$$

(We drop the terms involving the quadrupole for the moment.) Substituting in 4.1.13 gives

$$\begin{aligned} U_{\text{ind}}^B = -\sum_{n\neq 0} \langle 00|Tq^A q^B + T_\alpha(q^A \hat{\mu}_\alpha^B - \hat{\mu}_\alpha^A q^B) - T_{\alpha\beta}\hat{\mu}_\alpha^A \hat{\mu}_\beta^B + \cdots|0n\rangle \\ \times \langle 0n|Tq^A q^B + T_{\alpha'}(q^A \hat{\mu}_{\alpha'}^B - \hat{\mu}_{\alpha'}^A q^B) - T_{\alpha'\beta'}\hat{\mu}_{\alpha'}^A \hat{\mu}_{\beta'}^B + \cdots|00\rangle \\ \times (W_n^B - W_0^B)^{-1}. \end{aligned}$$

Now we can perform the implied integration over the coordinates of molecule A, which just yields the expectation values of the multipole moment operators. We note also that the matrix elements of q^B vanish, because the excited states are orthogonal to the ground state and q^B is just a constant. This gives

$$\begin{aligned} U_{\text{ind}}^B &= -\sum_{n\neq 0} \frac{\langle 0|T_\alpha q^A \hat{\mu}_\alpha^B - T_{\alpha\beta}\mu_\alpha^A \hat{\mu}_\beta^B + \cdots|n\rangle\langle n|T_{\alpha'} q^A \hat{\mu}_{\alpha'}^B - T_{\alpha'\beta'}\mu_{\alpha'}^A \hat{\mu}_{\beta'}^B + \cdots|0\rangle}{W_n^B - W_0^B} \\ &= -(q^A T_\alpha - \mu_\beta^A T_{\alpha\beta} + \cdots)\sum_{n\neq 0} \frac{\langle 0|\hat{\mu}_\alpha^B|n\rangle\langle n|\hat{\mu}_{\alpha'}^B|0\rangle}{W_n^B - W_0^B}(q^A T_{\alpha'} - \mu_{\beta'}^A T_{\alpha'\beta'} + \cdots). \end{aligned}$$

Now we can recognize here the sum-over-states expression for the polarizability $\alpha_{\alpha\alpha'}$ (see eqn (2.3.2)), so that the induction energy is

$$U_{\text{ind}}^B = -\tfrac{1}{2}(q^A T_\alpha - \mu_\beta^A T_{\alpha\beta} + \cdots)\alpha_{\alpha\alpha'}^B (q^A T_{\alpha'} - \mu_{\beta'}^A T_{\alpha'\beta'} + \cdots). \qquad (4.2.1)$$

But the expression $(q^A T_\alpha - \mu_\beta^A T_{\alpha\beta} + \cdots)$ is merely minus the electric field at B due to molecule A, so the induction energy is $-\frac{1}{2}F_\alpha^A(\mathbf{B})F_{\alpha'}^A(\mathbf{B})\alpha_{\alpha\alpha'}^B$, exactly as we would expect from a straightforward classical treatment of the field. The only part played by the quantum mechanical perturbation theory is to provide the formula for the polarizability.

In this derivation, we ignored all terms in the perturbation except for some of the ones involving the dipole operator for molecule B. It is clear by analogy or by explicit

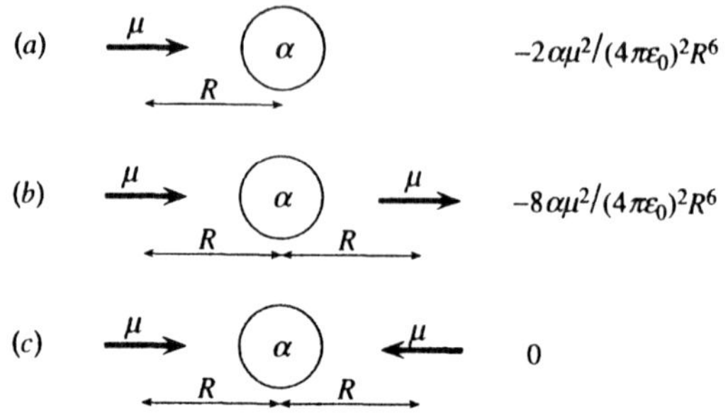

FIG. 4.1. Non-additivity of the induction energy

calculation that we shall have other terms in the induction energy, involving the dipole–quadrupole polarizability, the quadrupole–quadrupole polarizability, and so on, so that the induction energy takes the form

$$\begin{aligned} U^B_{\text{ind}} &= -\tfrac{1}{2}F^A_\alpha(\mathbf{B})F^A_{\alpha'}(\mathbf{B})\alpha^B_{\alpha\alpha'} - \tfrac{1}{3}F^A_\alpha(\mathbf{B})F^A_{\alpha'\beta'}(\mathbf{B})A^B_{\alpha,\alpha'\beta'} \\ &\quad - \tfrac{1}{6}F^A_{\alpha\beta}(\mathbf{B})F^A_{\alpha'\beta'}(\mathbf{B})C^B_{\alpha\beta,\alpha'\beta'} - \cdots. \end{aligned} \tag{4.2.2}$$

The simplest case arises when A is a spherical ion, like Na^+. Then the field at B is $-qT_\alpha = qR_\alpha/(4\pi\varepsilon_0 R^3)$. For the case where A is at the origin and B at $(0,0,z)$ the field is $F_z = q/(4\pi\varepsilon_0 z^2)$ and the induction energy is $q^2\alpha^B_{zz}/((4\pi\varepsilon_0)^2 z^4)$. In this case, therefore, the energy is proportional to R^{-4}. If A is neutral but polar ($\mu \neq 0$), the field is proportional to R^{-3} and the energy to R^{-6}. Notice that the induction energy is always negative.

4.2.1 *Non-additivity of the induction energy*

A very important feature of the induction energy emerges when we consider the case of a molecule surrounded by several others. The induction energy still takes the same form, i.e., $-\frac{1}{2}F_\alpha(\mathbf{B})F_{\alpha'}(\mathbf{B})\alpha^B_{\alpha\alpha'}$, but the field is now the total field due to the other molecules. (The simplest way to view this is to regard the rest of the system as comprising 'molecule A'.) However the same result can be obtained by working through the perturbation theory using as perturbation the sum of all intermolecular interactions, and again picking out the terms in which only molecule B is excited.) Consider two contrasting situations (Fig. 4.1). In Fig. 4.1a we have molecule B in the field of a single polar neighbour, so that $F(\mathbf{B}) = 2\mu/4\pi\varepsilon_0 R^3$, and the induction energy is $-2\alpha\mu^2/(4\pi\varepsilon_0)^2R^6$. In Fig. 4.1b there are two polar neighbours, aligned so that their fields are in the same direction at B. In this case the total field is twice as big as before, $F(\mathbf{B}) = 4\mu/4\pi\varepsilon_0 R^3$, and the induction energy is $-8\alpha\mu^2/(4\pi\varepsilon_0)^2R^6$, four times as big. In Fig. 4.1c there are again two polar neighbours, but aligned this time so that their fields are in opposite directions at B. Now the total field is zero and so is the induction energy.

This simple example illustrates dramatically the severe non-additivity of the induction energy. Both types of situation may occur. For example, the local environment of an ion in an ionic crystal is often centrosymmetric, so that the electric field has to vanish, and in such cases there is no dipole–dipole contribution to the induction energy. In molecular crystals and liquids, the opposite case is common; for example, in the tetrahedral structure found in ice and (approximately) in liquid water, there are two proton donors and two proton acceptors near each molecule, and the fields of these neighbours tend to add rather than cancel. In the case of induction energy, then, the assumption of pairwise additivity fails totally.

If we can simply add together the fields arising from the neighbouring molecules, the calculation of the induction energy would not present any difficulties, even though it is quadratic in the field. Unfortunately this also may fail. The simplest way to see this is to follow the derivation first given by Silberstein (1917) of the polarizability of a pair of neighbouring atoms. Two spherical atoms, with polarizabilities α^A and α^B are placed a distance R apart on the z axis. An external field F in the z direction polarizes both atoms, inducing a dipole moment in both of them. The induced dipole of each atom produces an additional field at the other, and this must be added to the applied field. So the equations for the induced moments are

$$\begin{aligned}\mu^A &= \alpha^A(F + 2\mu^B/(4\pi\varepsilon_0 R^3)),\\ \mu^B &= \alpha^B(F + 2\mu^A/(4\pi\varepsilon_0 R^3)).\end{aligned} \tag{4.2.3}$$

These equations are easily solved to give

$$\mu^A = \alpha^A F \frac{1 + 2\alpha^B/(4\pi\varepsilon_0 R^3)}{1 - 4\alpha^A\alpha^B/(4\pi\varepsilon_0)^2 R^6}, \quad \mu^B = \alpha^B F \frac{1 + 2\alpha^A/(4\pi\varepsilon_0 R^3)}{1 - 4\alpha^A\alpha^B/(4\pi\varepsilon_0)^2 R^6}. \tag{4.2.4}$$

So the effective field experienced by each atom is enhanced, and so is the induction energy. This is the case of a uniform external field, but much the same applies when the field is due to other molecules. It follows that we cannot in general expect to be able to add together the fields due to the static moments of the other molecules in order to calculate the induction energy. It may however be a reasonable *approximation* to do so, especially if the molecules are not very polar or not very polarizable.

In other geometries, the secondary effects of induction can reduce the magnitude of the induction energy. If the external field in the above example is in the x direction rather than the z direction, the equations for the moments become

$$\begin{aligned}\mu^A &= \alpha^A(F - \mu^B/(4\pi\varepsilon_0 R^3)),\\ \mu^B &= \alpha^B(F - \mu^A/(4\pi\varepsilon_0 R^3)),\end{aligned} \tag{4.2.5}$$

and the induced moments are

$$\mu^A = \alpha^A F \frac{1 - \alpha^B/(4\pi\varepsilon_0 R^3)}{1 - \alpha^A\alpha^B/(4\pi\varepsilon_0)^2 R^6}, \quad \mu^B = \alpha^B F \frac{1 - \alpha^A/(4\pi\varepsilon_0 R^3)}{1 - \alpha^A\alpha^B/(4\pi\varepsilon_0)^2 R^6}. \tag{4.2.6}$$

However the electrostatic energy often favours structures in which the induction energy is enhanced rather than reduced, and of course the induction energy itself favours such structures. We return to the question of cooperative induction effects in Chapter 9.

4.2.2 *Multipole expansion of the induction energy*

A more subtle characteristic of the induction energy emerges when we consider the multipole expansion in more detail. It is enough to consider the simplest case of a hydrogen-like atom A with nuclear charge Z in the field of a proton B. We take $Z > 1$ to avoid resonance effects (see Chapter 10). The perturbation is

$$\mathcal{H}' = \frac{Z}{R} - \frac{1}{|\mathbf{R} - \mathbf{r}|}, \tag{4.2.7}$$

where **R** is the position of the proton and **r** the position of the electron relative to an origin at the nucleus of atom A. Expressed as a multipole series using eqn (3.3.2) on p. 41, this becomes

$$\mathcal{H}' = \frac{1}{R}\left[Z - \sum_{n=0}^{\infty}\left(\frac{r}{R}\right)^n P_n(\cos\theta)\right], \tag{4.2.8}$$

where r and θ are polar coordinates for the electron in a coordinate system in which the proton is on the z axis.

Dalgarno and Lynn (1957*a*) were able to find an exact expression for the second-order energy of the atom under this perturbation. Their result is

$$W'' = -\sum_{n=1}^{\infty} \frac{(2n+2)!(n+2)}{n(n+1)(ZR)^{(2n+2)}}, \tag{4.2.9}$$

plus some terms that decrease exponentially as R increases; the latter describe the penetration term, which we shall discuss in Chapter 6.

Now we can use the standard ratio test to study the convergence of this series. The ratio of successive terms is $2n(n+3)(2n+3)/[(n+2)(ZR)^2]$, and it becomes infinite as $n \to \infty$, for any value of R. Consequently the multipole series expansion of the induction energy diverges for any R.

This is a very disturbing result. It suggests that in general the multipole expansion of the interaction energy may not converge. However Ahlrichs (1976), in a more rigorous treatment of earlier work by Brooks (1952), proved that it is always semi-convergent (asymptotically convergent). Specifically, he showed that the interaction energy can be expressed in the form

$$U_{AB}(R) = W_{AB}(R) - W_A - W_B = \sum_{\nu=0}^{N} U_\nu R^{-\nu} + O(R^{-N-1}), \tag{4.2.10}$$

so that the remainder left after truncating the multipole series at any term tends to zero as the intermolecular distance tends to infinity. Ahlrichs also showed that the exchange terms, which arise from overlap of the molecular wavefunctions and which will be discussed in Chapter 6, decrease faster with increasing R than any power of $1/R$. The upshot of all this is that although the multipole expansion cannot be shown to converge at any fixed value of R, and can be shown in some cases to diverge for all R, it gives results that can be made arbitrarily accurate if the intermolecular distance is large enough. In practice, even at distances that are not very large, useful results may be obtained by truncating the multipole series.

4.3 The dispersion energy

4.3.1 *Drude model*

We turn to the dispersion energy. Because this is a wholly non-classical phenomenon, it may be helpful in understanding it to contemplate a simple model, first introduced by

London (1930*b*) (see London (1937) for a version in English). We represent each atom by a Drude model: a one-dimensional harmonic oscillator in which the electron cloud, mass m, is bound to the nucleus by a harmonic potential with force constant k. The Hamiltonian for each atom takes the form $p^2/2m + \frac{1}{2}kx^2$. There is an atomic dipole proportional to the displacement x of the electrons from the nucleus, and for two adjacent atoms the interaction energy of these dipoles is proportional to the product of the dipole moments. Accordingly the Hamiltonian for the complete system takes the form

$$\mathcal{H} = \frac{1}{2m}(p_A^2 + p_B^2) + \frac{1}{2}k(x_A^2 + x_B^2 + 2cx_Ax_B),$$

where c is some coupling constant. This can be separated into two uncoupled oscillators corresponding to the normal modes $x_A \pm x_B$:

$$\begin{aligned}\mathcal{H} = &\frac{1}{4m}(p_A + p_B)^2 + \frac{1}{4}k(1+c)(x_A + x_B)^2 \\ &+ \frac{1}{4m}(p_A - p_B)^2 + \frac{1}{4}k(1-c)(x_A - x_B)^2.\end{aligned}$$

The normal mode frequencies are $\omega_\pm = \omega_0\sqrt{1 \pm c}$, where $\omega_0 = \sqrt{k/m}$ is the frequency of an isolated oscillator.

Now a classical system would have a minimum energy of zero, whether coupled or not. The quantum system however has zero-point energy, and the energy of the coupled system in its lowest state is

$$\tfrac{1}{2}\hbar(\omega_+ + \omega_-) = \tfrac{1}{2}\hbar\omega_0(\sqrt{1+c} + \sqrt{1-c}) = \tfrac{1}{2}\hbar\omega_0[2 - \tfrac{1}{4}c^2 - \tfrac{5}{64}c^4 - \cdots],$$

so that there is an energy lowering, compared with the zero-point energy $\hbar\omega_0$ of two uncoupled oscillators, of

$$U_{\text{disp}} = -\hbar\omega_0\left(\frac{1}{8}c^2 + \frac{5}{128}c^4 + \cdots\right).$$

The interaction between the atoms leads to a correlation between the motion of their electrons, and this manifests itself in a lowering of the energy. Notice that the energy depends on the square of the coupling constant c and is always negative; regardless of the sign of the coupling, the correlation between the motion of the electrons leads to a lowering of the energy. For interacting dipoles, the coupling c is proportional to R^{-3}, so the leading term in the dispersion energy is proportional to R^{-6}.

At very large separations, this picture changes slightly because the speed of light is finite. The correlation between the fluctuations in the two molecules becomes less effective, because the information about a fluctuation in the charge distribution of one molecule can only be transmitted to the other molecule at the speed of light. By the time that the other molecule has responded and the information about its response has reached the first molecule again, the electrons have moved, so that the fluctuations are no longer in phase. Detailed calculation (Casimir and Polder 1948) shows that this 'retardation' effect becomes important when the separation R is large compared with the wavelength λ_0

corresponding to the characteristic absorption frequency of the molecule, and the dispersion energy is then reduced by a factor of the order of λ_0/R and becomes proportional to R^{-7}. However this only occurs at much larger distances than we are concerned with, since λ_0 is typically several thousand Ångstrom.

4.3.2 *Quantum-mechanical formulation*

We return to the perturbation-theory expression, eqn (4.1.14). For simplicity we consider first only the dipole–dipole term in $\mathcal{H}'$:

$$
\begin{aligned}
U^{(6)}_{\text{disp}} &= -\sum_{m_A\neq 0}\sum_{n_B\neq 0} \frac{\langle 0_A 0_B|\hat{\mu}^A_\alpha T_{\alpha\beta}\hat{\mu}^B_\beta|m_A n_B\rangle\langle m_A n_B|\hat{\mu}^A_\gamma T_{\gamma\delta}\hat{\mu}^B_\delta|0_A 0_B\rangle}{W^A_{m0}+W^B_{n0}} \\
&= -T_{\alpha\beta}T_{\gamma\delta}\sum_{m_A\neq 0}\sum_{n_B\neq 0}\left(\frac{1}{W^A_{m0}+W^B_{n0}}\right. \\
&\qquad \left.\times\,\langle 0_A|\hat{\mu}^A_\alpha|m_A\rangle\langle m_A|\hat{\mu}^A_\gamma|0_A\rangle\langle 0_B|\hat{\mu}^B_\beta|n_B\rangle\langle n_B|\hat{\mu}^B_\delta|0_B\rangle\right),
\end{aligned}
\tag{4.3.1}
$$

where $W^A_{m0} = W^A_m - W^A_0$. This is an inconvenient expression to deal with, because although the matrix elements can be factorized into terms referring to A and terms referring to B, the denominator cannot. There are two commonly used ways to handle it. One was first introduced by London (1930*b*), and uses the *Unsöld* or *average-energy* approximation (Unsöld 1927). Here we first write (4.3.1) as

$$
\begin{aligned}
U^{(6)}_{\text{disp}} &= -T_{\alpha\beta}T_{\gamma\delta}\sum_{m_A\neq 0}\sum_{n_B\neq 0}\frac{W^A_{m0}W^B_{n0}}{W^A_{m0}+W^B_{n0}} \\
&\qquad\times\frac{\langle 0_A|\hat{\mu}^A_\alpha|m_A\rangle\langle m_A|\hat{\mu}^A_\gamma|0_A\rangle}{W^A_{m0}}\,\frac{\langle 0_B|\hat{\mu}^B_\beta|n_B\rangle\langle n_B|\hat{\mu}^B_\delta|0_B\rangle}{W^B_{n0}}.
\end{aligned}
\tag{4.3.2}
$$

If it were not for the inconvenient factor $W^A_{m0}W^B_{n0}/(W^A_{m0}+W^B_{n0})$, we should be able to factor out the sum-over-states expressions for $\alpha^A_{\alpha\gamma}$ and $\alpha^B_{\beta\delta}$. We therefore approximate this factor using average excitation energies U_A and U_B. Following Buckingham (1967), we use the Unsöld approximation in the form

$$
\frac{W^A_{m0}W^B_{n0}}{W^A_{m0}+W^B_{n0}} = \frac{U_A U_B}{(U_A+U_B)}(1+\Delta_{mn}),
\tag{4.3.3}
$$

where

$$
\Delta_{mn} = \frac{1/U_A - 1/W^A_{m0} + 1/U_B - 1/W^B_{n0}}{1/W^A_{m0}+1/W^B_{n0}}.
\tag{4.3.4}
$$

This is an identity for a particular m and n. If now we can choose U_A and U_B so that Δ_{mn} becomes negligible for all m and n (which requires that all the states $|m_A\rangle$ that make important contributions have excitation energies close to the average energy U_A, and likewise for the $|n_B\rangle$) then eqn (4.3.1) simplifies to

$$U_{\text{disp}}^{(6)} \approx -\frac{U_A U_B}{4(U_A+U_B)} T_{\alpha\beta} T_{\gamma\delta} \alpha_{\alpha\gamma}^A \alpha_{\beta\delta}^B. \tag{4.3.5}$$

For atoms, where $\alpha_{\alpha\gamma}$ reduces to $\overline{\alpha}\delta_{\alpha\gamma}$, this becomes

$$\begin{aligned} U_{\text{disp}}^{(6)} &\approx -\frac{U_A U_B}{4(U_A+U_B)} \overline{\alpha}^A \overline{\alpha}^B T_{\alpha\beta} T_{\alpha\beta} \\ &= -\frac{3U_A U_B}{2(U_A+U_B)} \frac{\overline{\alpha}^A \overline{\alpha}^B}{(4\pi\varepsilon_0)^2 R^6}. \end{aligned} \tag{4.3.6}$$

Here we have used the fact that

$$T_{\alpha\beta} T_{\alpha\beta} = \frac{(3R_\alpha R_\beta - R^2\delta_{\alpha\beta})(3R_\alpha R_\beta - R^2\delta_{\alpha\beta})}{(4\pi\varepsilon_0)^2 R^{10}} = \frac{6}{(4\pi\varepsilon_0)^2 R^6}. \tag{4.3.7}$$

Eqn (4.3.6) is the London formula for the dispersion energy between two atoms. The same formula can be used for molecules, when it gives the dispersion interaction averaged over relative orientations of the two molecules. It shows that the orientation-averaged dispersion energy takes the form* $-C_6 R^{-6}$, and that it is larger between more polarizable atoms, but it is evidently not a practical method for determining the dispersion coefficient C_6, since we have no way to determine U_A and U_B. It is usual to put them equal to the ionization energies, but this gives no more than a rough approximation to C_6. Alternatively, we can put U_A and U_B equal to the lowest excitation energies of A and B respectively to get an upper bound to the magnitude of the dispersion energy. Further possibilities include the Slater–Kirkwood formula, in which U_A is estimated as $(N_A/\alpha_A)^{1/2}$; here N_A is an effective number of valence electrons for molecule A, but as various approximations are built into this estimate the value of N_A often has to be much larger than one would expect (Pitzer 1959).

An alternative approach due to Casimir and Polder (1948) yields an exact formula which is much more useful. It depends on the identity

$$\frac{1}{A+B} = \frac{2}{\pi} \int_0^\infty \frac{AB}{(A^2+\nu^2)(B^2+\nu^2)} \mathrm{d}\nu, \tag{4.3.8}$$

which is valid for positive A and B, and which can be established by a simple contour integration. Applying this formula to the energy denominator in eqn (4.3.1), which we write as $\hbar(\omega_m^A + \omega_n^B)$, we get

$$\begin{aligned} U_{\text{disp}}^{(6)} &= -\frac{2\hbar}{\pi} T_{\alpha\beta} T_{\gamma\delta} \\ &\times \int_0^\infty {\sum_m}' \frac{\langle 0_A|\hat{\mu}_\alpha^A|m_A\rangle\langle m_A|\hat{\mu}_\gamma^A|0_A\rangle\omega_m^A}{\hbar((\omega_m^A)^2+\nu^2)} {\sum_n}' \frac{\omega_n^B\langle 0_B|\hat{\mu}_\beta^B|n_B\rangle\langle n_B|\hat{\mu}_\delta^B|0_B\rangle}{\hbar((\omega_n^B)^2+\nu^2)} \mathrm{d}\nu. \end{aligned} \tag{4.3.9}$$

*The sign of the dispersion coefficient C_6 is not a settled convention. Some authors use a convention in which C_6 is positive, in which case a minus sign appears in the expression for the dispersion energy, while others take C_6 to be negative. The definition of C_6 as a positive quantity avoids some irritating semantic difficulties over whether a 'larger' value for a negative quantity is larger in magnitude or less negative, i.e., smaller in magnitude.

Now we saw in Chapter 2 using time-dependent perturbation theory that the response to an oscillating electric field $F_\beta e^{-i\omega t}$ at a frequency ω that is far from any absorption frequencies is an oscillating dipole moment $\mu_\alpha = \alpha_{\alpha\beta}(\omega)F_\beta e^{-i\omega t}$, where

$$\alpha_{\alpha\beta}(\omega) = \sum_m{}' \frac{\omega_m\{\langle 0|\hat{\mu}_\alpha|m\rangle\langle m|\hat{\mu}_\beta|0\rangle + \langle 0|\hat{\mu}_\beta|m\rangle\langle m|\hat{\mu}_\alpha|0\rangle\}}{\hbar(\omega_m^2 - \omega^2)}. \tag{4.3.10}$$

We see that the dispersion energy can now be expressed in terms of the polarizability at the imaginary frequency iv:

$$U_{\text{disp}}^{(6)} = -\frac{\hbar}{2\pi}T_{\alpha\beta}T_{\gamma\delta}\int_0^\infty \alpha_{\alpha\gamma}^A(iv)\alpha_{\beta\delta}^B(iv)\,dv, \tag{4.3.11}$$

where

$$\alpha_{\alpha\gamma}^A(iv) = \sum_m{}' \frac{\omega_m\{\langle 0|\hat{\mu}_\alpha|m\rangle\langle m|\hat{\mu}_\gamma|0\rangle + \langle 0|\hat{\mu}_\gamma|m\rangle\langle m|\hat{\mu}_\alpha|0\rangle\}}{\hbar(\omega_m^2 + v^2)}. \tag{4.3.12}$$

For spherical atoms, eqn (4.3.11) reduces in the same way as before to give

$$\overline{U}_{\text{disp}}^{(6)} = -\frac{3\hbar}{(4\pi\varepsilon_0)^2\pi R^6}\int_0^\infty \bar{\alpha}^A(iv)\bar{\alpha}^B(iv)\,dv. \tag{4.3.13}$$

The concept of 'polarizability at imaginary frequency' is physically very bizarre. It can be thought of as describing the response to an exponentially increasing electric field, but this is stretching physical interpretation to perhaps unreasonable limits, and it is better to view it merely as a mathematical construct. At any rate, its mathematical properties are much more civilized than those of the polarizability at real frequencies, because the denominators have no zeros and increase monotonically with v. Accordingly, $\alpha_{\alpha\gamma}^A(iv)$ decreases monotonically from the static polarizability when $v = 0$, to zero as $v \to \infty$. This means that it can be determined quite accurately as a function of frequency, either by *ab initio* computation or from experimental data. In the case of *ab initio* computation, the values of $\alpha_{\alpha\gamma}^A(iv)$ and $\alpha_{\beta\delta}^B(iv)$ are computed at a number of suitable values of v, and the integral 4.3.11 is evaluated by numerical quadrature. Because the polarizability at imaginary frequency is mathematically well-behaved, the quadrature gives accurate results using a modest number of integration points—typically 20 or so. The polarizabilities themselves are unfortunately more difficult to calculate accurately; the calculations require large basis sets in order that the distortion due to the applied field can be represented accurately. The best calculations of dispersion energies by this method are probably those of Wormer and his colleagues (Rijks and Wormer 1989, Thakkar *et al.* 1992, Wormer and Hettema 1992).

Tang (1969) gave a modified version of the Slater–Kirkwood formula, based on a one-term approximation to the polarizability at imaginary frequency:

$$\tilde{\alpha}^A(iv) = \frac{\alpha^A(0)}{1 + (v/\eta_A)^2},$$

where $\alpha^A(0)$ is the static polarizability and η_A is a suitable parameter. η_A can be chosen so that $\tilde{\alpha}^A(iv)$ gives the correct value for C_6^{AA} when substituted into eqn (4.3.13); this requires that $\eta_A = \frac{4}{3}C_6^{AA}/(\alpha^A(0))^2$. Using values of $\tilde{\alpha}^A(\mathrm{i}v)$ and $\tilde{\alpha}^B(\mathrm{i}v)$ obtained in this way leads to an expression for C_6^{AB}:

$$C_6^{AB} = \frac{2\alpha^A(0)\alpha^B(0)C_6^{AA}C_6^{BB}}{(\alpha^A(0))^2C_6^{BB} + \alpha^B(0)^2C_6^{AA}}. \tag{4.3.14}$$

Note that although eqn (4.3.13) gives the orientational average of the R^{-6} term in the dispersion energy, the method can be used just as easily to obtain the orientation-dependent terms. For linear molecules the complete R^{-6} dispersion interaction takes the form (Stone and Tough 1984)

$$\begin{aligned} U_{\text{disp}}^6 = -\frac{C_6}{R^6}\{ & 1 + \gamma_{20}(\tfrac{3}{2}\cos^2\theta_A - \tfrac{1}{2}) + \gamma_{02}(\tfrac{3}{2}\cos^2\theta_B - \tfrac{1}{2}) \\ & + \gamma_{22}\cdot\tfrac{1}{2}[(2\cos\theta_A\cos\theta_B - \sin\theta_A\sin\theta_B\cos\varphi)^2 \\ & - \cos^2\theta_A - \cos^2\theta_B]\}, \end{aligned} \tag{4.3.15}$$

where the angles θ_A, θ_B and φ are defined as in §3.2 and

$$C_6 = \frac{3\hbar}{\pi(4\pi\varepsilon_0)^2}\int_0^\infty \overline{\alpha}^A(\mathrm{i}v)\overline{\alpha}^B(\mathrm{i}v)\,\mathrm{d}v, \tag{4.3.16}$$

$$\gamma_{20}C_6 = \frac{\hbar}{\pi(4\pi\varepsilon_0)^2}\int_0^\infty \Delta\alpha^A(\mathrm{i}v)\overline{\alpha}^B(\mathrm{i}v)\,\mathrm{d}v, \tag{4.3.17}$$

$$\gamma_{02}C_6 = \frac{\hbar}{\pi(4\pi\varepsilon_0)^2}\int_0^\infty \overline{\alpha}^A(\mathrm{i}v)\Delta\alpha^B(\mathrm{i}v)\,\mathrm{d}v, \tag{4.3.18}$$

$$\gamma_{22}C_6 = \frac{\hbar}{\pi(4\pi\varepsilon_0)^2}\int_0^\infty \Delta\alpha^A(\mathrm{i}v)\Delta\alpha^B(\mathrm{i}v)\,\mathrm{d}v. \tag{4.3.19}$$

As usual, $\overline{\alpha} = \frac{1}{3}(\alpha_{xx} + \alpha_{yy} + \alpha_{zz})$ and $\Delta\alpha = \alpha_{zz} - \frac{1}{2}(\alpha_{xx} + \alpha_{yy})$.

The important feature of this treatment is that except for a simple one-dimensional quadrature the expressions for the dispersion energy coefficients reduce to products of individual-molecule properties. These need only be calculated once, and the resulting dispersion coefficients do not depend on the intermolecular geometry.

4.3.3 *Spherical tensor formulation*

If we use the full spherical-tensor form of the interaction Hamiltonian, we obtain, instead of eqn (4.3.1), the expression

$$U_{\text{disp}}^{(6)} = -\sum_{m_A\neq 0}\sum_{n_B\neq 0}\frac{\langle 0_A0_B|\hat{Q}_t^A T_{tu}^{AB}\hat{Q}_u^B|m_An_B\rangle\langle m_An_B|\hat{Q}_{t'}^A T_{t'u'}^{AB}\hat{Q}_{u'}^B|0_A0_B\rangle}{W_{m0}^A + W_{n0}^B} \tag{4.3.20}$$

(where there is an implied summation over the repeated spherical-tensor suffixes t, t', u and u') and by applying the Casimir–Polder identity exactly as before, we obtain, instead of (4.3.9),

$$U_{\text{disp}}^{(6)} = -\frac{2\hbar}{\pi}T_{tu}^{AB}T_{t'u'}^{AB}\times$$

$$\int_0^\infty \sum_m{}' \frac{\langle 0_A|\hat{Q}_t^A|m_A\rangle\langle m_A|\hat{Q}_{t'}^A|0_A\rangle\omega_m^A}{\hbar((\omega_m^A)^2+v^2)} \sum_n{}' \frac{\omega_n^B\langle 0_B|\hat{Q}_u^B|n_B\rangle\langle n_B|\hat{Q}_{u'}^B|0_B\rangle}{\hbar((\omega_n^B)^2+v^2)}\,\mathrm{d}v, \tag{4.3.21}$$

and as before we identify components of the polarizability at imaginary frequency to obtain, instead of (4.3.11), the general form

$$U_{\text{disp}} = -\frac{\hbar}{\pi}T_{tu}^{AB}T_{t'u'}^{AB}\int_0^\infty \alpha_{tt'}^A(\mathrm{i}v)\alpha_{uu'}^B(\mathrm{i}v)\,\mathrm{d}v. \tag{4.3.22}$$

Remember that there is still a sum over t, t', u and u'.

When t, t', u and u' describe dipole components $10 = z$, $11c = x$ or $11s = y$, we have the dipole–dipole contribution to the dispersion energy, almost as before. There is a subtle but important difference: the components $\alpha_{tt'}^A$ and $\alpha_{uu'}^B$ are referred to *molecular* axes, so that the orientation dependence in (4.3.22) is all in the T_{tu}^{AB} and $T_{t'u'}^{AB}$, whereas $\alpha_{\alpha\gamma}^A$ and $\alpha_{\beta\delta}^B$ in eqn (4.3.11) are referred to *global* axes.

If t describes a quadrupole component and the rest are dipole components, then $\alpha_{tt'}^A$ becomes a quadrupole–dipole polarizability. In this case T_{tu}^{AB} is a quadrupole–dipole interaction function, proportional to R^{-4}, while $T_{t'u'}^{AB}$ is a dipole–dipole function as before, proportional to R^{-3}. Accordingly we now have a dispersion term proportional to R^{-7}. There are other R^{-7} terms involving dipole–quadrupole polarizabilities on B. If A and B are atoms or centrosymmetric molecules, these dipole–quadrupole polarizabilities are zero, so there is no R^{-7} term in the dispersion interaction in such cases. Even for non-centrosymmetric molecules, the coefficient C_7 of the R^{-7} term depends on the relative orientation of the molecules, and its average over orientations is zero, if all orientations have equal weight.

The next term, in R^{-8}, involves a quadrupole–quadrupole polarizability on one molecule and a dipole–dipole polarizability on the other, and this is non-zero for atoms as well as molecules. For non-centrosymmetric molecules there can also be an R^{-8} contribution from dipole–quadrupole polarizabilities on both A and B. The R^{-9} term is zero for atoms, like the R^{-7} term, because all the polarizabilities that might contribute to it are zero, so the next term for atoms is the R^{-10} term. Here there are two kinds of contribution, one involving a quadrupole–quadrupole polarizability on both atoms, and the other a dipole–dipole polarizability on one and an octopole–octopole polarizability on the other.

4.3.4 *Numerical values*

The numerical value of the dispersion coefficient C_6 increases sharply with molecular size—roughly as the square of the polarizability, as the London formula shows. Some values obtained by Meath and his collaborators using dipole oscillator strength distributions (see Chapter 12) are shown for illustration in Table 4.1. These are in atomic units and are believed to be accurate to about 1%. The atomic unit for the C_6 coefficient is $E_h a_0^6 = e^2 a_0^5/4\pi\varepsilon_0$, but a variety of other units is in use, commonly (energy unit)× Å^6.

The London formula shows that for interactions between unlike molecules, the combining rule $C_6^{AB} \approx (C_6^{AA}C_6^{BB})^{1/2}$ should be quite reliable. For Ar···Xe, for instance, the combining rule gives 135.6 a.u., and for CO···He it gives 10.9 a.u. These numbers are

Table 4.1 *Some values of dispersion coefficients.*

System	C_6/a.u.	C_8/a.u.	C_{10}/a.u.
He···He	1.46	13.9	182
Ne···Ne	6.6	57	700
Ar···Ar	64.3	1130	25000
Kr···Kr	133	2500	60000
Xe···Xe	286		
CO···CO	81.4		
SO_2···SO_2	294		
CS_2···CS_2	871		
HCCH···HCCH	204.1		
C_6H_6···C_6H_6	1723		
He···CO	10.7		
Ar···Xe	134.5		
He···HCCH	16.50		
HCCH···C_6H_6	593.0		

quite close to the accurate values in the Table. Similarly, the combining rule works very well for acetylene and benzene. Where the average excitation energy U is very different for the two molecules, it is likely to be less accurate; thus the combining rule gives $C_6 = 17.26$ a.u. for He···HCCH, whereas the accurate value is 16.50 a.u. In all these cases the Tang formula, eqn (4.3.14), gives rather better values: 134.1 a.u. for Ar···Xe, 10.55 a.u. for CO···He and 16.17 a.u. for He···HCCH, calculated using the static polarizabilities listed by Gray and Gubbins (1984).

The higher dispersion coefficients C_8, C_{10}, etc, may be calculated in much the same way as C_6. If the London approach is used, the result for the isotropically averaged C_8 is

$$C_8 = \frac{15U_AU_B(\bar{\alpha}^A\bar{\alpha}_2^B + \bar{\alpha}_2^A\bar{\alpha}^B)}{8(4\pi\varepsilon_0)^2(U_A+U_B)},$$

where $\bar{\alpha}_2^A$ and $\bar{\alpha}_2^B$ are the average quadrupole–quadrupole polarizabilities, $\frac{1}{5}\sum_\kappa \alpha_{2\kappa,2\kappa}$. If the same values of U_A and U_B are appropriate here as for C_6, the ratio $C_8/C_6 \approx 5\bar{\alpha}_2/2\bar{\alpha}$. Some numerical values are given in Table 4.1. For Ar, for instance, the ratio $C_8/C_6 = 17.6$, so the C_8/R^8 term becomes equal to C_6/R^6 when $R \approx 4.2\,a_0$. The equilibrium separation R_m for pairs of Ar atoms is $7.1\,a_0$, and at this distance the R^{-8} term is about one third the size of the R^{-6} term, while the R^{-10} term is about a factor of three smaller again. These contributions are by no means insignificant, and have to be included in accurate calculations. However they are reduced by damping effects, which will be discussed in Chapter 6. If higher values are required, an approximate recursion formula can be used (Tang and Toennies 1978):

$$C_{2n+6} = \left(\frac{C_{2n+4}}{C_{2n+2}}\right)^3 C_{2n}.$$

Thakkar (1988) found that this formula gave values within a few per cent of the accurate values for H atoms.

5

AB INITIO METHODS

5.1 Introduction

The term '*ab initio*' is used to describe methods that seek to calculate molecular properties from the beginning, that is, by solving Schrödinger's equation without using any empirical data. Although it is always necessary to make approximations, it is possible to organize them in such a way that the accuracy of the calculation can be systematically improved, up to the limit allowed by available computing resources.

We shall not attempt to review the principles of *ab initio* methods, but the reader will need to know something of these principles to understand the issues that arise, some of which are quite technical. We give an outline here of the points that affect the calculations of intermolecular forces; for a full account, see for example the book by Szabo and Ostlund (1989).

First, we note that the Born–Oppenheimer approximation is used almost universally. That is, the behaviour of the electrons is studied as if the nuclei were clamped in a given geometry. Solving the Schrödinger equation for the electrons then provides the electronic energy for that geometry, together with a wavefunction from which we can calculate properties such as the molecular multipole moments. The energy and the other properties are functions of the nuclear coordinates; in particular, we speak of the 'potential energy surface', which is the electronic energy as a function of nuclear coordinates. This is the potential energy term in the Schrödinger equation describing the nuclear motion. This separation of electronic and nuclear motion is a very good approximation except where there are near-degeneracies or crossings of electronic states.

Secondly, all standard *ab initio* programs ignore relativistic effects of all kinds. The neglect of relativistic effects (principally spin–orbit coupling) is usually a good approximation for calculations on molecules containing only light atoms, especially for closed-shell states. In the rare cases where it is important, it can be taken into account by perturbation theory.

Ab initio methods are used in the calculation of intermolecular forces in several ways. We have seen that forces at long range depend on molecular properties—multipole moments and polarizabilities—so we need accurate values of these quantities. Alternatively, we might seek the interaction energy as the difference of the energies of the interacting system and the separate molecules, all calculated *ab initio*. This is the so–called 'Supermolecule Method'. Finally, we can use perturbation theory for systems at distances where the long-range theory becomes invalid because the molecules overlap. In this case, it is no longer possible to write down analytical formulae for the energy, and *ab initio* methods have to be used to evaluate the energy terms. Perturbation methods of this sort are discussed in Chapter 6.

5.2 Basis sets

In any *ab initio* calculation it is necessary to select a set of basis functions. The choice is a compromise; generally one wishes to use the largest possible basis, but the cost of the calculation increases very sharply with the size of the basis. One reason for this—the main reason, in the case of self-consistent-field calculations—is the need to calculate the two-electron integrals. These have the form

$$(ij|kl) = \int\int \chi_i(1)^*\chi_j(1)\frac{1}{r_{12}}\chi_k(1)^*\chi_l(1)\,d\tau_1\,d\tau_2, \tag{5.2.1}$$

where the χ_i are the basis functions. Typically there are about $\frac{1}{8}N^4$ two-electron integrals for a calculation involving N basis functions, though the number is reduced if symmetry is taken into account, and for a large molecule, many of them may be negligibly small. If conventional methods are used, where the two-electron integrals are stored on disc and read from there as required, there is an absolute upper limit, set by the disc space available, to the size of the calculation. 'Direct' methods, in which the integrals are calculated anew every time they are needed, circumvent this limitation, but there is a corresponding increase in the computation time needed. This can be limited to some extent by avoiding the recalculation of integrals whose contribution to the calculation is negligible, and modern direct methods use sophisticated schemes for computing as few integrals as possible.

Early calculations used 'Slater-type orbitals' as basis functions. These have the form

$$\chi = R_{lk}(\mathbf{r})\exp(-\zeta r), \tag{5.2.2}$$

where R_{lk} is a solid harmonic, and the argument $\mathbf{r} \equiv (r, \theta, \varphi)$ is the position of the electron relative to a nucleus. The form of such functions is suggested by the hydrogen-atom wavefunctions; in particular, the *s*-type functions ($l = 0$) have a cusp at the nucleus. However they are very awkward to handle computationally, and modern basis sets use functions of the form

$$\chi = R_{lk}(\mathbf{r})\sum_i c_i N_i \exp(-\alpha_i r^2), \tag{5.2.3}$$

where R_{lk} is the appropriate solid harmonic. The individual $N_i \exp(-\alpha_i r^2)$ (where N_i is a normalizing factor) are called 'primitive gaussian functions' and the c_i are expansion coefficients, usually chosen to optimize the energy of the isolated atom but fixed for the molecule or supermolecule calculation.

The minimum quality of basis set that can be contemplated for useful calculations of intermolecular potentials is at about the 6–31G* or 6–31G** level (Hehre *et al.* 1971, Hariharan and Pople 1973, Francl *et al.* 1982). The 6–31G* basis uses a single basis function for the core atomic orbitals, represented by a linear combination of six primitive gaussians. For the valence orbitals, two sets of basis functions are used, one with three primitives and a more diffuse one consisting of a single primitive. The asterisk indicates that a set of *d* functions is added for each heavy (non-hydrogen) atom. These 'polarization functions' provide for distortion of the atomic orbitals in the molecular environment. The double asterisk in 6–31G** indicates that polarization functions—*p* orbitals in this case—are used on hydrogen atoms as well as on the heavy atoms. In calculations on intermolecular interactions, polarization functions may often be omitted for hydrogen atoms that are not involved in the interaction.

These basis sets use the same primitive exponents α_i for both $2s$ and $2p$ orbitals. The slightly better 'double-zeta plus polarization' or DZP basis sets use different exponents for $2s$ and $2p$, and also use two basis functions for each core orbital. For small molecules it may be possible to use a TZ2P basis: 'triple–zeta plus double polarization'. A possible next step beyond this involves introducing higher–rank polarization functions: f functions on heavy atoms (i.e., other than hydrogen) and d functions on hydrogen; or it may be possible to improve the valence basis further, by introducing additional s and p functions.

For calculations of molecular properties other than the energy itself, additional considerations enter. For the dipole–dipole polarizability, for instance, the sum-over-states expression (2.3.2) shows that matrix elements of the dipole operator between ground and excited states are involved. In the dipole-allowed excited states, p electrons may be excited to d orbitals (as well as $p \rightarrow s$ and $s \rightarrow p$) so it is essential to include d orbitals in the basis, even if (as in the Ne atom) they are not needed to describe the ground state SCF wavefunction. Werner and Meyer (1976) showed that three sets of d functions are needed to obtain reasonably accurate dipole–dipole polarizabilities for small molecules containing first-row atoms. For quadrupole–quadrupole polarizabilities, $p \rightarrow f$ excitations become important, so f functions are required, and so on. This means that high-rank polarizabilities are very difficult to calculate accurately, even for small molecules.

In calculations of intermolecular potentials it is necessary to carry out a large number of calculations at different values of the intermolecular coordinates, so it is worth spending some time optimizing the basis for the individual molecules. Other things being equal, the basis that minimizes the energy of the isolated molecule is to be preferred, because this is likely to reduce the effects of basis set superposition error (see §5.5.1 on p. 70). Feller (1992) found that the inclusion of diffuse basis functions was particularly important in this respect. At the same time, it is often necessary to include basis functions that have no effect on the energy of the isolated system, because molecular properties like the polarizability are involved. For example, in complexes of an inert gas atom such as argon with a polar molecule, d functions must be included for the argon atom in order that it can respond to the electric field of its partner; otherwise the induction energy will be seriously underestimated.

It is important to ensure that the basis is 'balanced'; that is, that the quality is comparable for all the atoms in the problem, and that the quality of the valence basis and the polarization basis are not too disparate. This is unfortunately an ill-defined criterion. The kind of problem that arises with unbalanced basis sets can be illustrated by the case, already mentioned, of an inert gas atom such as argon interacting with a polar molecule or ion. At the SCF level, the description of the isolated argon atom is unaffected by the addition of d orbitals, because they have the wrong symmetry. However they are needed in the complex in order to describe the polarizability of the argon atom and hence the induction energy. The problem with balance arises because if the d basis is good but the valence basis not very good, the electron distribution can tip too far into the d orbitals under the influence of an electric field, because the electronic energy is artificially high in the poor-quality valence basis. The effect is to exaggerate the induction energy. Evidently a good valence basis with a poor polarization basis leads to an underestimate of the induction energy. Getting the balance right is unfortunately difficult to achieve in a general way, and as with most questions involving basis sets, the choice generally has to be based on experience and intuition. Most of the widely available *ab initio* programs have libraries of basis sets which have been found satisfactory in a range of applications.

5.3 The supermolecule method

There are numerous 'black box' programs available for calculating the energy of a system of electrons and nuclei, either *ab initio*–by solving Schrödinger's equation, making some approximations but using no empirical data—or by semi-empirical methods, in which the calculation is further simplified by additional approximations and the use of some empirical data. This offers the possibility of treating a system of two or more interacting molecules as just another assembly of electrons and nuclei—a 'supermolecule'. The interaction energy U_{AB} of a pair of molecules A and B is then just the energy of the supermolecule less the energies of the isolated molecules:

$$U_{AB} = W_{AB} - W_A - W_B. \tag{5.3.1}$$

Nothing apparently could be simpler than to use any of the available programs to calculate W_{AB}, W_A, and W_B, and hence to obtain the interaction energy. In fact the approach is a very useful one, but its application is fraught with difficulties, which we shall now discuss.

5.4 Electron correlation and intermolecular interactions

The workhorse of *ab initio* calculation is the self-consistent field or SCF procedure, but this is an independent-electron model—the electrons are treated as if they move in an average field due to the other electrons, and no account is taken of the effect of the electron repulsion in keeping electrons apart. The motion of the electrons is said to be *uncorrelated.* The simplest example of the consequences is the H_2 molecule: in an SCF calculation each electron has probability $\frac{1}{2}$ of being on either atom, so there is a probability $\frac{1}{2} \times \frac{1}{2}$ that they are both on one particular atom. This is true for any internuclear distance, so we are led to conclude that the dissociation products are $H^+ \cdots H^-$ and $H^- \cdots H^+$ with probability $\frac{1}{4}$ each, and $H \cdots H$ with probability $\frac{1}{2}$. In the real molecule, the electrons repel each other, so the ionic configurations, in which both electrons are on the same atom, are less probable than this at all distances, and their probability tends to zero as the atoms separate.

Effects such as this are said to be due to *electron correlation.* A satisfactory *ab initio* calculation of intermolecular interactions has to include the effects of electron correlation in one way or another. One of the most important of these is the dispersion interaction, which is wholly a correlation effect, and is absent from any SCF calculation. Because calculations that include electron correlation explicitly are expensive, one treatment that has been quite widely used is the Hartree–Fock plus Dispersion or HFD method, in which an *ab initio* SCF calculation is supplemented by an empirical expression for the dispersion.* This has been quite successful for inert-gas atoms and non-polar molecules (Meath and Koulis 1991).

However, the SCF interaction itself is modified by the inclusion of electron correlation, and the HFD treatment cannot be regarded as a universal method. It is easier to see how this modification happens in the language of perturbation theory; although the *ab initio* calculation does not separate the contributions in the way that perturbation theory does, it describes the same effects.

The main effect of electron correlation is to modify the electrostatic interaction between the molecules, because the electron distribution is changed. Electron correlation tends to reduce charge separations, so in a distributed-multipole picture (discussed in Chapter 7) the atom and bond charges tend to be smaller. Higher moments are not usually affected so much. Since the electrostatic interaction is often large, the effect of electron correlation on it can be important in giving an accurate description of the interaction. The inclusion of electron correlation also affects properties such as polarizabilities, and consequently induction and dispersion energies. That is, although dispersion is itself a correlation effect, involving the correlated motion of electrons in different molecules, the motion of those electrons can themselves be affected by correlation between electrons in the same molecule.

At shorter range, electron correlation undoubtedly modifies the repulsion energy, but perturbation calculations of the repulsion energy that include the effects of correlation are very difficult, and only a few systems have been studied (Rijks *et al.* 1989, Jeziorski *et al.* 1993). These calculations suggest that the repulsion is quite strongly affected, in the sense that it is changed by 10–20% or more at any given distance. We return to this point in Chapter 6.

5.5 The variation principle

Ab initio methods are based on the variation principle: the energy of an approximate wavefunction (more precisely, the expectation value of the Hamiltonian, calculated exactly for the approximate wavefunction) cannot be lower than the lowest eigenvalue of the Hamiltonian. This principle holds when the Hamiltonian is the exact Schrödinger Hamiltonian, but also within certain approximations. In particular, it holds within the Born–Oppenheimer approximation. It also holds when the wavefunction is restricted to belong to a particular class of function; for example within the Hartree–Fock approximation, where the wavefunction is taken to be a single Slater determinant, or perhaps a specific combination of Slater determinants required to ensure that the wavefunction is an eigenfunction of spin. In this case (the Self-Consistent-Field or SCF method) the energy cannot be lower than the energy of the best wavefunction of this type, normally called the Hartree–Fock limit.

The variational principle also holds for certain approximate treatments of electron correlation, in particular for the widely used method of Configuration Interaction with single and double excitations (CISD), but not for all of them. It does not hold for any

* 'Hartree–Fock' is here used as a synonym for 'SCF', but it is more usually reserved as a term for the limit of the SCF treatment, i.e., in a complete basis.

semi-empirical method, essentially because the energy calculated by such methods is not the exact expectation value for some Hamiltonian, but only an approximation to the expectation value.

The variational principle is the benchmark that we use to judge the accuracy of most *ab initio* calculations. Although the exact lowest eigenvalue is hardly ever known, we can be confident that a calculation is in some sense better than another if it gives a lower energy. The wavefunction may not be better in all respects—it may for instance give worse results for some molecular properties—but we know that in the limit the energy can only coincide with the true energy if the wavefunction is correct, so that the process of seeking ever lower energies will in principle lead us to the right answer in the end.

In the calculation of interaction energies, however, the variation principle does not provide this kind of assurance. If we do a better calculation using a variational supermolecule method, then W_{AB}, W_A, and W_B must improve or at worst remain the same. But we have no way of knowing *a priori* whether the difference $U_{AB} = W_{AB} - W_A - W_B$ will increase or decrease. Moreover, although W_{AB}, W_A, and W_B will all approach their limiting values more closely whenever we improve the wavefunctions, there is no reason why U_{AB} should do the same. In the limit of exact wavefunctions, of course, U_{AB} will be correct, but in a sequence of calculations that approach this limit, U_{AB} need not approach its limit monotonically, but may get worse before it gets better.

This is a general feature of the use of *ab initio* methods to calculate energy differences, and is not confined to intermolecular energy calculations. A well-known example is the ammonia molecule, which has a pyramidal equilibrium geometry but which can invert via a planar configuration some $25\,\mathrm{kJ\,mol^{-1}}$ higher in energy. If a simple basis of atomic s and p functions is used in the calculation, the pyramidal geometry is found to have a lower energy than the planar one. If the sp basis is improved, however, the planar geometry becomes preferred, and remains the preferred geometry even with a complete sp basis (Rauk *et al.* 1970). Only if d functions are added to the N atom basis does the pyramidal geometry become preferred again—essentially because the d functions can all stabilize the pyramidal geometry, but the d_{xz} and d_{yz} functions have the wrong symmetry to affect the energy of the planar geometry.

This example illustrates both that our intuitions may be unreliable, and also that improvements of one sort or another may have very different effects on the energies whose difference we require. Putting the same statement the other way round: we have no reason to expect that the inevitable errors in W_{AB} and $W_A + W_B$ will cancel out. Since the error in a typical *ab initio* calculation is of the order of $1\,e\mathrm{V}$ per electron pair, the error in each of W_{AB} and $W_A + W_B$ can easily reach $10\,e\mathrm{V}$ or $1000\,\mathrm{kJ\,mol^{-1}}$ (Davidson and Chakravorty 1994). The interaction energy, on the other hand, is typically of the order of $10\,\mathrm{kJ\,mol^{-1}}$. It is remarkable that the errors cancel as well as they usually do, and the reasons for this are not fully understood, though one reason is that only a few electrons on each molecule are normally involved in the short-range interactions. However there are two well-known ways in which the errors are known not to cancel. These are *Basis Set Superposition Error* or BSSE, and lack of *size consistency*.

5.5.1 *Basis set superposition error*

Probably everyone who has ever thought of doing an *ab initio* calculation on a weakly bound complex or has read about one will have heard of Basis Set Superposition Error or BSSE. It is probably the subject of more boring papers than any other topic in quantum chemistry.

The principle of BSSE is not difficult to understand. When a calculation is done on an isolated molecule, the energy that can be achieved is limited by the available basis set, which is rarely as good as one would wish and is never complete. If we were to carry out a calculation on molecule A of a complex $A \cdots B$, using the basis functions of both molecules, in the same positions as they would have in the complex, but the electrons and nuclei only of molecule A, we would get an energy that is no higher, and usually lower, than we would find in a calculation using the basis functions of molecule A alone:

$$W_A(AB) \leq W_A(A).$$

We use the notation $W_A(AB)$ to describe a calculation of the energy W_A of molecule A using the joint (AB) basis, while $W_A(A)$ describes the calculation of the same energy using the A basis alone. When two molecules are brought together and a calculation is carried out on the supermolecule, the basis functions of each molecule become variationally available in this way to improve the description of the other molecule. This is quite separate from any actual physical interaction that may arise; it is an artefact, due to the inadequacy of the original basis sets for the isolated molecules. The result is that the energy of the supermolecule is spuriously low if we calculate it in the obvious way as the difference

$$U_{AB}^{0} = W_{AB}(AB) - W_A(A) - W_B(B). \tag{5.5.1}$$

A remedy for this defect was proposed by Boys and Bernardi (1970). It is called the 'counterpoise' procedure, and consists simply of using the supermolecule basis (AB) for all the calculations:

$$U_{AB}^{\mathrm{cp}} = W_{AB}(AB) - W_A(AB) - W_B(AB). \tag{5.5.2}$$

The difference between these calculations, $U_{AB}^{\mathrm{cp}} - U_{AB}^{0} = W_A(A) - W_A(AB) + W_B(B) - W_B(AB)$, is called the 'counterpoise correction' for the 'Basis Set Superposition Error' that is present in eqn (5.5.1).

An objection to this procedure has been made on the grounds that in $W_A(AB)$ the basis functions of molecule B are wholly available to molecule A, whereas in the calculation of $W_{AB}(AB)$ they are needed also to describe the wavefunction of molecule B. It is argued that the difference $W_A(A) - W_A(AB)$ is an overestimate of the correction that is needed for molecule A, and that a better formula would be $W_A(A) - W_A(AB')$, where the notation $W_A(AB')$ describes a calculation carried out for molecule A using the basis set of A augmented with only the *virtual* orbitals of B. There has been a great deal of passionate argument over this issue, with both sides claiming that only their method gives the true corrected interaction energy. Probably the most succinct argument in favour of the original Boys–Bernardi scheme is that of Gutowski *et al.* (1987), who remark that in the ideal calculation one would use a *complete* basis for the calculations on AB and on the isolated *A* and *B* molecules, and one would not then consider deleting the occupied orbitals of *B* from the basis for the calculation on *A*.

Despite the heat that has been generated by this matter, it is in truth a non-issue. The argument is based on the concept that there is in some sense a 'correct' interaction energy for any given basis. This is a quite meaningless concept for anything but a complete basis; an inadequate basis will give an inadequate result, and no 'correction' can eliminate all its deficiencies. The only way to get the correct result (at a given level of calculation—SCF, MP2 or whatever it may be) is to use a complete basis, and the only criterion for judging the usefulness of one correction procedure over another is to ask which of them reliably gives results that are closer to the basis set limit. Schwenke and Truhlar (1988) investigated this question, and concluded that neither of the procedures advocated as corrections for BSSE gives results that are consistently closer to the basis set limit than the uncorrected values. They suggest that it is better to use the counterpoise procedure not as a correction device but only as a way of estimating the inadequacy of the basis set. Frisch *et al.* (1986) came to a similar conclusion. The contrary view is held by Szalewicz *et al.* (1988), and by Gutowski *et al.* (1993), who found in a study of He_2 that counterpoise correction according to the original Boys–Bernardi prescription does improve the agreement between the corrected result and the Hartree–Fock limit, while the 'virtuals-only' prescription does not. More recent work on hydrogen-bonded complexes (Novoa *et al.* 1994) leads to the conclusion that the standard counterpoise procedure gives more consistent results as the basis set is improved than uncorrected SCF or MP2 calculations. However the corrections are larger for correlated wavefunctions than for SCF at a given level of basis set, and even for the best basis sets they are of the order of 1–3 $kJ\,mol^{-1}$.

Most workers currently believe that it is better to correct for BSSE, though few, if any, would see the correction as useful when the basis set is very poor. Indeed, the disagreement on this issue may just be a question of where the line is to be drawn between 'poor' and 'good' basis sets. It is possible to use much better basis sets for He_2 than for the larger systems that Schwenke and Truhlar studied. Probably a reasonable rule of thumb is to apply the counterpoise correction, and to treat the magnitude of the correction as a guide to the reliability of the result, except in cases where the basis set can be improved to the point where the corrected energy varies less with basis than the size of the correction.

One aspect of the BSSE correction process that may have led to the perception that the counterpoise procedure over-corrects is illustrated schematically in Fig. 5.1. The customary procedure is to find the minimum of the uncorrected curve and apply the correction there, as shown schematically by (a) in Fig. 5.1. However, if the error varies with separation, then the minimum of the uncorrected potential curve will occur at a different separation from the minimum of the corrected curve ((b) in Fig. 5.1), and the BSSE correction at the minimum of the uncorrected curve will inevitably be larger than the correction at the minimum of the corrected curve. The correct procedure is evidently to find the minimum of the corrected curve, but this is more difficult. Recent work has shown that in some cases the customary procedure can lead to significant errors in the binding energy and the position of the minimum. Clearly the errors are worst when the BSSE depends strongly on the separation, which seems to happen when the interaction

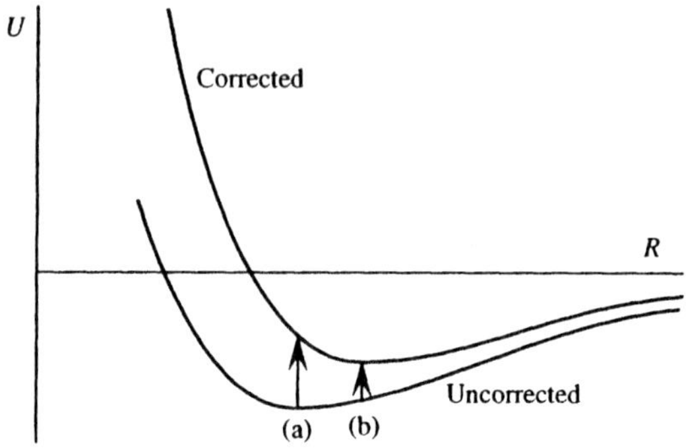

FIG. 5.1. The BSSE correction at the minimum of the uncorrected surface (a) is larger than the correction at the minimum of the corrected surface (b).

is dominated by dispersion (Eggenberger *et al.* 1991).

It is unfortunate that so much effort has gone into the subject of BSSE, because it has obscured the fact that there are other errors in a supermolecule calculation that are very imperfectly understood. Indeed, there is very little understanding of what those errors might be. However a study by Szalewicz *et al.* (1988) of the binding energy of the water dimer suggested that the uncertainty remaining in the best calculations is still of the order of 2.5 kJ mol^{-1}, or 10% of the binding energy in what is a particularly strongly bound complex.

5.6 Electron correlation and size consistency

The correlation of the electrons that results from their mutual repulsion can be described in many ways, all of them expensive in computer time. This is not the place for a detailed account of all the available methods, but it is necessary to understand the basic principles of some of them. See Szabo and Ostlund (1989) for a fuller discussion. The oldest method is 'configuration interaction', so called because the dominant SCF determinant, or 'root configuration', is allowed to mix with states in which the electrons are assigned to different orbitals, i.e., in different configurations. Again, the H_2 molecule provides the simplest example. The SCF wavefunction (ignoring spin) takes the form

$$\begin{aligned}\psi^{\mathrm{SCF}} &= \tfrac{1}{2}[s_A(1)+s_B(1)][s_A(2)+s_B(2)]\\ &= \tfrac{1}{2}[s_A(1)s_A(2)+s_A(1)s_B(2)+s_B(1)s_A(2)+s_B(1)s_B(2)],\end{aligned}$$

in which we see the coefficient $\frac{1}{2}$ for the ionic functions $s_A(1)s_A(2)$ and $s_B(1)s_B(2)$ (probability $(\frac{1}{2})^2 = \frac{1}{4}$). The functions s_A and s_B are atomic orbitals; in an accurate SCF treatment they would not be pure hydrogen 1s orbitals but would be modified by bond formation. In the CI method this state is mixed with others, in particular with the configuration σ_u^2 in which both electrons are in the σ_u orbital: $\frac{1}{2}[s_A(1)-s_B(1)][s_A(2)-s_B(2)]$. If we subtract a multiple of this from the original σ_g^2 configuration we get

$$\begin{aligned}\psi^{\mathrm{CI}} &= \sigma_g^2 - \lambda\sigma_u^2 \\ &= \tfrac{1}{2}(1-\lambda)s_A(1)s_A(2) + \tfrac{1}{2}(1+\lambda)s_A(1)s_B(2) \\ &\qquad + \tfrac{1}{2}(1+\lambda)s_B(1)s_A(2) + \tfrac{1}{2}(1-\lambda)s_B(1)s_B(2),\end{aligned}$$

so for positive λ the weight of the ionic terms is reduced. λ is a variational parameter which can be chosen to minimize the energy.* Its optimum value depends on the internuclear distance, and at large separations $\lambda \to 1$ and the ionic terms disappear altogether.

It can be shown that singly excited configurations alone do not introduce any electron correlation, though they can improve the wavefunction in conjunction with more highly excited configurations. Doubly excited configurations are the most important; higher excitations have a significant effect but their inclusion is very expensive. Accordingly the standard method of this type is Configuration Interaction with Single and Double excitations, or CISD.

The CISD method is variational. Unfortunately it is not *size consistent.* To explain this technical concept we suppose that we now wish to calculate the energy of a pair of H_2 molecules. At distances where they interact, we use a supermolecule calculation. The reference calculation corresponding to infinite separation involves a counterpoise-corrected calculation for each isolated molecule, using the supermolecule basis in each case. We normally need to do these single-molecule calculations separately, because the supermolecule basis is often the largest basis for which the calculation is feasible, and we cannot do a calculation with a basis that is twice as big. However we need to be sure that the result we would get by adding the single-molecule energies from the separate calculations is the same as we would get from a supermolecule calculation on the two separated molecules simultaneously. If this is the case, the calculation is said to be size consistent.

The H_2 example allows us to see why the CISD method fails this test. For a single molecule, the CISD wavefunction (in the simplified version discussed above) is $\sigma_g^2 - \lambda\sigma_u^2$. If we add the energies for two separate calculations, that is equivalent to using a simple product of a wavefunction like this for molecule A and another for molecule B, i.e., $(\sigma_g^{A^2} - \lambda\sigma_u^{A^2})(\sigma_g^{B^2} - \lambda\sigma_u^{B^2})$. However, when we multiply this out, we find a term $\lambda^2\sigma_u^{A^2}\sigma_u^{B^2}$, in which *four* electrons have been excited out of the ground configuration $\sigma_g^{A^2}\sigma_g^{B^2}$ of the two-molecule system. If we were to carry out a CISD calculation on the two-molecule system, however, we would include single and double excitations only, so this configuration would not appear and the answer would be different.

This effect is quite large—much too large to ignore. For a pair of water molecules separated by 10,000 a.u., an SCF calculation using a DZP basis gives almost exactly twice the energy calculated for a single molecule using the same basis set. The small difference, less than 10^{-10} a.u., can be attributed to rounding errors; the physical interaction between the molecules at this distance is dominated by the dipole–dipole interaction, of order $R^{-3} = 10^{-12}$ a.u. However the correlation energy calculated by CISD using a DZP

*Some people find it difficult to understand how the energy can be lowered by mixing excited states into the wavefunction. There is no difficulty if one understands that the original wavefunction is contaminated by contributions from excited states. The purpose of the CI procedure is to cancel out these contaminating terms.

basis is -0.375207 a.u., while the correlation energy calculated for a single molecule by the same method is -0.197067 a.u.. (The 'correlation energy' is the difference between the energy calculated with correlation effects included, and the energy obtained from the SCF calculation, and is negative.) The difference between the first value and twice the second is 0.018927 a.u. = 49.7 kJ mol^{-1}. That is, if we calculate the interaction energy by the supermolecule method, subtracting twice the energy of a single molecule from the energy of the supermolecule, the result will be in error by an amount of the order of 50 kJ mol^{-1} if we use CISD.

It is clear from this example that the calculation of intermolecular interactions must be carried out in a size-consistent way. There are two main classes of electron correlation method that are size consistent. Many-body perturbation theory or MBPT treats the electron repulsion, or more precisely the difference between the electron repulsion and the average of it that appears in the SCF method, as a perturbation. Methods of this type include Møller–Plesset perturbation theory (MPn for perturbation theory to order n). The zeroth-order Hamiltonian is the Fock operator $\mathcal{F}$, and the perturbation is $\mathcal{H} - \mathcal{F}$. The zeroth-order wavefunction for a closed-shell system is the SCF Slater determinant, and the zeroth-order energy is the sum of the occupied orbital energies. The energy to first order is the expectation value of the full Hamiltonian for the SCF wavefunction, i.e., the SCF energy. Correlation corrections appear at second order and beyond. The MP2 method, for example, gives a correction to the energy that is a sum over all doubly excited configurations. The MP2 energy for an assembly of two or more non-interacting systems is just the sum of the MP2 energies for the individual systems, as required for size consistency.

The other main class of size-consistent treatment is the coupled-cluster (CC) scheme (Čižek and Paldus 1980). In the CCSD treatment the correlated wavefunction can be written as $\exp(\hat{T}_1 + \hat{T}_2)\Psi_0$, where Ψ_0 is the SCF wavefunction and $\hat{T}_1$ and $\hat{T}_2$ are operators that transform Ψ_0 into a linear combination of singly excited and doubly excited configurations respectively. The first terms in the expansion of the exponential, $1 + \hat{T}_1 + \hat{T}_2$, give the CISD form of wavefunction, though the coefficients of the terms in the expansion would be different. The CCSD treatment is size consistent because the product of two CCSD wavefunctions is $\exp(\hat{T}_1^A + \hat{T}_2^A)\Psi_0^A \exp(\hat{T}_1^B + \hat{T}_2^B)\Psi_0^B = \exp(\hat{T}_1^A + \hat{T}_2^A + \hat{T}_1^B + \hat{T}_2^B)\Psi_0^A\Psi_0^B$, which is of CCSD form.

Neither of these methods is variational; in principle the CCSD wavefunction could be optimized variationally, but that is impractical and a different approach is used. However size consistency is more important than the variational property; we have seen that a variational calculation of energies provides no particular guarantees as to the quality of energy differences.

5.7 Density functional theory

A popular *ab initio* method in recent years has been the use of Density Functional Theory or DFT. This is based on a theorem due to Hohenberg and Kohn (1964), which states that a knowledge of the electron density of a system in its ground state is enough to determine the energy: 'the energy is a functional of the density'. An informal understanding of the theorem was provided by Wilson (1968), who remarked that the exact density has cusps only at the positions of the nuclei, and the gradient of the density at the cusp depends on the nuclear charge. Moreover the integral of the density over all space gives the number of electrons. A knowledge of the density therefore allows one to write down the Hamiltonian, from which everything can be determined. This also illustrates that although the theorem is rigorous, it does not in itself provide a practical way to calculate the energy from a given density. The literature contains many proposed functionals which may be used for this purpose, but at the present time there is no fundamental theory for determining the one true functional. The functionals in current use are based on the theory of the uniform electron gas, together with various limiting properties that the functional must satisfy.

An overview of the present state of the art in this field may be found in Handy (1994). As far as intermolecular potentials are concerned, the position may be summed up very briefly: the method is currently of very little practical use. Xantheas (1995) found that the O···O distance in the water dimer varied from 2.6 to 3.3 Å, depending on the functional used, while the binding energy varied from 2.10 to 12.2 kcal/mol, but concluded that the B–LYP functional gave results in good agreement with MP2 calculations. del Bene *et al.* (1995) on the other hand used DFT to study eight hydrogen-bonded complexes for which experimental data were available, and concluded that 'density functional calculations fail to yield reliable binding energies, intermolecular distances and hydrogen-bonded X–H frequency shifts'. An attempt to apply the method to inert gases is described by Pérez-Jordá and Becke (1995), and here the results are very poor. One of the important problems which has not yet been overcome is that none of the functionals currently in use gives an interaction energy between molecules that behaves like R^{-6} at long range, so that they completely fail to describe the dispersion energy. A further problem in some methods is that the practical calculation of the energy involves numerical quadratures whose errors are comparable with intermolecular interaction energies, and fluctuate in an unpredictable fashion as the intermolecular distance varies. These limitations are not fundamental features of the method, and since density functional theory is a very active field of research at present, they may be overcome sooner rather than later.

5.8 Semi-empirical methods

Although semi-empirical methods cannot be described as *ab initio*, a brief mention is appropriate here. The most time-consuming step of a true *ab initio* SCF calculation is the computation of the two-electron integrals (see eqn (5.2.1) on p. 65). Semi-empirical methods avoid this computational cost by neglecting most of them and approximating most of the rest. Many of the one-electron integrals are also approximated. The approximations involve empirical parameters which are determined by requiring the calculations to reproduce experimental data, for example on heats of formation of a range of small molecules. The AM1 (Dewar *et al.* 1985) and PM3 (Stewart 1989) methods are two popular examples. As they are calibrated against heats of formation, that is, against enthalpies of reactions involving the making and breaking of large numbers of chemical bonds, they cannot be expected to be reliable for calculating small energy differences such as the binding energies of molecular complexes. However they are widely used to calculate structures and charge distributions of molecules that are too large for standard *ab initio* methods to be used. They are also sometimes used to obtain a first guess at the structure of a moderately large molecule, to save time in a subsequent geometry optimization using a true *ab initio* method. Semi-empirical methods have reached a high level of sophistication, and are quite reliable provided that they are used for the purposes for which they were designed.

5.9 Choice of geometries

It takes up to six coordinates to describe the relative configuration of two molecules, and some thought should be given to the set of coordinate values to be used. Even for a pair of linear molecules, four coordinates are needed. There are many calculations in the literature for homonuclear diatomics where a wide range of intermolecular separations was studied, but the three orientational coordinates were represented by only three or four configurations—commonly linear ($\theta_A = \theta_B = \varphi = 0$), parallel ($\theta_A = \theta_B = \pi/2$, $\varphi = 0$), crossed ($\theta_A = \theta_B = \pi/2$, $\varphi = \pi/2$), and T ($\theta_A = 0$, $\theta_B = \pi/2$, $\varphi = 0$). These configurations have the advantage that their symmetry makes the calculation faster, but that in itself limits very severely the information that they provide about the potential surface. For example, the energy derivatives with respect to the angular coordinates are all zero by symmetry for these configurations.

A better strategy, if it is required to characterize the potential energy surface as efficiently as possible, is to select perhaps five values for each angular coordinate (fewer for φ near $\theta_A = 0$ or $\theta_B = 0$) and a somewhat larger number of R values. A reasonable choice for the θ values is a set of Gauss–Legendre integration points (regarded as values of $\cos\theta$) (Abramowitz and Stegun 1965). This would yield 1000 or so points for a pair of linear molecules, but the number will be reduced by symmetry—by a factor of 8 for homonuclear diatomics. If this is still too many, a proportion can be selected at random. In this way it is possible to cover the intermolecular coordinate space reasonably efficiently and economically.

The reason for this type of approach is that it is almost always necessary to fit some kind of analytic function to the computed points, and this procedure makes the fitted function much more representative of the surface as a whole than when only a few angular geometries are used. Moreover some of the points omitted from a random selection can be used for additional test calculations to check the accuracy of the fitted function.

5.10 Morokuma analysis

One of the limitations of the supermolecule approach is that it provides just a single number, the total interaction energy, at each configuration, in contrast to perturbation methods, which give a number of terms of well-defined physical significance. Morokuma (1971) proposed a scheme for decomposing the supermolecule energy to obtain similar information. The initial scheme involved carrying out the following calculations, after first obtaining the unperturbed wavefunctions Ψ_0^A and Ψ_0^B for the two molecules. The unperturbed energy for the supersystem is the sum of the energies of the isolated molecules: $E_0 = E_0^A + E_0^B$.

1. In the first step, the wavefunction is a simple product $\Psi_0^A\Psi_0^B$ of the unperturbed wavefunctions. The expectation value of the energy is $E_1 = E_0 + E_{\rm es}$; that is, the

change in energy yields the electrostatic energy of interaction of the unperturbed charge distributions.

2. In the next step, the wavefunction is still a simple product, antisymmetric only with respect to permutations of electrons within A and B but not with respect to permutations that exchange electrons between the molecules. This is optimized with all integrals deleted that contain overlap between the basis functions of molecule A and those of molecule B. That is, integrals involving products of the form $\chi_i^A(1)\chi_j^B(1)$ are deleted. However intermolecular Coulomb integrals of the form $\int \chi_i^A(1)\chi_j^A(1)(r_{12})^{-1}\chi_k^B(2)\chi_l^B(2)\,d\tau_1\,d\tau_2$ survive, so each molecule's wavefunction distorts under the influence of the other's electrostatic field, but with no mixing of the basis functions of one molecule into the orbitals of the other. The resulting energy contains the induction energy: $E_2 = E_0 + E_{es} + E_{ind}$.

3. In the third step, the wavefunction is an antisymmetrized product of the unperturbed wavefunctions: $\Psi_3 = \mathcal{A}\Psi_0^A\Psi_0^B$. As we shall see in Chapter 6, the effect of the antisymmetrization is to introduce the exchange repulsion: $E_3 = E_0 + E_{es} + E_{er}$.

4. Finally, the antisymmetrized product is optimized using the full Fock matrix for the supersystem. In the original scheme, the resulting energy was written as $E_4 = E_0 + E_{es} + E_{ind} + E_{er} + E_{ct}$, so that the remaining part of the interaction energy was ascribed to charge transfer.

Kitaura & Morokuma (Kitaura and Morokuma 1976, Morokuma and Kitaura 1981) later suggested a modified approach. The Fock matrix is expressed in the basis of molecular orbitals for the separated molecules, and can be partitioned according to whether the orbitals belong to molecule A or molecule B, and according to whether they are occupied or virtual:

$$F = \begin{array}{c} \\ A_{occ} \\ \\ A_{virt} \\ \\ B_{occ} \\ \\ B_{virt} \\ \\ \end{array} \begin{array}{cccc} A_{occ} & A_{virt} & B_{occ} & B_{virt} \end{array} \\ \begin{pmatrix} F_{oo}^{AA} & F_{ov}^{AA} & F_{oo}^{AB} & F_{ov}^{AB} \\ (\mathrm{esx}) & (\mathrm{ind}) & (\mathrm{ex}') & (\mathrm{ct}) \\ F_{vo}^{AA} & F_{vv}^{AA} & F_{vo}^{AB} & F_{vv}^{AB} \\ (\mathrm{ind}) & (\mathrm{esx}) & (\mathrm{ct}) & (\mathrm{ex}') \\ F_{oo}^{BA} & F_{ov}^{BA} & F_{oo}^{BB} & F_{ov}^{BB} \\ (\mathrm{ex}') & (\mathrm{ct}) & (\mathrm{esx}) & (\mathrm{ind}) \\ F_{vo}^{BA} & F_{vv}^{BA} & F_{vo}^{BB} & F_{vv}^{BB} \\ (\mathrm{ct}) & (\mathrm{ex}') & (\mathrm{ind}) & (\mathrm{esx}) \end{pmatrix} \tag{5.10.1}$$

Note that although the basis functions for this description are those for the separated molecules, the Fock matrix is constructed for the complete system, i.e., including all the intermolecular interactions. If now the normal SCF procedure is followed, we get the usual supermolecule wavefunction and energy. Kitaura & Morokuma, however, pointed out that parts of the Fock matrix can be suppressed, at every iteration, and the resulting energy corresponds to the omission of certain terms in the interaction. Specifically:

1. If all the off-diagonal blocks are suppressed, leaving only the 'esx' blocks, the resulting energy contains the electrostatic term plus part of the exchange–repulsion;
2. If the off-diagonal blocks labelled 'ind' are retained, the induction energy is obtained as well;
3. If the 'ex′' blocks are retained, the rest of the exchange energy is obtained;
4. If the 'ct' blocks are retained, the result contains the charge transfer energy 'ct'.
5. If all the off-diagonal terms are retained, the result is the complete supermolecule energy, which contains all the above effects together with an assortment of cross-terms.

These schemes for energy decomposition have been very widely used. However it will be apparent that as described, they suffer from the same disadvantage as any other supermolecule calculation: that is, they are affected by Basis Set Superposition Error. This manifests itself especially in the charge transfer terms, in which virtual orbitals of one molecule are allowed to mix with occupied orbitals of the other. The cure for this, as in the ordinary supermolecule treatment, is to perform the reference calculations for the isolated molecules in the dimer basis. Such a method was suggested by Sokalski *et al.* (Sokalski, Hariharan and Kaufman 1983, Sokalski, Roszak, Hariharan and Kaufman 1983, Cammi *et al.* 1985), though Gutowski and Piela (1988) claim that the form of the resulting decomposition is incorrect. The original scheme (Morokuma 1971) has also been criticised on the grounds that in step 2 of the calculation, when the wavefunction of each molecule is optimized in the field of the other, the Pauli principle is not satisfied because the wavefunction is not antisymmetrized. The result is an over-estimate of the induction energy, which can be quite severe at short range (Gutowski and Piela 1988, Frey and Davidson 1989, Cybulski and Scheiner 1990). The modified scheme always uses an antisymmetrized wavefunction, so this problem does not then arise.

6

PERTURBATION THEORY OF INTERMOLECULAR FORCES AT SHORT RANGE

6.1 Introduction

The perturbation theory for well-separated molecules described in Chapter 4 (the 'long-range approximation' or 'polarization approximation') is very successful if the molecules are a long distance apart, but at short range it fails completely. Part of the reason for the failure of the theory as usually formulated is that the multipole expansion breaks down, but as we have seen, the multipole expansion is not fundamental to the theory, since it is possible to use a 'non-expanded' treatment. We shall see in Chapter 7 how to overcome the failure of the multipole expansion to converge without resorting to this device.

A more fundamental failure is that the repulsion between molecules that occurs at short range is completely missing from the long-range theory. This failure arises from the fact that if the molecules are close enough for their wavefunctions to overlap, we can no longer ignore exchange, as we did in Chapter 4. To see the effect that this has on the energy, it will suffice initially to study one-electron atoms, and to compare the expectation value of the energy for an antisymmetrized wavefunction with the expectation value for a simple product. We shall see that at long range the result is the same, but at short range new terms appear that can be identified as the repulsion energy.

For two hydrogen atoms in $1s$ states, the zeroth-order wavefunction in the long-range treatment is $\psi = \big|a(1)b(2)\big\rangle$, a simple product of the normalized $1s$ wavefunction a for atom A and the $1s$ wavefunction b for atom B. Note that in this treatment, electron 1 is associated with atom A and electron 2 with atom B, and exchange is ignored. We obtain the energy of the system to first order in the usual way, by evaluating $\langle\psi|\mathcal{H}|\psi\rangle/\langle\psi|\psi\rangle$. The Hamiltonian is

$$\begin{aligned}\mathcal{H} &= -\tfrac{1}{2}\nabla_1^2 - \tfrac{1}{2}\nabla_2^2 - \frac{1}{r_{A1}} - \frac{1}{r_{A2}} - \frac{1}{r_{B1}} - \frac{1}{r_{B2}} + \frac{1}{r_{12}} + \frac{1}{r_{AB}} \\ &= \mathcal{H}^A(1) + \mathcal{H}^B(2) - \frac{1}{r_{A2}} - \frac{1}{r_{B1}} + \frac{1}{r_{12}} + \frac{1}{r_{AB}},\end{aligned}$$

so the energy in the long-range approximation is

$$\begin{aligned}W_{\text{PA}} &= \langle\psi|\mathcal{H}|\psi\rangle \\ &= \big\langle a(1)b(2)\big|\mathcal{H}\big|a(1)b(2)\big\rangle \\ &= \Big\langle a(1)b(2)\Big|W_a + W_b - \frac{1}{r_{A2}} - \frac{1}{r_{B1}} + \frac{1}{r_{12}} + \frac{1}{r_{AB}}\Big|a(1)b(2)\Big\rangle \\ &= W_a + W_b + U_{\text{es}}.\end{aligned}$$

Here U_{es} is the classical electrostatic interaction energy between the unperturbed charge densities:

$$
\begin{aligned}
U_{\text{es}} &= -\int \rho^b(2)\frac{1}{r_{A2}}\,\mathrm{d}^3\mathbf{r}_2 - \int \rho^a(1)\frac{1}{r_{B1}}\,\mathrm{d}^3\mathbf{r}_1 \\
&\quad + \int \rho^a(1)\rho^b(2)\frac{1}{r_{12}}\,\mathrm{d}^3\mathbf{r}_1\,\mathrm{d}^3\mathbf{r}_2 + \frac{1}{r_{AB}},
\end{aligned} \tag{6.1.1}
$$

where $\rho^a(1) = |a(1)|^2$ is the electron density for the unperturbed atom A and likewise $\rho^b(1) = |b(1)|^2$.

If the charge densities overlap, however, we should antisymmetrize, so that the wavefunction for the system becomes $\Psi = \sqrt{\frac{1}{2}}|a(1)b(2) - a(2)b(1)\rangle$. This is not normalized; the normalization integral is

$$
\int |\Psi|^2\,\mathrm{d}\tau_1\,\mathrm{d}\tau_2 = \tfrac{1}{2}\langle a(1)b(2) - a(2)b(1)|a(1)b(2) - a(2)b(1)\rangle = 1 - S^2, \tag{6.1.2}
$$

where $S = \langle a(1)|b(1)\rangle$. The Hamiltonian is just the same as before, but note that by reorganizing the terms we can also write it as

$$
\mathcal{H} = \mathcal{H}^A(2) + \mathcal{H}^B(1) - \frac{1}{r_{A1}} - \frac{1}{r_{B2}} + \frac{1}{r_{12}} + \frac{1}{r_{AB}},
$$

so for the 'symmetric' treatment the energy to first order is

$$
\begin{aligned}
W_S &= \frac{\langle\Psi|\mathcal{H}|\Psi\rangle}{\langle\Psi|\Psi\rangle} \\
&= \tfrac{1}{2}\langle a(1)b(2) - a(2)b(1)|\mathcal{H}|a(1)b(2) - a(2)b(1)\rangle/(1 - S^2) \\
&= \langle a(1)b(2)|\mathcal{H}|a(1)b(2) - a(2)b(1)\rangle/(1 - S^2) \\
&= \frac{1}{1 - S^2}\Big[\Big\langle a(1)b(2)\Big|W_a + W_b - \frac{1}{r_{A2}} - \frac{1}{r_{B1}} + \frac{1}{r_{12}} + \frac{1}{r_{AB}}\Big|a(1)b(2)\Big\rangle \\
&\qquad - \Big\langle a(1)b(2)\Big|W_a + W_b - \frac{1}{r_{A1}} - \frac{1}{r_{B2}} + \frac{1}{r_{12}} + \frac{1}{r_{AB}}\Big|a(2)b(1)\Big\rangle\Big] \\
&= (W_a + W_b) + \frac{1}{1 - S^2}\Big[\Big\langle a(1)b(2)\Big|-\frac{1}{r_{A2}} - \frac{1}{r_{B1}} + \frac{1}{r_{12}} + \frac{1}{r_{AB}}\Big|a(1)b(2)\Big\rangle \\
&\qquad - \Big\langle a(1)b(2)\Big|-\frac{1}{r_{A2}} - \frac{1}{r_{B1}} + \frac{1}{r_{12}} + \frac{1}{r_{AB}}\Big|a(2)b(1)\Big\rangle\Big].
\end{aligned}
$$

There has been some juggling of labels here; we can exchange electron labels 1 and 2 provided we do it throughout an integral.

We can express this result as a sum of the unperturbed energy $W_a + W_b$, the classical electrostatic interaction U_{es} as in the long-range approximation, and some new terms, which are

$$
U_{\text{er}} = \frac{S^2}{1 - S^2}\Big\langle a(1)b(2)\Big|-\frac{1}{r_{A2}} - \frac{1}{r_{B1}} + \frac{1}{r_{12}}\Big|a(1)b(2)\Big\rangle
$$

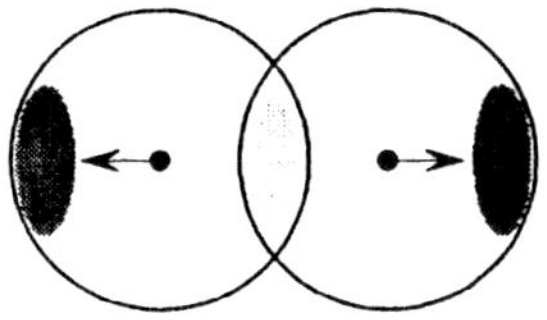

FIG. 6.1. The electron density of a pair of closed-shell atoms is less than the sum of the atomic densities between the nuclei (light shading) and greater in the region remote from the other atom (darker shading).

$$\begin{aligned} &-\frac{1}{1-S^2}\left\langle a(1)b(2)\middle|-\frac{1}{r_{A2}}-\frac{1}{r_{B1}}+\frac{1}{r_{12}}\middle|a(2)b(1)\right\rangle \\ &= -\frac{S^2}{1-S^2}\left[\left\langle b\middle|\frac{1}{r_A}|b\right\rangle+\left\langle a\middle|\frac{1}{r_B}|a\right\rangle-\left\langle a(1)b(2)\middle|\frac{1}{r_{12}}\middle|a(1)b(2)\right\rangle\right] \\ &+\frac{1}{(1-S^2)}\left[S\left\langle a\middle|\frac{1}{r_A}+\frac{1}{r_B}\middle|b\right\rangle-K_{ab}\right]. \end{aligned} \tag{6.1.3}$$

$K_{ab} \equiv (ab|ba) \equiv \langle a(1)b(2)|1/r_{12}|a(2)b(1)\rangle$ is the exchange integral between the orbitals a and b, and gives a negative, i.e. attractive, contribution to the energy, while the term $\langle a|1/r_A + 1/r_B|b\rangle$ describes a repulsion arising from the overlap of the electrons from the two molecules. The other terms can be thought of as corrections to the electrostatic energy. The net effect of the whole expression is repulsive, and it is called the *exchange–repulsion* energy. Note that if there is no overlap ($S = 0$) the antisymmetrized wavefunction Ψ gives the same result as the simple product ψ.

It is illuminating to think about these effects in terms of the charge density. For the simple product wavefunction $|a(1)b(2)\rangle$, the charge density is merely $\rho^a + \rho^b$, the superposition of the unperturbed atomic charge distributions. For the antisymmetrized wavefunction Ψ, however, it is

$$\begin{aligned} \rho(\mathbf{r}) &= \frac{1}{1-S^2}\int(\delta(\mathbf{r}-\mathbf{r}_1)+\delta(\mathbf{r}-\mathbf{r}_1)\Psi^2\,d^3\mathbf{r}_1\,d^3\mathbf{r}_2 \\ &= \frac{1}{1-S^2}((\rho^a(\mathbf{r})+\rho^b(\mathbf{r})-2Sa(\mathbf{r})b(\mathbf{r})). \end{aligned}$$

Now on the plane between the atoms and equidistant from them, the two atomic wavefunctions have equal values, and $\rho^a(\mathbf{r}) = \rho^b(\mathbf{r}) = a(\mathbf{r})b(\mathbf{r})$. Consequently the density is $\rho(\mathbf{r}) = 2\rho^a(\mathbf{r})/(1+S)$, which is *less* than the density $\rho^a(\mathbf{r})+\rho^b(\mathbf{r}) = 2\rho(\mathbf{r})$ on the same plane for the simple product wavefunction. At a point on the side of atom A remote from atom B, on the other hand, we have $b(\mathbf{r}) \ll a(\mathbf{r})$ and $\rho^b(\mathbf{r}) \ll \rho^a(\mathbf{r})$, so $\rho(\mathbf{r}) \approx \rho^a(\mathbf{r})/(1-S^2)$, which is *greater* than the density for the product wavefunction.

For the antisymmetrized wavefunction, then, as compared with the product wavefunction, the electron density is decreased between the atoms and increased at the side of each atom remote from the other. We can now apply the Hellman–Feynman theorem, which tells us that the force on a nucleus can be evaluated from the charge distribution as if it were classical, and we see that the change in the charge distribution produces an additional force on each nucleus in the direction away from the other. In other words, the effect of the antisymmetrization is to modify the electron density in such a way as to cause a repulsive force on the nuclei. (See Fig. 6.1.) This is the force corresponding to the exchange repulsion energy U_{er}.

Nothing has been said so far about spin. If the spins associated with orbitals a and b are the same (both α or both β, giving in either case a component of the triplet state ${}^3\Sigma_u$) nothing changes; the spin functions can simply be integrated out. If they are different, however, the overlap integral S becomes zero, and U_{er} also vanishes. Thus there is no exchange–repulsion for electrons of opposite spin. In the more usual closed-shell case, therefore, we can regard the repulsion as arising purely between electrons of the same spin, while electrons of opposite spin merely behave like classical charge distributions.

If the spins are different in the hydrogen-atom case, of course, chemical bonding occurs. This arises because there are two states with opposite spin, $|a^\alpha b^\beta\rangle$ and $|a^\beta b^\alpha\rangle$, and they are strongly coupled. The matrix element of the Hamiltonian between them is

$$\begin{aligned}
&\langle a^\alpha b^\beta|\mathcal{H}|a^\beta b^\alpha\rangle \\
&\quad= \tfrac{1}{2}\langle a^\alpha(1)b^\beta(2) - a^\alpha(2)b^\beta(1)|\mathcal{H}|a^\beta(1)b^\alpha(2) - a^\beta(2)b^\alpha(1)\rangle \\
&\quad= \langle a^\alpha(1)b^\beta(2)|\mathcal{H}|a^\beta(1)b^\alpha(2) - a^\beta(2)b^\alpha(1)\rangle \\
&\quad= \left\langle a^\alpha(1)b^\beta(2)\middle|W_a + W_b - \frac{1}{r_{A2}} - \frac{1}{r_{B1}} + \frac{1}{r_{12}} + \frac{1}{r_{ab}}\middle|a^\beta(1)b^\alpha(2)\right\rangle \\
&\qquad - \left\langle a^\alpha(1)b^\beta(2)\middle|W_a + W_b - \frac{1}{r_{A1}} - \frac{1}{r_{B2}} + \frac{1}{r_{12}} + \frac{1}{r_{AB}}\middle|a^\beta(2)b^\alpha(1)\right\rangle \\
&\quad= -S^2(W_a + W_b) - \left\langle a(1)b(2)\middle|-\frac{1}{r_{A2}} - \frac{1}{r_{B1}} + \frac{1}{r_{12}} + \frac{1}{r_{AB}}\middle|a(2)b(1)\right\rangle.
\end{aligned}$$

The mixing of these two states produces new states that are the sum and difference: $|a^\alpha b^\beta\rangle \pm |a^\beta b^\alpha\rangle$. The reader may care to verify that one of these is the third component of the ${}^3\Sigma_u$ state, and has the same energy as the other two components, so it suffers exchange repulsion just as they do; while the other state is substantially lowered in energy and becomes the ${}^1\Sigma_g$ bonding state of the H_2 molecule.

Thus we see that antisymmetrization introduces new effects that are completely missing when we use a simple product wavefunction. The most important of these for our purposes is that there is a strong repulsion between closed-shell molecules when their wavefunctions overlap significantly. We have to modify the perturbation treatment so as to take this into account correctly.

6.1.1 *Short-range perturbation theory*

Unfortunately the need for antisymmetrization complicates the theory very considerably. The problems that arise are twofold. First, if we assign electrons to individual molecules in order to separate the Hamiltonian into an unperturbed part $\mathcal{H}^0$ and a perturbation $\lambda\mathcal{H}'$, we obtain a zeroth-order Hamiltonian and a perturbation that are not symmetric with respect to permutations that exchange electrons between the molecules:

$$[P, \mathcal{H}^0] \neq 0, \qquad [P, \lambda\mathcal{H}'] \neq 0, \tag{6.1.4}$$

but the total Hamiltonian is, as always, symmetric:

$$[P,(\mathcal{H}^0+\lambda\mathcal{H}')]=0. \tag{6.1.5}$$

Combining these equations we find

$$[P,\mathcal{H}^0]=-[P,\lambda\mathcal{H}']\neq 0, \tag{6.1.6}$$

so that we have a non-zero first-order quantity that is equal to a zeroth-order quantity. This contradicts the basic assumption that we make when we separate the perturbation equations into their different orders, i.e., that we can collect terms in different powers of λ and equate them separately to zero. It means that there is no unique definition of the order of a term in the perturbation expansion.

A closely related problem has to do with the symmetry of the wavefunctions. If the wavefunctions of the two molecules overlap, then, as we have seen, we should antisymmetrize the product functions that we use as zeroth-order wavefunctions, in order to satisfy the Pauli principle. That is,

$$|m_A n_B\rangle = \mathcal{A}|m_A\rangle|n_B\rangle, \tag{6.1.7}$$

where the notation is now to be understood as implying that $|m_A n_B\rangle$ is antisymmetric with respect to all electron permutations, whereas $|m_A\rangle|n_B\rangle$ is antisymmetric with respect only to permutations of the A electrons among themselves and of the B electrons among themselves. But whereas the simple products are orthogonal, even at short range (because $|m_A\rangle$ and $|n_B\rangle$ refer to different electrons) the antisymmetrized products are not. The easiest way to see this is to look at an example. Consider the case of two hydrogen atoms, so that the ground states $|0_A\rangle$ and $|0_B\rangle$ are $1s$ states, and take an excited state $|1_B\rangle$ in which atom B is excited to a $2p$ state. Then

$$\begin{aligned}\langle 0_A 0_B\,|0_A 1_B\rangle &= \langle \mathcal{A}s_A(1)s_B(2)|\mathcal{A}s_A(1)p_B(2)\rangle\\ &= \tfrac{1}{2}\langle s_A(1)s_B(2)-s_A(2)s_B(1)|s_A(1)p_B(2)-s_A(2)p_B(1)\rangle\\ &= \tfrac{1}{2}[\langle s_A(1)|s_A(1)\rangle\langle s_B(2)|p_B(2)\rangle-\langle s_A(1)|p_B(1)\rangle\langle s_B(2)|s_A(2)\rangle\\ &\quad -\langle s_B(1)|s_A(1)\rangle\langle s_A(2)|p_B(2)\rangle+\langle s_B(1)|p_B(1)\rangle\langle s_A(2)|s_A(2)\rangle],\\ &= -\langle s_A|p_B\rangle\langle s_A|s_B\rangle,\end{aligned} \tag{6.1.8}$$

and this need not be zero.

Now, as we know from elementary quantum mechanics, the eigenstates of any hermitian operator are orthogonal. If the states $|m_A n_B\rangle$ do not form an orthogonal set they cannot be eigenstates of any hermitian operator. To put it another way, there can be no hermitian zeroth-order Hamiltonian that has these antisymmetrized products as eigenstates.

This is a serious difficulty. It does not mean that we can no longer use perturbation theory, but it does mean that we cannot use the standard Rayleigh–Schrödinger treatment. In its place we have to put something else, but there is no unique approach, and many different methods have been used. Comparisons between the different methods are difficult, because they use the same terminology for quantities that may be differently defined. For example, all methods yield a quantity that at long range becomes the induction energy. At short range, however, this becomes modified, and the modifications are different in different methods. A short-range term such as the charge-transfer energy may be viewed as a separate effect or included as part of the induction energy. Furthermore, because there is no natural and unambiguous separation into orders of perturbation theory, a term that is second-order in one method may be viewed as including third-order contributions in another.

It might be thought possible in principle to carry out ordinary Rayleigh–Schrödinger perturbation theory, starting with the simple unsymmetrized product functions, and using the same functions throughout the calculation, with certain electrons assigned to molecule A and the rest to molecule B. The zeroth-order function for two hydrogen atoms would then be $|a(1)b(2)\rangle$, and the result of the perturbation theory, if it converges, would be close to the antisymmetrized function $\sqrt{\frac{1}{2}}|a(1)b(2)-a(2)b(1)\rangle$. If we use a complete basis to describe atom A, we could express $b(1)$ in terms of these basis functions, and conversely $a(2)$ could be expressed in terms of the basis set used for atom B. We can see, however, that in practical applications, where the size of the basis set is limited, there would be no possibility of approximating the $a(2)b(1)$ term accurately by this technique. Moreover the convergence of such a treatment would be slow, because the difference between the unperturbed function $|a(1)b(2)\rangle$ and the perturbed function $\sqrt{\frac{1}{2}}|a(1)b(2)-a(2)b(1)\rangle$ is not in any sense small. In fact the overlap between them tends to $\sqrt{\frac{1}{2}}$ as the separation increases. In the many-electron case, with n_A electrons on A and n_B on B, there are $(n_A+n_B)!/n_A!n_B!$ ways of assigning the electrons, and the overlap between any particular one of these and the antisymmetrized function, which is a linear combination of all of them, tends to $\sqrt{n_A!n_B!/(n_A+n_B)!}$ at long range, and this becomes smaller and smaller as the number of electrons increases.

Even for the H$\cdots$H case, Kutzelnigg (1992) showed analytically that the radius of convergence of the Rayleigh–Schrödinger perturbation expansion, viewed as a power series in λ, is only slightly greater than 1—in fact it exceeds 1 by an amount of order e^{-2R}, and so approaches 1 as $R\to\infty$. Cwiok *et al.* (1992) confirmed this result numerically, and found that to calculate the interaction energy accurately for H$\cdots$H at 8 bohr, it was necessary to sum the Rayleigh–Schrödinger expansion approximately to infinite order. Clearly this is not possible in practical applications.

However there is another, more serious, problem with the Rayleigh–Schrödinger expansion. If we work with wavefunctions that are antisymmetrized with respect to the electrons of A, and with respect to those of B, but we do not impose antisymmetry with respect to exchanges of electrons between A and B, then there are solutions of the Schrödinger equation that do not satisfy the Pauli principle, and that lie lower in energy than all the antisymmetric ones. For the He$\cdots$H system, for instance, it is possible to write down a wavefunction in which all the electrons, including the one formally assigned to H as well as those assigned to He, occupy the He 1s orbital. This Pauli-forbidden, or 'unphysical', state is clearly lower in energy than any correctly antisymmetric physical state. Claverie (1971) was the first to point out this problem; he argued that if the Rayleigh–Schrödinger expansion starting from the lowest-energy unperturbed state converged at all, it would converge to the lowest-energy unphysical state, which he called the 'mathematical ground state', and not to the lowest physical state. Morgan and Simon (1980) showed that there is an infinite number of these non-physical states below the physical ground state for any system containing at least one atom with atomic number greater than 2.

Consequently it is necessary to find new versions of perturbation theory that overcome these problems. Such methods—methods that deal with intermolecular perturbation theory in the region where exchange cannot be ignored—are generally called 'exchange perturbation theories'. Claverie (1978) has given a very thorough account both of the mathematical issues and of the attempts up to that time to overcome them. The methods so far proposed fall into two main groups. The 'symmetric' methods take as their unperturbed functions the set of antisymmetrized products (6.1.7). These are not orthogonal, as we have seen, and various schemes have been proposed to handle the non-orthogonality. 'Symmetry-adapted perturbation theories', on the other hand, start from the simple product functions $|m_A\rangle|n_B\rangle$, without antisymmetrization, so that in the unperturbed problem each electron is assigned to one or other molecule. These unperturbed functions are orthogonal, but because the perturbed wavefunction must be antisymmetric it is necessary to carry out an antisymmetrization at each order of perturbation theory, or to adopt some other technique to ensure that even if the wavefunction is not antisymmetric the energy is nevertheless correct.

6.2 Symmetric perturbation methods

6.2.1 *Expansion in powers of overlap*

Since it is to a large degree the non-orthogonality of the antisymmetrized products that causes the difficulties, an approach that has been adopted by some workers is simply to neglect the non-orthogonality at zeroth order. That is, the overlap matrix $\mathbf{S}$ is written in the form $\mathbf{S} = \mathbf{I} + \mathbf{t}$, and the assumption made that the off-diagonal part $\mathbf{t}$ of the overlap matrix is small. (Note that the antisymmetrized product functions require normalization first.) It is then possible to develop a treatment involving expansion in powers of the overlap as well as in powers of the perturbation, and because the unperturbed functions are orthogonal to zeroth order in the overlap, a hermitian unperturbed Hamiltonian can be found. Several methods of this kind have been proposed (Basilevsky and Berenfeld 1972*a*, 1972*b*, Kvasnicka *et al.* 1974, Stone and Erskine 1980, Mayer and Surjan 1993).

Unfortunately this approach does not give good results. The assumption of small overlap becomes increasingly untenable as the quality of the calculation is improved. Even with basis sets that are far from complete, diffuse basis functions on one molecule may overlap strongly with similar functions on the other, so that the series in powers of overlap converges poorly.

The methods of this type were for the most part proposed at a time when the capacity of computers was limited, and the problems that arise from the introduction of diffuse basis functions were not apparent. Even without diffuse basis functions, however, the results are not very good. In a calculation using a minimal basis for two H_2 molecules, for example, where the exact result can be calculated analytically, the energy to second order was in error by 50%, even when the intermolecular overlap integrals were only about 0.1 (Hayes and Stone 1984*a*). Since the formulae of perturbation theory become increasingly complicated after second order, any method that is to be useful in practice must give good results at second order.

6.2.2 *Orthogonalization of the antisymmetrized-product basis*

An alternative way to avoid the difficulties that arise from a non-orthogonal basis might be to construct an orthogonal basis by Schmidt or Löwdin orthogonalization. Unfortunately this suffers from similar difficulties. It is easier to think about it in terms of molecular orbitals rather than many-electron states. We have a set of orbitals ψ_k^A for molecule A, orthogonal among themselves, and a similar set $\psi_{k'}^B$ for molecule B, also orthogonal among themselves but not to the ψ_k^A. In each case there is a set of occupied orbitals and a set of vacant (virtual) orbitals. In order to make $\psi_{k'}^B$ orthogonal to the ψ_k^A it is necessary to add a little of each ψ_k^A to it. In a Schmidt treatment, we might have

$$\tilde{\psi}_{k'}^B = \psi_{k'}^B - \sum_k \psi_k^A \langle \psi_k^A | \psi_{k'}^B \rangle.$$

Inevitably, there is a contamination of the orbitals of B by those of A. This particular formulation is unsymmetrical between A and B, so it is not a satisfactory method. We might seek to improve it by a procedure such as the following. Take a Slater determinant describing the SCF ground state of the unperturbed $A+B$ system. First orthogonalize the occupied orbitals of A with those of B, using a symmetrical method such as Löwdin orthogonalization. This ensures that the SCF description of the ground state is unchanged, because it merely mixes the columns of the Slater determinant. Then Schmidt orthogonalize the virtual orbitals of A and B to all the occupied orbitals. This will modify the virtual orbitals but leave the occupied orbitals unchanged. Finally, orthogonalise the virtual orbitals of A symmetrically with those of B. This procedure achieves orthogonality of the orbitals, and hence of the states constructed from them, while changing the low-lying states as little as possible, and the ground state not at all. Reed *et al.* (1986) described a variant of this procedure suitable for correlated wavefunctions, in which the natural orbitals for the unperturbed molecules were used as a starting-point, and their procedure preserved the form of the orbitals with occupation numbers near 2 as far as possible consistent with achieving orthogonality. Nevertheless all such schemes show the same feature: there is inevitably a contamination of the orbitals of each molecule by those of the other. This means that it becomes difficult to regard molecular states constructed from these orbitals as unperturbed states in any natural sense, and this makes the physical interpretation of numerical results difficult and very misleading. In particular, it seems to lead to much larger estimates of the charge transfer energy than other methods (Reed *et al.* 1986, 1988, King and Weinhold 1995).

6.2.3 *Non-hermitian zeroth-order Hamiltonian*

If we require the antisymmetrized products to be eigenfunctions of a zeroth-order operator $\mathcal{H}^0$, it follows, as we have seen, that this operator cannot be hermitian. However it is possible to construct a suitable non-hermitian operator. Since the complete Hamiltonian is hermitian, the perturbation $\mathcal{H}'$ is also necessarily non-hermitian.

Jansen (1967) proposed one method for constructing $\mathcal{H}^0$. It is most easily understood with the help of an example. If we consider the interaction of a helium atom with a hydrogen atom, then at long range we can allocate particular electrons to each atom. The electron on the hydrogen atom may be either electron 1, or electron 2, or electron 3. In each case the other two electrons are on the He atom. If electron 1 is on the hydrogen, the Hamiltonian for the H atom is $\mathcal{H}^{\mathrm{H}}(1)$, and that for the He atom is $\mathcal{H}^{\mathrm{He}}(2,3)$. If we have wavefunctions $\psi^{\mathrm{H}}(1)$ and $\psi^{\mathrm{He}}(2,3)$ for the isolated atoms, satisfying $\mathcal{H}^{\mathrm{H}}(1)\psi^{\mathrm{H}}(1) = W^{\mathrm{H}}\psi^{\mathrm{H}}(1)$ and $\mathcal{H}^{\mathrm{He}}(2,3)\psi^{\mathrm{He}}(2,3) = W^{\mathrm{He}}\psi^{\mathrm{He}}(2,3)$ respectively, the zeroth-order wavefunction for the separated atoms is the product $\psi^{\mathrm{H}}(1)\psi^{\mathrm{He}}(2,3)$ of the wavefunctions for the two atoms. Similarly, in the case when electron 2 is on the H atom, the unperturbed Hamiltonian is $\mathcal{H}^{\mathrm{H}}(2)+\mathcal{H}^{\mathrm{He}}(1,3)$ and the wavefunction $\psi^{\mathrm{H}}(2)\psi^{\mathrm{He}}(1,3)$, and when electron 3 is on the H atom they are $\mathcal{H}^{\mathrm{H}}(3)+\mathcal{H}^{\mathrm{He}}(1,2)$ and $\psi^{\mathrm{H}}(3)\psi^{\mathrm{He}}(1,2)$.

The antisymmetrized unperturbed wavefunction for the whole system is a linear combination of these functions:

$$\Psi(1,2,3) = \sqrt{\tfrac{1}{3}}\left(\psi^{\mathrm{H}}(1)\psi^{\mathrm{He}}(2,3) - \psi^{\mathrm{H}}(2)\psi^{\mathrm{He}}(1,3) + \psi^{\mathrm{H}}(3)\psi^{\mathrm{He}}(1,2)\right). \tag{6.2.1}$$

(Remember that $\psi^{\mathrm{He}}(2,1) = -\psi^{\mathrm{He}}(1,2)$.)

Now we define a projection operator Λ_1 which annihilates any function which does not have electron 1 assigned to the H atom, and Λ_2 and Λ_3 similarly. Thus $\Lambda_1\Psi(1,2,3) = \sqrt{\frac{1}{3}}\psi^{\mathrm{H}}(1)\psi^{\mathrm{He}}(2,3)$, for example. Then the zeroth-order Hamiltonian is

$$\begin{aligned}\mathcal{H}^0 &= (\mathcal{H}^{\mathrm{H}}(1)+\mathcal{H}^{\mathrm{He}}(2,3))\Lambda_1 + (\mathcal{H}^{\mathrm{H}}(2)+\mathcal{H}^{\mathrm{He}}(1,3))\Lambda_2 \\ &\quad + (\mathcal{H}^{\mathrm{H}}(3)+\mathcal{H}^{\mathrm{He}}(1,2))\Lambda_3.\end{aligned} \tag{6.2.2}$$

When this operates on $\psi^{\mathrm{H}}(1)\psi^{\mathrm{He}}(2,3)$, the first term gives

$$(\mathcal{H}^{\mathrm{H}}(1)+\mathcal{H}^{\mathrm{He}}(2,3))\psi^{\mathrm{H}}(1)\psi^{\mathrm{He}}(2,3) = (W^{\mathrm{H}}+W^{\mathrm{He}})\psi^{\mathrm{H}}(1)\psi^{\mathrm{He}}(2,3),$$

and the other two terms give zero because $\psi^{\mathrm{H}}(1)\psi^{\mathrm{He}}(2,3)$ is annihilated by Λ_2 and Λ_3. In the same way, the perturbation is

$$\mathcal{H}' = \mathcal{H}'(1;2,3)\Lambda_1 + \mathcal{H}'(2;1,3)\Lambda_2 + \mathcal{H}'(3;1,2)\Lambda_3, \tag{6.2.3}$$

where $\mathcal{H}'(1;2,3)$ is the form that the perturbation would take at long range if electron 1 were on the H atom and the other two electrons on He. Since the operators $\mathcal{H}^{\mathrm{H}}(1)+\mathcal{H}^{\mathrm{He}}(2,3))+\mathcal{H}'(1;2,3)$, $\mathcal{H}^{\mathrm{H}}(2)+\mathcal{H}^{\mathrm{He}}(1,3))+\mathcal{H}'(2;1,3)$ and $\mathcal{H}^{\mathrm{H}}(3)+\mathcal{H}^{\mathrm{He}}(1,2))+\mathcal{H}'(3;1,2)$ are all equal to the complete Hamiltonian for the system, and $\Lambda_1+\Lambda_2+\Lambda_3 = 1$, it follows that $\mathcal{H}^0+\mathcal{H}'$ is also the complete Hamiltonian. However neither $\mathcal{H}^0$ nor $\mathcal{H}'$ is hermitian.

The perturbation theory developed using this formalism (Jansen 1967) has not been widely used, perhaps because many people are uncomfortable with operators that are not hermitian. Juanós i Timoneda and Hunt (1986) used it successfully to calculate the collision-induced dipole moment of the He···H system, and Polymeropoulos *et al.* (1984) applied it to the calculation of three-body exchange interactions for inert gases.

6.2.4 *Biorthogonal functions*

An alternative approach has been proposed by Surjan *et al.* (1985), following earlier work by Gouyet (1973). Denoting a member $|m_A n_B\rangle = \mathcal{A}|m_A\rangle|n_B\rangle$ of the set of antisymmetrized products by $|I\rangle$, we can construct the overlap matrix, with elements $S_{IJ} = \langle I|J\rangle$. We define a new 'biorthogonal' set of functions $|\bar{I}\rangle = \sum_J |J\rangle(S^{-1})_{JI}$. By construction these satisfy $\langle\bar{K}|I\rangle = \delta_{KI}$, as can easily be verified. We now define an operator $\mathcal{H}^0$ whose eigenvectors are the $|I\rangle$: $\mathcal{H}^0|I\rangle = W_I^0|I\rangle$. (Jansen's approach shows that such an operator can be constructed.) $\mathcal{H}^0$ is non-hermitian, but the *matrix* $\mathbf{H}$ with elements $H_{JI} = \langle\bar{J}|\mathcal{H}^0|I\rangle$ is not only hermitian but diagonal. (If this appears to be contradictory, note that the matrix elements are constructed with the biorthogonal functions $\langle\bar{J}|$ on the left and the original functions $|I\rangle$ on the right. If $\mathcal{H}^0$ were hermitian, we would have $\langle\bar{J}|\mathcal{H}^0|I\rangle^* = \langle I|\mathcal{H}^0|\bar{J}\rangle$, but the elements of $\mathbf{H}$ satisfy $\langle\bar{J}|\mathcal{H}^0|I\rangle^* = \langle\bar{I}|\mathcal{H}^0|J\rangle$.)

It is now possible to work through the standard Rayleigh–Schrödinger perturbation theory formalism, using the $|I\rangle$ as ket vectors in every matrix element and the $\langle\bar{I}|$ as bra vectors, to obtain expressions for the perturbed energy (Surjan *et al.* 1985). The method was applied to a number of examples (Kochanski and Gouyet 1975*a*, 1975*b*, Surjan and Poirier 1986), and gave good results in some cases, but not in all. It was thought at one time that the poor results were a consequence of the biorthogonal formalism, since the overlap matrix S may become singular or ill-conditioned when the basis set becomes large, but in a recent paper, Surjan and Mayer (1991) argue that they arise not from overlap problems but from an incorrect formulation of the perturbation operator. Neverthless the energy expressions involve matrix elements in which the $\langle\bar{I}|$ appear on the left, and the problems of contamination of the orbitals of each molecule by those of the other are just as severe as they are in procedures that involve orthogonalization.

6.2.5 *Non-orthogonal perturbation theory*

The use of orthogonal expansion functions is a convenience in carrying out calculations, and is a natural consequence of conventional Rayleigh–Schrödinger perturbation theory, where the functions used as a basis for expanding the first-order wavefunction are eigenfunctions of the zeroth-order Hamiltonian. However it is not necessary to use an orthogonal expansion set, and at the cost of somewhat greater complexity in the calculations, a non-orthogonal set can be used. It is also possible to formulate the problem in such a way that no separation into zeroth-order Hamiltonian and perturbation is necessary. In the case of intermolecular interactions, the natural basis is the set of antisymmetrized product functions. Taking the pragmatic view that we can only ever handle a finite number of them in numerical calculations, we regard them as a basis for a matrix treatment of the problem. That is, we seek to solve the matrix equation

$$(\mathbf{H} - W\mathbf{S})\mathbf{c} = 0, \tag{6.2.4}$$

where $\mathbf{H}$ is the matrix of the Hamiltonian between the unperturbed functions, $\mathbf{S}$ is the overlap matrix, W the perturbed energy, and $\mathbf{c}$ the vector of coefficients. This is then solved perturbatively by splitting $\mathbf{H}$ into a zeroth-order part $\mathbf{H}^0$ and a perturbation $\mathbf{H}'$. It is possible to choose both $\mathbf{H}^0$ and $\mathbf{H}'$ to be hermitian, but the $|m_A n_B\rangle$ cannot all be eigenfunctions of $\mathbf{H}^0$ for the reasons already discussed. If $\mathbf{H}^0$ is chosen to comprise the diagonal part of **H**, the following expression emerges for the perturbed energy of the system (Hayes and Stone 1984*a*):

$$W = H_{00} - \sum_k{}' \frac{(H_{0k} - H_{00}S_{0k})(H_{k0} - H_{00}S_{k0})}{H_{kk} - H_{00}} + \cdots. \tag{6.2.5}$$

The choice of $\mathbf{H}^0$ as the diagonal part of **H** is not the only possibility. Any diagonal matrix can be used, and if $\mathbf{H}^0$ has diagonal elements E_k the energy to second order has the same form, i.e.,

$$W = H_{00} - \sum_k{}' \frac{(H_{0k} - E_0S_{0k})(H_{k0} - E_0S_{k0})}{E_k - E_0} + \cdots. \tag{6.2.6}$$

Higher-order terms will be different, and evidently the convergence of the perturbation series may be improved or (more likely) worsened by such a change.

If SCF wavefunctions, i.e., single Slater determinants in the closed-shell case, are used as the single-molecule basis, the antisymmetrized product functions are simply Slater determinants constructed from the molecular orbitals of the two molecules. In this case a choice for the E_k which does not appear to affect the convergence adversely is the Møller–Plesset formula: $E_k = \sum_i n_{ik}\varepsilon_i$, with n_{ik} the occupation number of orbital i in determinant k and ε_i the orbital energy. The formulae needed for manipulating these determinantal functions in the context of perturbation theory have been worked out (Hayes and Stone 1984*b*), and a perturbation theory using these formulae has been developed (Hayes and Stone 1984*a*) and implemented to second order. Subsequently, Magnasco and Figari (1986) described a similar method, also to second order, and also showed how to generalize the manipulation of the determinantal functions (Figari and Magnasco 1985), so that the perturbation theory could in principle be extended to higher order.

6.2.6 *Expansion in orders of exchange*

A further approximation which has sometimes been used is based on the premise that if exchange is negligible at long range its effects will be small at shorter range. The overlap between a product function $|m_A\rangle|n_B\rangle$ and the function obtained from it by exchanging one of the electrons of A with one of the electrons of B is of order S^2, where S represents the magnitude of overlap integrals between orbitals on A and orbitals on B. (See eqn (6.1.2), for example.) The magnitude of this overlap decreases exponentially with distance, say like $e^{-\alpha R}$. The overlap between $|m_A\rangle|n_B\rangle$ and a function obtained from it by two electron exchanges is of order S^4, so if S^2 is small, the contributions of these double exchanges may be negligible. The contributions of triple exchanges will be even smaller, and so on. Consequently the approximation is made that only single exchanges need be considered. Such a method has been described, for example, by Hess *et al.* (1990). An approximation of this sort may be necessary when correlated zeroth-order wavefunctions are used, because the formulae become intractable when exchanges of more than one pair of electrons are included, but it is not necessary for SCF zeroth-order wavefunctions, where the standard procedures for handling Slater determinants can be used, modified to cope with the non-orthogonality of the orbitals. An obvious limitation of the approximation is that when the molecules are close together, S is no longer small, and the neglect of terms of order S^4 may no longer be justifiable. This problem becomes worse as the molecules approach, so the effects of exchange on the potential energy curve are increasingly underestimated.

6.3 Symmetry-adapted perturbation theories

The difficulties that appear in formulating the symmetric perturbation theories arise from the fact that there can be no hermitian zeroth-order Hamiltonian which has the antisymmetrized products as eigenfunctions, because of their non-orthogonality. Symmetry-adapted perturbation theories avoid these difficulties by taking as their expansion functions the simple products $|m_A\rangle|n_B\rangle$, without antisymmetrization. These are eigenfunctions of a well-defined zeroth-order Hamiltonian in which some of the electrons are assigned to molecule A and the rest to molecule B. However, because the true wavefunction for the system must be antisymmetric, it is necessary to work with the antisymmetrized functions $|m_A n_B\rangle = \mathcal{A}|m_A\rangle|n_B\rangle$ alongside the unsymmetrized ones. The first theory of this sort was given by Eisenschitz and London (1930) in the early days of quantum mechanics. Van der Avoird (1967*a*, 1967*b*) gave a version of this treatment in more modern notation. Although very elegant, it suffers from a lack of rigour, and Van der Avoird (1967*c*) also gave a more rigorous treatment. To second order, the results are the same: in a notation using ψ_k to denote a member of the set of unperturbed simple product functions $|m_A\rangle|n_B\rangle$, with ψ_0 for the unperturbed ground-state function $|0_A\rangle|0_B\rangle$, the first-order energy is

$$W' = \frac{\langle \mathcal{A}\psi_0|\mathcal{A}\mathcal{H}'\psi_0\rangle}{\langle \mathcal{A}\psi_0|\mathcal{A}\psi_0\rangle}, \tag{6.3.1}$$

and the second-order energy is

$$W'' = \frac{1}{\langle \mathcal{A}\psi_0|\mathcal{A}\psi_0\rangle}\sum_k{}' \frac{\langle \mathcal{A}(\mathcal{H}'-W')\psi_0|\mathcal{A}\psi_k\rangle\langle \mathcal{A}\psi_k|\mathcal{A}(\mathcal{H}'-W')\psi_0\rangle}{W_0^0 - W_k^0}. \tag{6.3.2}$$

It is instructive to compare these formulae with those of the symmetric non-orthogonal treatment, eqn (6.2.5). At first sight they look different, but in fact they are the same. From (6.3.1) the energy to first order is

$$W^0 + W' = \frac{\langle \mathcal{A}\psi_0|\mathcal{A}(W^0+\mathcal{H}')\psi_0\rangle}{\langle \mathcal{A}\psi_0|\mathcal{A}\psi_0\rangle} = \frac{\langle \mathcal{A}\psi_0|\mathcal{A}\mathcal{H}\psi_0\rangle}{\langle \mathcal{A}\psi_0|\mathcal{A}\psi_0\rangle}, \tag{6.3.3}$$

and this is the same as H_{00} if ψ_0 is normalized and antisymmetric, as it is in the symmetric non-orthogonal treatment. In the second-order energy, eqn (6.3.2), note that $(\mathcal{H}^0 - W_0^0)\psi_0 = 0$, so we can replace $(\mathcal{H}' - W')\psi_0$ by $(\mathcal{H}^0 + \mathcal{H}' - W_0^0 - W')\psi_0 = (\mathcal{H} - H_{00})\psi_0$ to give the second-order term of (6.2.5).

6.3.1 *Iterative symmetry-forcing procedures*

There is a family of related perturbation methods of the symmetry-adapted type. One method of deriving them is to seek a primitive function ψ which is not symmetric, but from which the true perturbed wavefunction Ψ may be derived by antisymmetrization: $\Psi = \mathcal{A}\psi$. The primitive function ψ is not uniquely defined, and by imposing different conditions on it one obtains different perturbation expansions (Amos 1970). These include the Van der Avoird version of the Eisenschitz–London theory mentioned above, and the equivalent treatment due to Hirschfelder (1967) (these methods being usually referred to together as EL–HAV), the HS method of Hirschfelder and Silbey (1966), and the MS–MA method, derived independently by Murrell and Shaw (1967) and Musher and Amos (1967), their two methods being later shown to be equivalent (Johnson and Epstein 1968). The convergence of these methods was compared for a simple model problem by Epstein and Johnson (1968); a discouraging result of this comparison was that the radius of convergence of the power series in the perturbation parameter λ was in many cases much smaller than the value of 1 needed to describe the perturbed problem. Jeziorski *et al.* (1978) also compared the convergence, using the more realistic example of H_2^+, and showed that the MS–MA method diverges.

Jeziorski and Kołos (1977) proposed a different approach to the derivation of these perturbation methods. We start with the unperturbed Hamiltonian $\mathcal{H}^0 = \mathcal{H}^A + \mathcal{H}^B$, with some electrons assigned to molecule A and the rest to molecule B. Assume that we have a set of zeroth-order eigenfunctions ψ_k and eigenvalues W_k:

$$\mathcal{H}^0 \psi_k = W_k \psi_k, \tag{6.3.4}$$

where as before ψ_k is an unsymmetrized product of unperturbed functions for the two isolated molecules. Now define the 'reduced resolvent' R_0 (see Appendix C), which acts like the inverse of the operator $\mathcal{H}^0 - W_0$ for functions orthogonal to ψ_0 but gives zero when operating on ψ_0:

$$R_0 \psi_k = \begin{cases} (W_k - W_0)^{-1} \psi_k, & k \neq 0, \\ 0, & k = 0. \end{cases} \tag{6.3.5}$$

In operator notation, this can be expressed in the form

$$\begin{aligned} R_0(\mathcal{H}^0 - W_0) &= 1 - P_0, \\ R_0 P_0 &= 0, \end{aligned} \tag{6.3.6}$$

where $P_0 = |\psi_0\rangle\langle\psi_0|$. In terms of the complete set of eigenvalues and eigenvectors of $\mathcal{H}^0$, R_0 may be expressed in the form

$$R_0 = \sum_{k \neq 0} |\psi_k\rangle (W_k - W_0)^{-1} \langle \psi_k|. \tag{6.3.7}$$

Now we look for a solution Ψ to the perturbed problem:

$$\mathcal{H}\Psi = W\Psi, \tag{6.3.8}$$

subject to the usual intermediate normalization condition $\langle\psi_0|\Psi\rangle = 1$. Writing $W = W_0 + \Delta W$ and $\mathcal{H} = \mathcal{H}^0 + \mathcal{H}'$, we have

$$(\mathcal{H}^0 + \mathcal{H}' - W_0 - \Delta W)\Psi = 0, \tag{6.3.9}$$

and when we operate on this equation with R_0 and use (6.3.6) we get

$$(1 - P_0 + R_0(\mathcal{H}' - \Delta W))\Psi = 0, \tag{6.3.10}$$

or

$$\Psi = \psi_0 + R_0(\Delta W - \mathcal{H}')\Psi. \tag{6.3.11}$$

Also from (6.3.9) we see that

$$\Delta W = \langle \psi_0 | \mathcal{H}' | \Psi \rangle. \tag{6.3.12}$$

Now we may attempt to solve these equations iteratively, through the scheme

$$\Delta W_n = \langle \psi_0 | \mathcal{H}' | \Psi_{n-1} \rangle, \tag{6.3.13}$$
$$\Psi_n = \psi_0 + R_0(\Delta W_n - \mathcal{H}')\Psi_{n-1}. \tag{6.3.14}$$

This scheme may or may not converge; whether it does or not will depend on the function that we use to start the iterations. If we choose $\Psi_0 = \psi_0$, the first iterations yield

$$\begin{aligned}
\Delta W_1 &= \langle \psi_0 | \mathcal{H}' | \psi_0 \rangle, \\
\Psi_1 &= \psi_0 + R_0(\Delta W_1 - \mathcal{H}')\psi_0, \\
&= \psi_0 + \sum_k{}' \psi_k \frac{\langle \psi_k | \mathcal{H}' | \psi_0 \rangle}{W_0 - W_k}, \\
\Delta W_2 &= \langle \psi_0 | \mathcal{H}' | \psi_0 \rangle + \langle \psi_0 | \mathcal{H}' R_0 \mathcal{H}' | \psi_0 \rangle \\
&= \langle \psi_0 | \mathcal{H}' | \psi_0 \rangle + \sum_k{}' \frac{\langle \psi_0 | \mathcal{H}' | \psi_k \rangle \langle \psi_k | \mathcal{H}' | \psi_0 \rangle}{W_0 - W_k},
\end{aligned}$$

which are the standard results of Rayleigh–Schrödinger perturbation theory. Clearly this initial choice is a bad one, because Ψ is antisymmetric and is very different from ψ_0, which is not. We have already seen that Rayleigh–Schrödinger perturbation theory converges badly for just this reason. If we start instead from the antisymmetrized function $\mathcal{A}\psi_0$ we should do better. Because of the normalization condition we have to set $\Psi_0 = \langle \psi_0 | \mathcal{A}\psi_0 \rangle^{-1} \mathcal{A}\psi_0$. Jeziorski and Kołos (1977) showed that this generates the MS–MA perturbation scheme.

Even if we start from an antisymmetric Ψ_0, eqn (6.3.14) does not maintain the antisymmetry, so Ψ_n is not in general antisymmetric for finite n, and indeed it need not be antisymmetric in the limit $n \to \infty$, even if the process converges. This does not matter so long as the energy expression is right, but it seems intuitively that it would be more satisfactory if the antisymmetry could be maintained. Jeziorski and Kołos (1977) proposed to improve the convergence by forcing the correct symmetry on Ψ_{n-1} before using it in the calculation of ΔW_n or Ψ_n or both. The iteration scheme becomes

$$\Delta W_n = \langle \psi_0 | \mathcal{H}' | \mathcal{G}\Psi_{n-1} \rangle, \tag{6.3.15}$$
$$\Psi_n = \psi_0 + R_0(\Delta W_n - \mathcal{H}')\mathcal{F}\Psi_{n-1}, \tag{6.3.16}$$

where $\mathcal{F}$ and $\mathcal{G}$ are the symmetry-forcing operators. They must satisfy $\mathcal{G}\Psi = \mathcal{F}\Psi = \Psi$, but otherwise they are arbitrary. In practice the available choices are, for $\mathcal{F}$, either 1 or $\mathcal{A}$; and for $\mathcal{G}$, either 1 or a modified antisymmetrizer defined by $\mathcal{G}\chi = \langle \psi_0 | \mathcal{A}\chi \rangle^{-1} \mathcal{A}\chi$, this form being needed in order to preserve the intermediate normalization of the antisymmetrized function. A wide variety of perturbation schemes, including the MS–MA, HS and EL–HAV schemes, can be generated using suitable choices for the symmetry-forcing (which does not have to be the same at every iteration) and for the starting function. In a subsequent paper, Jeziorski *et al.* (1978) found that the choice $\mathcal{G} = 1$, $\mathcal{F} = \mathcal{A}$, $\Psi_0 = \langle\psi_0|\mathcal{A}\psi_0\rangle^{-1}\mathcal{A}\psi_0$, suggested by Jeziorski and Kołos (1977) and called the JK scheme, was the most satisfactory for strong interactions, while a scheme in which the symmetry is forced only in the evaluation of the energy is most successful at longer range where the interaction is weaker. This scheme, called the symmetrized Rayleigh–Schrödinger (SRS) scheme, takes the form

$$\Delta W_n = \langle\psi_0|\mathcal{H}'|\mathcal{G}\Psi_{n-1}\rangle, \tag{6.3.17}$$

$$\Psi_n = \psi_0 + R_0(\langle\psi_0|\mathcal{H}'|\Psi_{n-1}\rangle - \mathcal{H}')\Psi_{n-1}, \tag{6.3.18}$$

with $\Psi_0 = \psi_0$ and $\mathcal{G}$ defined by $\mathcal{G}\chi = \langle\psi_0|\mathcal{A}\chi\rangle^{-1}\mathcal{A}\chi$, as above. Here the corrections to the wavefunction have precisely the same form as in the ordinary Rayleigh–Schrödinger scheme, and it is only in the formula for the energy that the antisymmetrizer appears. This method has been used successfully for $(H_2O)_2$ and $(HF)_2$ (Rybak *et al.* 1991) and $Ar\cdots H_2$ (Williams *et al.* 1993), but the success of this method is not yet fully understood (Adams 1994). One may obtain a power series expansion of the interaction energy by replacing $\mathcal{H}'$ in any of these schemes by $\lambda\mathcal{H}'$ and picking out powers of λ from the resulting expressions. If this is done for the SRS scheme, the resulting power series for the ground state of H_2 has a radius of convergence of 1.0000000031 at the Van der Waals minimum (Cwiok *et al.* 1992), and so converges extremely slowly for $\lambda = 1$. Moreover the radius of convergence can be shown to be less than 1 if either of the monomers has more than 2 electrons (Kutzelnigg 1980). It seems likely that the SRS scheme works because at second order (which is the best that can be achieved in practice for most systems) it gives results that differ hardly at all from the HS scheme, which converges much better but is computationally impracticable (Cwiok *et al.* 1994).

The importance of the symmetry-forcing approach lies in the fact that it is possible to develop a satisfactory perturbative treatment of intermolecular forces by means of an iterative scheme rather than a conventional power series expansion. In this way, the difficulties that arise from the ambiguities in the definition of perturbation order can be circumvented, and the arbitrariness that remains in the formulation of the iterative scheme is seen to be not a fundamental problem but merely a flexibility that can be exploited to achieve the most effective convergence of the process.

6.3.2 *The treatment of electron correlation*

It must be said that the existence of this flexibility is to a large degree academic, since practical calculations on systems of any size cannot be carried out beyond second order, and even then considerable practical difficulties remain. The most important of these is that the iterative scheme is formulated in terms of the reduced resolvent R_0, which is defined in terms of a sum over states, and these states are products of the exact eigenstates of the isolated molecules, which we do not know.

Accordingly it is necessary, if we are to make any progress, to start from a simplified treatment of the isolated molecules. The natural choice is the Self-Consistent-Field approximation, in which the effects of electron correlation are ignored (see Chapter 5). It is a practicable treatment for molecules of significant size, and although it too is not exactly soluble, it is possible to get quite close to the exact solution (the 'Hartree–Fock limit') using modern basis sets.

The procedure then is to separate the complete Hamiltonian in the following way:

$$\mathcal{H} = \mathcal{F}^A + \mathcal{F}^B + \zeta^A \mathcal{G}^A + \zeta^B \mathcal{G}^B + \lambda \mathcal{H}', \tag{6.3.19}$$

where $\mathcal{F}^A$ and $\mathcal{F}^B$ are the Fock operators for the isolated molecules A and B, respectively, $\mathcal{G}^A = \mathcal{H}^A - \mathcal{F}^A$ describes the effects of electron correlation in molecule A and similarly $\mathcal{G}^B = \mathcal{H}^B - \mathcal{F}^B$, and $\mathcal{H}'$ is the intermolecular interaction, as before. It is now necessary to carry out a triple perturbation expansion, the coefficients ζ_A, ζ_B and λ being used to identify orders of perturbation. The theory becomes extremely complicated, and has been described in detail by Jeziorski *et al.* (1993, 1994).

It seems at the moment that following a long period of neglect by most people, the perturbation approach to the calculation of intermolecular forces at short range may be due for a revival. The symmetric method of Hayes and Stone (1984*a*) has been applied to systems as large as the benzene-tetracyanoethylene complex, and while it does not handle correlation corrections (except for the dispersion energy itself) it provides an efficient route to the uncorrelated terms. The symmetry-forcing methods have for many years been confined to small systems of little interest to chemists, but they have now been developed to the point where results of high accuracy are emerging for larger systems such as $(H_2O)_2$ and $(HF)_2$ (Rybak *et al.* 1991). Even larger systems, such as uracil-water, have been studied, though only with a single-zeta-plus-polarization basis (Rybak *et al.* 1992).

6.4 The first-order energy at short range

In spite of the plethora of perturbation treatments, the first-order energy is essentially the same in all methods: it is the difference between the expectation value of the full Hamiltonian $\mathcal{H}$ for the antisymmetrized product function $|0_A 0_B\rangle$ and the zeroth-order energy. The latter is unambiguously $W_A^0 + W_B^0$ without any need to specify a zeroth-order Hamiltonian. We have seen that there are two contributions to the first-order energy: the electrostatic energy and the exchange–repulsion energy.

6.4.1 *The electrostatic energy at short range: penetration*

The electrostatic energy can be described at any distance by the expression 4.1.3; it is defined in terms of the unperturbed charge densities of the two molecules. At long range, the multipole expansion provides a satisfactory alternative, but at short range, even when the multipole expansion still converges, it is in error by an amount called the *penetration energy*. Consider as an illustration the interaction between a proton and a hydrogen-like atom of nuclear charge Z. The wavefunction of the latter is $\sqrt{Z^3/\pi}\exp(-Zr)$, so that the electron charge density is $-e(Z^3/\pi)\exp(-2Zr)$. The potential at the proton due to this electron density is the solution of Poisson's equation:

$$\nabla^2 V = -\frac{\rho}{\varepsilon_0} = 4Z^3 \exp(-2Zr), \tag{6.4.1}$$

since $4\pi\varepsilon_0 = 1$ in atomic units. Since the charge density is spherically symmetric, the potential is also spherically symmetric, so this equation becomes

$$\frac{1}{r^2}\frac{\partial}{\partial r}r^2\frac{\partial V}{\partial r} = 4Z^3\exp(-2Zr),$$

which can be solved by standard methods to give

$$V(r) = -\frac{1}{r} + \exp(-2Zr)\left(Z + \frac{1}{r}\right). \tag{6.4.2}$$

Note that this does not include the potential Z/r due to the nucleus. We see that there are two terms. One is proportional to $-1/r$ and is the multipole expansion of the potential due to the spherical electron cloud. We note that this is trivially convergent at all non-zero values of r. The other term contains a factor $\exp(-2Zr)$ and describes the correction to the potential at short distances that arises from the finite extent of the charge distribution. This is the penetration correction. Note that when $r \to 0$ the potential remains finite, as it should; there is no singularity in the electron charge distribution, and therefore no singularity in the potential. In the general case, the complete penetration correction cancels out all the r^{-n} singularities that arise from the multipole expansion of the potential due to the electrons, leaving only the Z/r terms from the nuclei.

Note also that the term in $\exp(-2Zr)$ does not contribute to the multipole expansion. Viewed as a function of $z = 1/r$, $\exp(-2Zr) = \exp(-2Z/z)$ has an essential singularity at $z = 0$: it tends to zero as $z \to 0$ from positive values, but to ∞ as $z \to 0$ from negative real values, so it has no Taylor expansion in powers of z.

At moderate distances the penetration correction is approximately proportional to $\exp(-2Zr)$, i.e., to the charge density. It is a positive correction to the potential, so it yields a negative correction to the interaction energy in the usual case where the electron distribution of one molecule penetrates into another. If we may assume that the penetration correction to the potential of molecule A, ΔV^A_{pen} is proportional to the charge density ρ^A in the general case, then the correction to the energy is

$$\Delta U_{\text{pen}} \approx \int \Delta V^A_{\text{pen}}(\mathbf{r})\rho^B(\mathbf{r})\,\mathrm{d}^3\mathbf{r} \propto \int \rho^A(\mathbf{r})\rho^B(\mathbf{r})\,\mathrm{d}^3\mathbf{r}.$$

Note that the 'constant' of proportionality includes the nuclear charge Z, so this is unlikely to be very accurate as a general formula. Nevertheless it provides some guidance to the behaviour of the effect, and moreover suggests that the penetration correction can be modelled for practical purposes as part of the repulsion energy, which has a similar behaviour, as we shall see.

It is possible to write 6.4.2 in the form

$$\begin{aligned} V(r) &= -\frac{1}{r}[1 - \exp(-2Zr)(1 + rZ)] \\ &= -\frac{1}{r}f_1(2Zr), \end{aligned} \tag{6.4.3}$$

in which the quantity in square brackets is treated as a *damping function* $f_1(2Zr)$ that is 1 at long range and tends to zero at short range in such a way as to suppress the singularity in the multipole term. The dispersion and induction terms, where similar effects arise (see §6.5.1) are commonly treated this way, but the treatment of electrostatic penetration by means of damping functions does not seem to have been explored.

6.4.2 *The exchange–repulsion energy*

The main part of the first-order energy is normally described as 'exchange–repulsion', and is a combination of two effects, as we noted above. The 'exchange' energy is a consequence of the fact that the electron motions can extend over both molecules, and is an attractive term, while the 'repulsion', as its name suggests, is a repulsive term that arises when the electrons attempt to occupy the same region of space, and are forced to redistribute because the Pauli principle forbids electrons of the same spin to be in the same place. The exchange–repulsion is termed 'exchange' by some authors.

For single-determinant (SCF) wavefunctions, we can write the exchange repulsion energy in the form (Hayes and Stone 1984*a*)

$$U_{\mathrm{er}} = U_{\mathrm{exch}} + U_{\mathrm{rep}},$$

where

$$\begin{aligned} U_{\mathrm{exch}} &= -\tfrac{1}{2}\sum_{k\in A}\sum_{l\in B}(kl|lk), \\ U_{\mathrm{rep}} &= \sum_{kl}\langle k|h_A + h_B - \tfrac{1}{2}\nabla^2|l\rangle(T_{kl} - \delta_{kl}) \\ &\quad + \tfrac{1}{2}\sum_{klmn}(kl|mn)[T_{kl}T_{mn} - \delta_{kl}\delta_{mn} - T_{kn}T_{ml} + \delta_{kn}\delta_{ml}]. \end{aligned}$$

Here h_A is the electron–nuclear attraction for molecule A, and $(kl|mn)$ is a two-electron integral $\int k(1)l(2)r_{12}^{-1}m(1)n(2)\,d\tau_1\,d\tau_2$. T is the inverse of the overlap matrix between the occupied spin-orbitals of the system. The sums are taken over the occupied spin-orbitals. Notice that T_{kl} is zero if orbitals k and l have different spin. In the interaction of two hydrogen atoms that we discussed above, exchange–repulsion arises only if the two electrons have the same spin. If they have opposite spins, chemical bonding could occur. In an interaction between two closed-shell molecules, this is not a possibility, and in this case there is an exchange repulsion between pairs of electrons of the same spin, while pairs of electrons of opposite spin merely interact like classical charge densities.

6.5 The second-order energy at short range

The detailed predictions of the theory at second order depend on the form of perturbation theory that is used, but there are common features that are generally agreed to be physically valid. We recall that at long range there are two contributions to the second-order energy, namely the induction energy and the dispersion energy, and that these are normally expanded as power series in $1/R$. The features that we have discussed for the electrostatic energy also arise here, but in a form that is very much less amenable to calculation.

These two features are penetration and poor convergence. In the case of the electrostatic interaction, it was quite easy to show that there is a region where the multipole expansion converges, though convergence may be slow at short range; and that there is an error in the multipole expansion, even when it does converge—this is the penetration effect. In the case of the induction energy, it can be shown, both in general (Brooks 1952, Ahlrichs 1976) and by example (Dalgarno and Lynn 1957*a*, Kreek and Meath 1969) that the R^{-n} expansion of the induction energy never converges, as we saw in Chapter 4. It is asymptotically convergent, which means that the terms in the series decrease to a minimum magnitude and then increase again, and in series of this type it is often the case that a good estimate can be obtained by truncating the series at its smallest term, the truncation error being of the order of the first neglected term.

This feature means that the convergence characteristics of the multipole expansion of the induction energy are much less well defined than they are for the electrostatic interaction, and the same is true for the dispersion energy.

The second-order energy at short range is given by eqn (6.3.2). We can rewrite this in the following form, using the fact that $\mathcal{A}^2 = \mathcal{A}$

$$W'' = \frac{1}{\langle\psi_0|\mathcal{A}\psi_0\rangle}\sum_k{}' \frac{|\langle\psi_k|\mathcal{A}(\mathcal{H}' - W')\psi_0\rangle|^2}{W_0^0 - W_k^0}. \qquad (6.5.1)$$

Now the antisymmetrizer $\mathcal{A}$ can be written in the form

$$\mathcal{A} = N^{-1}(1 - P^{(1)} + P^{(2)} - \cdots), \qquad (6.5.2)$$

where $P^{(1)}$ is the sum of the $N_A N_B$ permutations that exchange one of the N_A electrons of A with one of the N_B electrons of B, $P^{(2)}$ is the sum of all the permutations that exchange two electrons from A with two electrons from B, and so on, and N is the total number $(N_A + N_B)!/N_A!N_B!$ of all these permutations, including the identity permutation that exchanges no electrons.

If we ignore the effect of all these permutations except the identity, we recover the formula of the long-range approximation. If this is evaluated without using the multipole expansion, we obtain non-expanded expressions for the induction and dispersion energy in the long-range approximation. (Remember that the induction energy and the dispersion energy are separated according to the nature of the excited state ψ_k in the sum over states: if ψ_k describes a state in which A alone is excited, it contributes to the induction energy of A, and so on.) The difference between each of these expressions and the corresponding multipole approximation is a penetration term, analogous to the penetration term in the electrostatic energy.

When we take the electron exchanges into account, both the normalization integral $\langle\psi_0|\mathcal{A}\psi_0\rangle$ and the matrix elements in the sum over states change. If the overlap between the molecules is small, the changes will be of order S^2, where S is a measure of the overlap of molecular obitals of A with those of B. The difference between the induction energy when exchange is taken into account, and the induction energy calculated using the (non-expanded) long-range approximation, is the 'exchange–induction' energy. Similarly, there is an 'exchange–dispersion' energy.

In the triple perturbation theory developed by Jeziorski *et al.* (1993), further corrections arise from the effects of intramolecular correlation. These corrections modify both the long-range approximation results and the exchange terms. The triple perturbation theory yields an expansion of the interaction energy in powers of the parameters λ,

Table 6.1 *Summary of perturbation terms in symmetry-adapted perturbation theory*

Symbol		Name	Physical Interpretation
$U_{\mathrm{es}}^{(0)}$	$E_{\mathrm{pol}}^{(10)}$	Electrostatic energy	Electrostatic energy of interaction, calculated classically from the unperturbed Hartree–Fock charge distributions.
$U_{\mathrm{es}}^{(l)}$	$E_{\mathrm{pol}}^{(1l)}$	Electrostatic energy: correlation corrections	Correction to the electrostatic energy arising from lth-order intramolecular correlation corrections to the charge distributions.
$U_{\mathrm{er}}^{(0)}$	$E_{\mathrm{exch}}^{(10)}$	Exchange repulsion	'Closed-shell repulsion': the modification to the first-order energy of the Hartree–Fock monomers arising from exchange of electrons between them.
$U_{\mathrm{er}}^{(l)}$	$E_{\mathrm{exch}}^{(1l)}$	Exchange repulsion: correlation corrections	Intramolecular correlation corrections to the closed-shell repulsion
$U_{\mathrm{ind}}^{(0)}$	$E_{\mathrm{ind}}^{(20)}$	Induction energy	Energy arising from the distortion of each molecule in the field due to the unperturbed Hartree–Fock charge distribution of the other.
$U_{\mathrm{exch\text{-}ind}}^{(0)}$	$E_{\mathrm{exch\text{-}ind}}^{(20)}$	Exchange induction	Modification to the induction energy when exchange effects are taken into account.
$U_{\mathrm{disp}}^{(0)}$	$E_{\mathrm{disp}}^{(20)}$	Dispersion energy	Energy arising from the correlated fluctuations of the Hartree–Fock electron distributions of each molecule.
$U_{\mathrm{exch\text{-}disp}}^{(0)}$	$E_{\mathrm{exch\text{-}disp}}^{(20)}$	Exchange dispersion	Modification to the dispersion energy when exchange effects are taken into account.
$U_{\mathrm{disp}}^{(l)}$	$E_{\mathrm{disp}}^{(2l)}$	Dispersion energy: correlation corrections	Correction to the dispersion energy arising from intramolecular correlation.

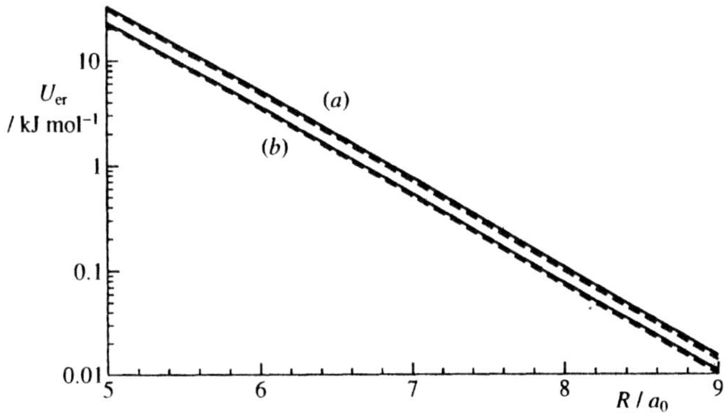

FIG. 6.2. The dependence of the exchange–repulsion energy U_{er} on separation for (a) $Ar\cdots H_2$ in a linear configuration, (b) $Ar\cdots H_2$ in a T configuration. In each case, the two adjacent lines show the repulsion with correlation corrections (continuous line), and without (dashed line).

ζ_1 and ζ_2 appearing in eq.(6.3.19). We can classify the interaction terms accordingly. Jeziorski *et al.* use a notation in which $E^{(ijk)}$ is the term of order i in the interaction, j in the intramolecular correlation for A and k in the intramolecular correlation for B. A subscript 'pol' is used for the non-expanded expression from the long-range approximation, and a subscript 'exch' for the exchange correction. They also use a notation with only two superscripts, where $E^{(il)}$ is the sum of the terms $E^{(ijk)}$ with $j+k=l$, so that l describes the total order of intramolecular correlation. In fact the label i is superfluous, since the second-order terms in the interaction are separated into induction and dispersion, and labelled accordingly, while the first-order terms are the electrostatic energy U_{es} and the exchange–repulsion U_{er}. A summary of all the terms, giving both notations, is given in Table 6.1.

One of the advantages of using perturbation theory in this way is that it is only necessary to carry out the correlation corrections to the order that is appropriate for each term separately. The electrostatic interaction is very sensitive to correlation effects, and Jeziorski *et al.* recommend using correlation corrections up to $j+k=4$. The induction energy, on the other hand, is often a small term, and if that is so the correlation corrections may be negligible. Dispersion is dominated by the Hartree–Fock term, but correlation corrections up to $j+k=2$ may be significant*, and the exchange–dispersion is a small but not negligible repulsive contribution. However these are generalizations from calculations on a rather small range of systems, and further calculations are needed to establish whether they are generally valid.

The exchange–repulsion is also affected by correlation. Calculations on small systems such as $Ar\cdots H_2$ (Williams *et al.* 1993) show that the correlation correction up to $j+k=2$ is about 10–15% of the uncorrelated exchange–repulsion, and is usually posi-

*It is difficult to find a satisfactory terminology here. Dispersion is a correlation effect, so it seems contradictory to speak of a 'Hartree–Fock' dispersion term. What we mean by this expression is the contribution to the dispersion energy that is obtained using the long-range approximation with Hartree–Fock unperturbed wavefunctions.

tive, i.e., it increases the repulsion. Figure 6.2 shows the exchange–repulsion on a semi-logarithmic plot for $Ar \cdots H_2$, with and without correlation corrections, and from this we can draw some important conclusions. First, the variation of $\log U_{er}$ with distance is very nearly linear. Secondly, the slope of the line is very nearly the same for the cases shown, though the intercept of the line with any given energy value depends on the system and on its orientation. These observations show that the exchange–repulsion can be accurately represented by a function of the form

$$U_{er} = C \exp[-\alpha(R - \rho)], \tag{6.5.3}$$

where α describes the slope of the logarithmic plot and varies only slightly between different systems, or different orientations of the same system, while ρ, which describes the separation at which the repulsion energy has the fixed value C, varies rather more. In fact ρ provides a description of the shape of the molecules concerned: it is the value of R at which U_{er} is equal to the constant C.

The third conclusion from the data shown in Fig. 6.2 is that the correlation correction changes α hardly at all, and increases ρ by an amount of the order of 0.05 bohr. Indeed, it is easy to check that if U_{er} is increased by 10% at all distances, and $\alpha \approx 2a_0{}^{-1}$, which is a typical value, then the value of ρ in eqn (6.5.3) must increase by 0.048 bohr. This is probably a reasonable guide to the behaviour of larger molecules, where the correlation corrections to U_{er} are beyond the scope of current calculations, but the uncorrected values can be computed routinely, for example by the method of Hayes and Stone (1984*a*). (See, for example, the calculations on chlorinated hydrocarbon dimers by Price *et al.* (1994).) It would be an over-simplification to suppose that this is all we need to do to take account of correlation corrections; the percentage change in U_{er} is not the same at all distances, and is unlikely to be the same for all systems. Indeed, Moszynski *et al.* (1995) found that for $HCCH \cdots He$, it was necessary to include correlation terms up to $j + k = 4$ to achieve satisfactory convergence, and although the correlation correction is repulsive in the linear geometries, where it increases the repulsion energy by about 20%, it is attractive for the T geometry, where the He atom interacts with the π bond; here the correction decreases the repulsion by about 10%. It is clear that further work is needed here.

6.5.1 *Damping functions*

The short-range perturbation theories discussed in this chapter do not make use of the multipole expansion, so the terms listed in Table 6.1 are in 'non-expanded' form. Consequently they remain finite as the molecular separation tends to zero; they are components of the electronic energy of the whole system, which remains finite for any nuclear configuration. (The only singularities in the total energy are terms like $Z_a Z_b e^2 / 4\pi\varepsilon_0 R$, arising from electrostatic repulsion between the nuclei.) However, if we use the long-range theory and the multipole expansion, the dispersion energy appears as a power series in $1/R$, with leading term $-C_6 R^{-6}$. At short distances this cannot be correct. The proper behaviour can be achieved by expressing the dispersion energy in the form (for atoms)

$$U_{disp} = -f_6(R)\frac{C_6}{R^6} - f_8(R)\frac{C_8}{R^8} - f_{10}(R)\frac{C_{10}}{R^{10}} - \cdots, \tag{6.5.4}$$

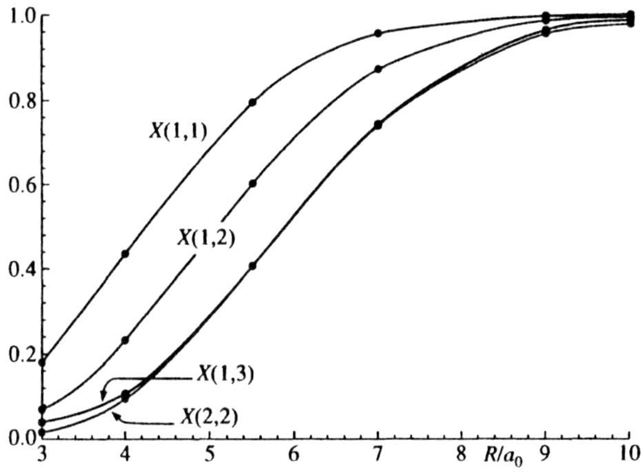

FIG. 6.3. Dispersion damping functions for Ar···Ar. $X(1,1)$: dipole–dipole, $X(1,2)$: dipole–quadrupole, $X(2,2)$: quadrupole–quadrupole and $X(1,3)$: dipole–octopole

where $f_6(R)$, $f_8(R)$, etc, are 'damping functions' that tend to 1 as $R \to \infty$, and to zero as $R \to 0$. Each damping function $f_n(R)$ must suppress the corresponding R^{-n} singularity, so it must tend to zero like R^n as $R \to 0$. For non-centrosymmetric molecules there will be similar terms in odd powers of $1/R$.

Knowles and Meath (1986*b*, 1986*a*, 1987) have carried out *ab initio* calculations of the dispersion interaction U^0_{disp} between atoms and small molecules, and have determined the form of the damping functions. Figure 6.3 shows the results for Ar···Ar. The notation $X(1,2)$ indicates that the damping function shown is for the dispersion term involving the dipole–dipole polarizability (rank 1) on one atom and the quadrupole–quadrupole polarizability (rank 2) on the other. These curves all have a broadly similar shape, but the higher-rank damping functions begin to take effect at greater distances. For molecules, the C_n coefficients are orientation-dependent, and it is clear that the damping factors too must have some orientation dependence. At present there is no evidence as to what this should be.

The induction energy behaves similarly; for ions the leading term in the multipole expansion behaves like R^{-4}, while for neutral polar molecules it is proportional to R^{-6}. Again, satisfactory behaviour can in principle be achieved by multiplying each term in the multipole expansion by a damping function that suppresses the singularity.

The damped dispersion energy calculated by Knowles and Meath is a representation of the dispersion energy in the long-range theory, without including any exchange effects, and without allowing for electron correlation in the separated molecules. In the multipole-expanded treatment, electron correlation just modifies the polarizabilities at imaginary frequency that appear in the Casimir–Polder formulae for the dispersion coefficients, eqns (4.3.16–4.3.19). Consequently, the form of eqn (6.5.4) should be equally valid for the correlation-corrected dispersion energy, but the damping functions as well as the dispersion coefficients will probably be different in the correlated case. Exchange effects, on the other hand, do not arise at long range, so they will modify only the damping functions and not the dispersion coefficients. No attempt seems to have been made to explore any of these effects.

6.6 Charge transfer and basis set superposition error

Perturbation theories of intermolecular interactions commonly identify a further contribution to the interaction energy. This is the charge transfer term. It is the source of a good deal of error and confusion, and it may be as well to make it plain at the outset that first, it is not a new term but is part of the short-range induction energy; and second, as commonly formulated it is a serious source of basis set superposition error.

The charge-transfer phenomenon was first identified by Mulliken (1952) in the electronic spectra of molecular complexes. A 'charge-transfer complex' comprises an electron acceptor A with a high electron affinity A_A and an electron donor B with a low ionization potential I_B. The ionic system $A^- \cdots B^+$ then has an energy $I_B - A_A$ at infinite separation that is not much higher than the energy of the neutral system $A \cdots B$, and at finite separations its energy is further lowered by the electrostatic interaction, which is $-e^2/4\pi\varepsilon_0 R$ plus higher-rank terms. When the molecules are close enough for their wavefunctions to overlap, these states can mix, so that the ground state becomes $|AB\rangle + \varepsilon|A^-B^+\rangle$, in an obvious notation, and the excited state becomes approximately $|A^-B^+\rangle - \varepsilon|AB\rangle$. (We are not concerned here with cases such as NaCl, where the ionic state becomes the ground state at short distances.)

There is an allowed electronic transition between these two states. Its intensity is proportional to the square of the matrix element of the dipole operator, which is $\mu_{\rm ct} = (\langle AB| + \varepsilon\langle A^-B^+|)\hat{\mu}(|A^-B^+\rangle - \varepsilon|AB\rangle)$. The dominant term in this expression is $\varepsilon\langle A^-B^+|\hat{\mu}|A^-B^+\rangle$, i.e., ε times the dipole moment of the ionic state, which is approximately eR if the molecules are separated by R. If the mixing coefficient ε becomes significant, therefore, there is an intense electronic transition between the ground state $|AB\rangle$ of the complex and the ionic state $|A^-B^+\rangle$. A very characteristic feature of a charge-transfer transition is that it does not correspond to any electronic transition in either isolated molecule, because it involves the state $|A^-B^+\rangle$, which is unique to the complex.

This simple idea of Mulliken's was very successful in explaining the phenomenon of charge-transfer spectra, and many examples are known. For instance, tetracyanoquinodimethane (TCNQ) and anthracene are both colourless, but when mixed they yield an intensely coloured deep green complex. Unfortunately Mulliken also noted that if the neutral and charge-transfer states became mixed, then there must be a stabilization associated with this mixing, and it was assumed that it was this stabilization that was responsible for binding the complex together. At the time, it was not possible to carry out reliable calculations to confirm this assumption. Dewar and Thompson (1966) warned against it, pointing out that it was not consistent with observations of binding energies and charge-transfer transition frequencies, and that the electrostatic interaction would be big enough in many of these systems to provide strong bonding. Nevertheless, the concept of 'charge-transfer stabilization' has entered the collective conciousness of the chemical community and has become firmly established. Now that it has become possible to carry out accurate calculations of the terms in the interaction energy, it can be shown that the charge-transfer component of the binding energy is usually small compared with other contributions (Stone 1993), and this is in line with recent experimental results (Cozzi *et al.* 1993).

However, it is crucial in carrying out these calculations to be aware of the way in which the charge-transfer energy appears in the calculations. We have seen that the induction energy arises from states in which one of the molecules, say B, is excited. It is also possible to identify states in which one of the electrons of B becomes excited into an orbital that is localized not on B but on the electron acceptor A, and if the molecules are close enough together, such states will contribute to the interaction. The sum of contributions of this sort may then be identified as the charge-transfer energy.

There are serious errors in this view. First of all, observe that the effect on the wavefunction of mixing in such excited states is to mix virtual orbitals of A into the occupied orbitals of B. This is precisely the process of using the basis functions of A to improve the description of B that leads to basis set superposition error. Accordingly, the charge-transfer energy will be heavily contaminated with BSSE unless steps are taken to compensate for it or to prevent it occurring. Secondly, notice that as we improve the basis for B, it will eventually become possible to describe any orbital of A in terms of basis functions on B, so that the charge-transfer state, with an electron from B occupying an orbital of A, can be described entirely in terms of the B basis set. Consequently, the genuine charge-transfer effect, as distinct from the BSSE component, will eventually be described as part of the induction energy when the basis set is good enough.

In practice, any basis set chosen for the purpose of describing B is unlikely to be good enough to describe orbitals for A at all adequately. If however we use the dimer basis to determine the unperturbed wavefunctions for both A and B, in the style of Boys and Bernardi (1970), we provide enough flexibility in the basis set of B to describe the charge-transfer states and the charge-transfer energy with good accuracy. At the same time, we ensure that no purely variational improvement can occur as a side-effect of the perturbation calculation, so that BSSE is absent. But the charge-transfer states will then formally be excited states of B, and their contributions to the interaction will appear as part of the induction energy. Within the perturbation scheme of Hayes and Stone (1984*a*), it is possible to use the dimer basis in this way, and to separate out the 'charge-transfer' terms from the total induction energy (Stone 1993). The separation is basis-set dependent, but the charge-transfer component turns out to be not very sensitive to basis set. The remaining part of the induction energy is more sensitive to basis, since it describes the distortion of the wavefunction in the electric field of the other molecule, and requires high-rank polarization functions if it is to be described accurately, but the charge-transfer energy describes the occupation of a low-lying valence orbital on the other molecule, and a reasonably good account of that can be given without polarization functions.

When the charge-transfer energy is calculated in this way, it turns out to be relatively small, and indeed is often negligible. An implication of this is that the importance attached to 'frontier–orbital' effects in organic reactions is probably misplaced. In this approach, due to Klopman and Hudson (1967, 1968), Salem (1968), and Fukui and Fujimoto (1968), the energy change in the early stages of an organic reaction is expressed as a sum of three parts: the electrostatic interaction, the exchange–repulsion, and a second-order term which is wholly charge-transfer. The charge-transfer term is commonly approximated by the term with the lowest energy denominator, describing the transfer of an electron from the highest occupied molecular orbital, or 'HOMO' of the electron donor (or nucleophile) to the lowest unoccupied molecular orbital, or 'LUMO' of the electron acceptor (electrophile). The HOMO and LUMO are the 'frontier orbitals'. This idea has been a valuable unifying concept in understanding a wide range of organic reactions (Fleming 1976), but numerical calculations of the frontier–orbital component of the interaction energy, which is part of the charge-transfer term, show it to be small (Stone and Erskine 1980; Craig and Stone 1994). At present, it is not clear why frontier–orbital theory is so successful. In one particularly important case where the theory is commonly used, namely in justifying the Woodward–Hoffmann rules for electrocyclic reactions (Woodward and Hoffmann 1970), there are many quite different explanations that do not invoke frontier–orbital ideas (Stone 1978), but no alternative has been found for predicting stereoselectivity and regioselectivity effects.

7

DISTRIBUTED MULTIPOLE EXPANSIONS

7.1 Convergence of the multipole expansion

The multipole expansion of the electrostatic interaction is based on a power series expansion of $1/(\mathbf{B}+\mathbf{b}-\mathbf{A}-\mathbf{a}) = 1/(\mathbf{R}+\mathbf{b}-\mathbf{a})$, initially in powers of $|\mathbf{b}-\mathbf{a}|/R$ (eqn (3.3.2)). If this is to converge absolutely, at a particular R, it must converge for arbitrary orientations of both molecules. This means that the series must converge for the worst-case orientation of $\mathbf{a}$ and $\mathbf{b}$, i.e., for the largest value of $|\mathbf{b}-\mathbf{a}|$, which is $a+b$. Since the angular factors multiplying $|\mathbf{b}-\mathbf{a}|^k$ and R^{-k-1} are Racah spherical harmonics, with maximum magnitude 1, the series will converge absolutely if $a+b < R$ and will diverge if $a+b > R$ (Amos and Crispin 1976*a*).

For classical assemblies of point charges, this result can easily be expressed in pictorial form. We construct a sphere with its centre at $\mathbf{A}$, just large enough to enclose all the charges belonging to A, and a similar sphere for B. Then we enquire whether the spheres intersect for a given R. If they do, it is possible to find particles a and b, on the surface of each sphere, such that $a+b > R$, and the multipole expansion will diverge. (See Fig. 7.1.) If we increase R so that the spheres no longer intersect, then $a+b < R$ for all pairs of particles, and the series will converge.

For those readers accustomed to the concept of radius of convergence as it applies to a series in powers of a complex variable, it is worth emphasising that we are dealing here with a series in powers of $1/R$. For this reason, the interior of the sphere describes the region where the series *diverges*, and we shall use the term 'divergence sphere' to emphasize this fact. In contrast to the case of a circle of convergence in the theory of complex variables, which one wishes to be as large as possible, we require the spheres of divergence to be as *small* as possible for optimum convergence.

For atoms and molecules the implication at first sight is that the multipole series must always diverge, because the charge distribution of any molecule extends formally to infinity, even though it dies away exponentially with distance. Consider, however, the in-

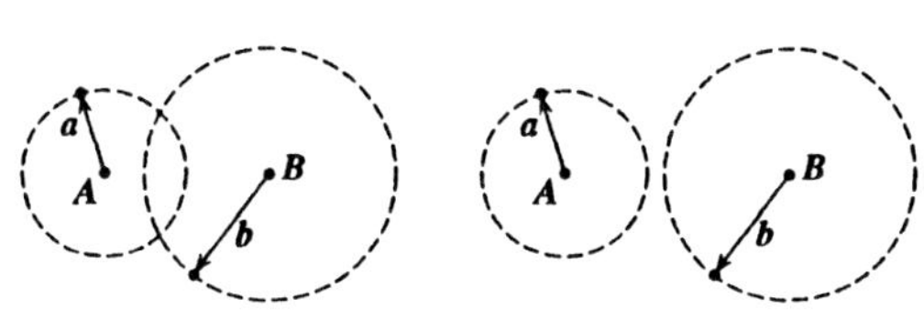

FIG. 7.1. Convergence of the multipole series for the interaction of two charges displaced by **a** and **b** from the respective local origins **A** and **B**. We construct spheres with centres at **A** and **B** that are just large enough to enclose the two charges. The series diverges if the spheres intersect (*left*), and converges if they do not (*right*).

teraction between two Li^+ ions. Both particles have charge +1, but they are spherical (^1S) atoms, and all their higher multipole moments vanish. Consequently the multipole expansion comprises the single term $q^A q^B/R$ and converges trivially.*

The electronic wavefunction can be represented by an expansion in terms of gaussian functions of the form $R_{lm}(\mathbf{r}-\mathbf{p})\exp(-\alpha|\mathbf{r}-\mathbf{p}|^2)$, centred at one or more arbitrary points $\mathbf{p}$, usually at the nuclei. When the wavefunction takes this form, it can be shown that the divergence sphere for a molecule is the sphere enclosing just the *nuclei* (Stone and Alderton 1985, Vigné-Maeder and Claverie 1988), a much less demanding condition. The proof of this important result will be given below (p. 109).

Nevertheless convergence problems remain. It is still possible to find cases where the multipole series will certainly diverge. For example, the radius of a sphere enclosing all the nuclei of benzene has a radius of about 2.4 Å. For hexafluorobenzene the radius needs to be about 2.8 Å. The multipole expansion of the electrostatic interaction between benzene and hexafluorobenzene will therefore converge only if their centres are more than 5.2 Å apart. In the $C_6H_6 \cdots C_6F_6$ complex, however, the molecular planes are about 3.5 Å apart. This means that the multipole expansion cannot be used to describe the interaction.

Even when the divergence spheres do not overlap, the multipole expansion will converge slowly if they are not widely separated, so we have a problem even for molecules smaller than benzene.

7.2 Distributed multipole expansions

The solution to this problem is to use, not a molecular or single-site multipole expansion in which we describe the entire molecular charge distribution by an expansion about a single point, but a *distributed multipole expansion*, in which the molecule is divided into regions, each described by its own multipole moments. A region will usually be an atom or a small group of atoms like a methyl group. Sometimes, for accurate work, it is helpful to have separate regions associated with the bonds. Each region has its own origin, usually at the nucleus for an atomic region. We use the term 'site' for the origin of a region.

The interaction between two molecules A and B then takes the form

$$\mathcal{H}' = \sum_{a\in A}\sum_{b\in B}\Big[\, T^{ab}\hat{q}^a\hat{q}^b + T^{ab}_{\alpha}(\hat{q}^a\hat{\mu}^b_{\alpha} - \hat{\mu}^a_{\alpha}\hat{q}^b) + T^{ab}_{\alpha\beta}(\tfrac{1}{3}\hat{q}^a\hat{\Theta}^b_{\alpha\beta} - \hat{\mu}^a_{\alpha}\hat{\mu}^b_{\beta} + \tfrac{1}{3}\hat{\Theta}^a_{\alpha\beta}\hat{q}^b) + \cdots\Big] \qquad (7.2.1)$$

(compare eqn (3.2.1) on p. 39). The sum runs over the sites a of molecule A and the sites b of molecule B, and $\hat{q}^a$, $\hat{\mu}^a_{\alpha}$, etc, are the operators for the charge, dipole moment, etc, of region a.

There are several ways to obtain this kind of expansion.

*If the ions are close enough for their charge distributions to interpenetrate significantly, this expression for the electrostatic interaction is in fact incorrect, but that is a separate issue which we discussed in Chapter 5. The 'penetration error' in the multipole expansion decreases exponentially as the distance between the ions increases.

7.2.1 *Empirical models*

At sufficiently long range, the central multipole expansion is valid, and in this case the distributed multipole description and the central multipole description must agree. If we use eqn (2.7.6) to change the origin of all the distributed moments to the molecular centre, the resulting multipoles must be the central moments. This provides a constraint on the distributed moments that may be used to help determine their values.

The difficulty with this approach is that the information contained in the central moments is not usually sufficient to determine the distributed moments, or to discriminate between alternative models. Consider, for example, the quadrupole moment of the N_2 molecule, which has a value of -1.09 ± 0.05 a.u. (Buckingham *et al.* 1983, Huot and Bose 1991) We may contemplate a number of distributed-multipole models that reproduce this value. A point-charge model, with a charge of q at the N nuclei and a balancing charge of $-2q$ at the centre of the bond, has a quadrupole moment of $2q(\frac{1}{2}l)^2$, where l is the bond-length, so we require $q = -0.51\,e$. (The bond length is $2.0674\,a_0$.) Alternatively, we might assign a point dipole of magnitude μ to each N atom. In this case, eqn (2.7.2) on p. 33 shows that the quadrupole moment is $2 \times 2\mu(\frac{1}{2}l)$ (if the positive direction of the dipole moment is outwards) so we need $\mu = -0.26\,ea_0$. A third possibility is to assign a point quadrupole Θ to each atom, in which case the total quadrupole is just 2Θ, so we need $\Theta = -0.55\,ea_0^2$.

How can we tell which, if any, of these is right? Occam's razor might favour the simplest model, the point-charge picture, if it were not for the fact that everything we know about chemical bonding contradicts a model in which a charge of $+1$ is assigned to the centre of a triple bond. Most chemists have a less well-developed intuition about atomic dipole and quadrupole moments, so it is less easy to assess the merits of the other two models. One possibility is to examine the hexadecapole moment. It is not very well determined experimentally; a value of -10.4 ± 1 a.u. has been deduced from collision-induced infrared spectra of gaseous N_2 (Poll and Hunt 1981), but this is a rather indirect method and depends on an assumed intermolecular potential for N_2. *Ab initio* calculation yields a value of -6.76 a.u. (Amos 1985). The charge, dipole and quadrupole models give central hexadecapole moments of -1.2, -2.4 and -7.2 respectively. This provides some support for the quadrupole model, but is far from conclusive. In fact we shall see that all of these models are wrong. The experimental data are just not complete enough to distinguish between these or the many other models that we might contemplate. This is by no means an unusual case; it is usually possible to measure only the first one or two non-zero moments, and usually only the first with any accuracy.

7.3 Distributed multipole analysis

We therefore need to make some use of computational results. One possibility would be to fit a model to computed central moments, and some workers have done this, but a much better approach is to compute the distributed moments directly. The method of *distributed multipole analysis* (Stone 1981, Stone and Alderton 1985) will be described here, but similar procedures have been used by Rein (1973), Sokalski and Poirier (1983) and by Vigné-Maeder and Claverie (1988). An alternative procedure based on the use of localized molecular orbitals has been proposed by Amos and Crispin (1976*b*).

It is customary nowadays to express computed wavefunctions in terms of gaussian functions of the form $\chi = R_{lm}(\mathbf{r}-\mathbf{a})\exp[-\zeta(\mathbf{r}-\mathbf{a})^2]$, centred at one or more arbitrary points $\mathbf{a}$, usually at the nuclei, though additional functions are sometimes used, for example at the centres of bonds. Then a one-electron wavefunction such as a molecular orbital takes the form

$$\psi(\mathbf{r}) = \sum_s c_s \chi_s(\mathbf{r}-\mathbf{a}_s). \tag{7.3.1}$$

The charge density then takes the form of a sum of products, with coefficients which are elements of the density matrix P_{st}:

$$\rho(\mathbf{r}) = \sum_{st} P_{st}\chi_s(\mathbf{r}-\mathbf{p}_s)\chi_t(\mathbf{r}-\mathbf{p}_t). \tag{7.3.2}$$

This form for the charge density does not depend on any particular assumptions about the form of the electronic wavefunction; for example it may include electron correlation.

We need to examine the form of a typical product of basis functions, say

$$\chi_a\chi_b = R_{lm}(\mathbf{r}-\mathbf{a})\exp[-\alpha(\mathbf{r}-\mathbf{a})^2] \times R_{l'm'}(\mathbf{r}-\mathbf{b})\exp[-\beta(\mathbf{r}-\mathbf{b})^2].$$

The product of two spherical gaussian functions $\exp[-\alpha(\mathbf{r}-\mathbf{a})^2] \times \exp[-\beta(\mathbf{r}-\mathbf{b})^2]$ can be written as a single gaussian centred at a point on the line from $\mathbf{a}$ to $\mathbf{b}$:

$$\begin{aligned}&\exp[-\alpha(\mathbf{r}-\mathbf{a})^2]\exp[-\beta(\mathbf{r}-\mathbf{b})^2]\\ &\qquad = \exp\left[-\frac{\alpha\beta}{\alpha+\beta}(\mathbf{a}-\mathbf{b})^2\right]\exp[-(\alpha+\beta)(\mathbf{r}-\mathbf{p})^2],\end{aligned} \tag{7.3.3}$$

where $\mathbf{p}$ is the point $(\alpha\mathbf{a}+\beta\mathbf{b})/(\alpha+\beta)$. This formula was first derived by Boys (1950), and used by him as the basis for the practical application of gaussian wavefunctions in *ab initio* calculations. The basis function χ_a also contains regular spherical harmonics $R_{lm}(\mathbf{r}-\mathbf{a})$. These can be moved to $\mathbf{p}$ using eqn (2.7.5):

$$R_{lm}(\mathbf{r}-\mathbf{a}) = \sum_{k=0}^{l}\sum_{q=-k}^{k}\left[\binom{l+m}{k+q}\binom{l-m}{k-q}\right]^{\frac{1}{2}} R_{kq}(\mathbf{r}-\mathbf{p})R_{l-k,m-q}(\mathbf{p}-\mathbf{a}),$$

giving a linear combination of regular spherical harmonics of ranks up to l. We do the same with the $R_{l'm'}(\mathbf{r}-\mathbf{b})$ in χ_b, to give a linear combination of spherical harmonics of ranks up to l'. The product of two spherical harmonics of ranks k and k' is a linear combination of spherical harmonics of ranks from $|k-k'|$ to $k+k'$ (Clebsch–Gordan series; see Appendix B). When we multiply all of these expressions together, we obtain a rather complicated linear combination of terms of the form $R_{kq}(\mathbf{r}-\mathbf{p})\exp[-\zeta(\mathbf{r}-\mathbf{p})^2]$, involving spherical harmonics R_{kq} of ranks from 0 to $l+l'$.

Now the term $R_{kq}(\mathbf{r}-\mathbf{p})\exp[-\zeta(\mathbf{r}-\mathbf{p})^2]$ in the charge distribution has multipole moments, evaluated with origin at $\mathbf{p}$, that are all zero except for Q_{kq}, because of the orthogonality of the spherical harmonics. Accordingly the product distribution $\chi_a\chi_b$ of basis functions of angular momentum l and l' has multipole moments, with respect to an origin at $\mathbf{p}$, of ranks from 0 to $l+l'$ only. Thus an exact multipole representation of the charge distribution consists of a set of multipoles like this for each product of primitive gaussians appearing in the expansion of the charge density.

This may seem extremely complicated at first sight. Fortunately there are two features which make the practical application of this result very much simpler than might be expected. The first of these is that for most molecules of interest, the angular momentum of the basis functions is quite small. The valence orbitals of small molecules are constructed principally from s and p orbitals, and although d, f, and higher angular momentum functions may be needed to describe polarization effects, their contributions are relatively small. We see that the overlap of two s functions ($l = l' = 0$) leads only to a point charge at the overlap centre $\mathbf{p}$; the overlap of an s function with a p function generates a charge and a dipole; and the overlap of two p functions generates charge, dipole, and quadrupole. For a qualitative understanding of distributed multipoles, nothing further is needed, though the moments that arise from the polarization functions have to be included in accurate calculations.

The second simplifying feature is that in the numerical computation of the distributed multipole moments it is not necessary to carry out the algebraic manipulations explicitly. If we know that the multipole moments relative to $\mathbf{p}$ are limited in rank to $l+l'$, we can evaluate all of them by standard techniques, using Gauss–Hermite quadrature. Standard theory shows that $(l+l'+1)$-point Gauss–Hermite quadrature gives exact results for all the moments that arise.

However one complication remains: the formula for the position of the overlap centre $\mathbf{p}$ shows that it depends on the exponents α and β of the two gaussian functions, and will in general be different for every pair of gaussians. We deal with this problem by choosing a small number of *sites* at suitable positions in the molecule, and using eqn (2.7.6) to move the moments at the overlap centre to one or other of the chosen sites. The way in which this is done—the allocation algorithm—varies between different implementations of the theory.

Before exploring allocation algorithms, we note that the point $\mathbf{p} = (\alpha\mathbf{a}+\beta\mathbf{b})/(\alpha+\beta)$ lies on the line between $\mathbf{a}$ and $\mathbf{b}$. If we use atom-centred gaussian basis functions, as is usual, we see that all the multipoles in the exact multipole representation lie on lines joining the nuclei. Accordingly, they all lie within any sphere that encloses the nuclei. The convergence sphere for a single-site multipole description of an assembly of point multipoles has its centre at the single site, and a radius just large enough to enclose all the multipoles. The required sphere therefore just encloses all the nuclei, as previously stated. The observant reader will notice that this begs the question of whether the atom-centred gaussian expansion of the wavefunction is itself convergent. We do not attempt to answer this question, but merely observe that in practice the expansion appears to converge quite quickly.

7.3.1 *Allocation algorithms for distributed multipole analysis*

The importance of a correct choice of allocation algorithm arises from the fact that, as explained in §2.7, there is a penalty for shifting the origin for multipole moments too far. If a moment Q_{km} at one origin is described by a multipole expansion about a new origin at a distance s from the old one, the new expansion contains moments of all ranks $l \geq k$, multiplied by a numerical factor proportional to s^{l-k}, so the higher moments rapidly become unacceptably large if s is too big.

To quantify this effect, we can define a divergence sphere for each site in a distributed multipole expansion, just as we did previously for the complete molecule. The sphere is centred at the site origin, and has to be just large enough to enclose all the multipoles that are to be transferred there.

There are two main allocation algorithms in current use. The *Mulliken* algorithm, which is used in the Cumulative Atomic Multipole Moment (CAMM) method (Sokalski and Poirier 1983), is a generalization of the method used to calculate Mulliken populations (Mulliken charges), and involves allocating half of the multipoles to each of the sites **a** and **b** from which the basis functions came. The *nearest-site* algorithm involves transferring all of the multipoles to the nearest site, which may be **a** or **b** or some other site entirely. The Mulliken algorithm has the merit of simplicity, but there is nothing else to be said in its favour. Its disadvantage is that it may involve relatively large movements of the multipoles, especially if the two sites concerned are fairly far apart. If one of the basis function exponents α and β is much larger than the other, the overlap centre **p** will be very close to either **a** or **b**, and it is much more satisfactory in this case to move all the multipoles to the nearest site rather than half of them to the other. In terms of divergence spheres, we can easily see that the divergence sphere of each site for the Mulliken algorithm will reach nearly to the most distant atom with which there is a significant degree of basis-function overlap. Since modern high-quality basis sets include a number of very diffuse functions on each atom, the divergence spheres are typically quite large. Even for a small molecule such as N_2, the divergence spheres for each site have a radius equal to the bond-length when sites are placed at each atom. This may be compared with the central multipole expansion, where the divergence sphere has only to enclose the two nuclei and has a radius equal to half the bond-length. We have to conclude that the Mulliken algorithm can actually be worse than the central-multipole description, and this is confirmed by detailed calculation (Stone and Alderton 1985).

For the nearest-site algorithm, on the other hand, each divergence sphere extends just half-way to the nearest other site. If for N_2 we take two sites, one at each nucleus, the divergence spheres have radius equal to half the bond-length. In this case the distributed-multipole treatment is no better in terms of convergence than the single-site multipole expansion. However we can do much better if we add a further site at the centre of the bond. Now the divergence spheres have a radius of only one quarter of the bond-length, and the convergence is very much better (Stone and Alderton 1985).

One minor problem with the nearest-site algorithm arises in the case where $\alpha = \beta$, which is common in symmetrical molecules. In this case sites **a** and **b** are equidistant from the overlap centre, and to preserve the symmetry it is necessary to divide the multipoles equally between the two sites, but this causes no practical difficulty, and in particular has an insignificant effect on the radius of the divergence spheres.

A more serious problem is that the values for the distributed multipoles are very sensitive to the choice of basis set. This happens because a change of basis set involves changes in the exponents of the basis functions (α and β in eqn (7.3.3), so the overlap centres **p** are different for different basis sets. The allocation of a particular set of overlap multipoles to one site or another depends on whether the corresponding **p** falls to one side or the other of the dividing surface between the sites, so small changes in the positions of the overlap centres can have dramatic effects on the multipole moments. In one sense this is not a serious problem, because the convergence properties of the nearest-site algorithm ensure that the distributed multipoles always give the optimum description of the electrostatic interactions, and sets of distributed multipoles that look very different may in fact yield similar interaction energies. It is however a practical inconvenience in two ways. First, it makes it much more difficult to compare distributed multipoles from different basis sets and to assess whether improvements in the basis are making a significant difference; and secondly, the effect of truncating the description, for example by discarding all moments above quadrupole, may be different for different basis sets. Indeed, the errors may be greater for better basis sets containing functions with high angular momentum, because these make larger contributions to the higher moments.

Wheatley has suggested a different allocation algorithm that appears to overcome these problems to some extent (Wheatley 1993*b*), but it only applies to linear molecules. According to this prescription, the maximum rank of multipole that is to be retained, $l_{\max}$, is chosen in advance. The z axis of coordinates is chosen to lie along the molecular axis, so that only the Q_{lm} with $m = 0$ are non-zero (see §2.6); we abbreviate Q_{l0} by Q_l. Then an overlap multipole Q_l at position $z = p$ is represented by moments at each of the n sites, with the moment $Q_{l'}^{\text{new}}(a)$ for the site at a being given by

$$Q_{l'}^{\text{new}}(a) = \frac{(l_{\max}-l)!\,(l_{\max}+n-l'-1)!}{(l_{\max}-l')!\,(l_{\max}+n-l-1)!}\binom{l'}{l}(p-a)^{l'-l}Q_l\frac{\prod_{a'\neq a}(p-a')}{\prod_{a'\neq a}(a-a')}. \quad (7.3.4)$$

This applies for $l \le l' \le l_{\max}$. In addition, there are correction terms that account for overlap multipoles of ranks from $l_{\max}+1$ to $l_{\max}+n-1$ by modifying the site multipoles with rank $l_{\max}$.

This procedure has not been used extensively yet, but early results (Wheatley 1993*b*) suggest that the sensitivity to basis set is greatly reduced, and the accuracy of the multipole description is enhanced.

7.4 Other distributed-multipole methods

For completeness, we mention two other ways of calculating distributed multipoles, though neither is suitable for application to intermolecular forces because they have poor convergence properties. One is to associate a region of physical space with each site using Voronoi polyhedra. The Voronoi polyhedron of a given site is the region of space that is closer to that site than to any other. Then distributed multipoles can be defined for each site:

$$Q_{lm}^{a} = \int_{V_a} R_{lm}(\mathbf{r}-\mathbf{a})\rho(\mathbf{r})\,\mathrm{d}^3\mathbf{r}. \quad (7.4.1)$$

Here a labels the site, with its centre at $\mathbf{a}$, and V_a denotes the volume enclosed by the Voronoi polyhedron. The trouble with this approach is that although it seems logical to divide space up in this way, it does not lead to an efficient (that is, rapidly convergent) multipole description. For N_2, for example, with a site at each N nucleus, the dividing surface between the two regions is the plane through the centre of the molecule perpendicular to the axis. The nitrogen molecule as a whole is not far from spherical, but the

Table 7.1 *Distributed multipoles for N_2. 5s4p2d basis.*

	Total moments	Distributed multipole analysis		
Site:	centre	N	centre	N
Position:	0.0	−1.0337	0.0	1.0337
SCF:				
Q_0	–	0.60	−1.20	0.60
Q_1	–	0.81	–	−0.81
Q_2	−0.88	−0.09	1.36	−0.09
CASSCF:				
Q_0	–	0.50	−1.00	0.50
Q_1	–	0.80	–	−0.80
Q_2	−0.92	−0.07	1.27	−0.07
MP2:				
Q_0	–	0.43	−0.86	0.43
Q_1	–	0.73	–	−0.73
Q_2	−1.12	−0.12	1.21	−0.12

two halves obtained by this bisection are very non-spherical, and their multipole moments increase rapidly with increasing rank (Stone and Alderton 1985). Even the central-multipole model provides a very much better description.

Another approach which suffers from similar problems is Bader's 'Atoms in Molecules' treatment. In this method, the dividing surface between two atoms is the 'zero-flux' surface, which is determined directly from the charge density (Bader 1990). As a method for analysing molecular charge densities it has many attractive features. Once again, however, it leads to atomic regions that are highly non-spherical, and although the multipole moments for these atomic regions can be evaluated (Cooper and Stutchbury 1985) (at considerable computational expense, though there are ingenious techniques for improving the efficiency of the calculation) they suffer from the same disadvantages as those for the polyhedral dissection. In the case of N_2, in fact, the dividing surface is the same in both methods, for symmetry reasons.

7.5 Examples

We return to the N_2 molecule. The distributed multipoles, obtained using three sites, with the nearest-site allocation algorithm, are shown in Table 7.1. The sites are the two N atoms, at $z = \pm 1.0337a_0$ in the local coordinate system, and the centre of the bond. The values were calculated using a $5s4p2d$ basis set. The CASSCF or Complete Active Space Self-Consistent Field method is a way of taking account of electron correlation; see Szabo and Ostlund (1989). The left-hand column of numbers shows the central moments, while the other three columns show the distributed multipole analysis. The different levels of calculation give somewhat different results—in particular, the SCF calculation exaggerates the charge separation—but the general features are clear. Evidently none of the

Table 7.2 *Distributed multipoles for CO. 5s4p2d basis.*

	Total moments	Distributed multipole analysis		
Site:	centre	C	centre	O
Position:	0.0	−1.218	−0.152	0.914
SCF:				
Q_0	–	0.63	−0.72	0.09
Q_1	−0.10	0.86	0.10	−0.49
Q_2	−1.50	−0.57	0.74	0.35
MP2:				
Q_0	–	0.30	−0.40	0.10
Q_1	+0.17	0.66	0.14	−0.40
Q_2	−1.51	−0.55	0.68	0.21
Experiment:				
Q_1	+0.043			
Q_2	−1.4			

simple models discussed earlier comes anywhere near being an accurate description. The atomic charges are positive, in agreement with chemical expectations, not negative, as would be required by the 3-charge model. The dipole moments on the atoms are three times as big as were required for the simple model involving dipoles only. The value of 0.8 a.u. is quite large in chemical terms—it is about 2 D, a value somewhat larger than the molecular dipole moment of HF. These dipoles have their positive ends directed towards the centre of the molecule and the negative ends outwards, so they can be regarded as describing the σ lone pair orbitals on each N atom. Finally, the atomic quadrupoles are quite small, but there is a significant quadrupole moment, around 1.2 a.u., associated with the bond centre. It is easy to understand this as being associated with the π bond orbitals. Note that it is positive; that is, it has the opposite sign, though a similar magnitude, to the overall molecular quadrupole moment.

The CO molecule provides an interesting comparison. If the numbers in Table 7.2 are compared with those in Table 7.1, a broadly similar pattern can be seen. However the negative charge in the bond is balanced by a positive charge that is mostly on the C atom, in agreement with electronegativity considerations. This effect alone would give a fairly large dipole with the wrong sign, though significantly smaller when MP2 corrections are included, in line with the general observation that SCF calculations tend to exaggerate charge separations. However there is an atomic dipole on each atom (with negative end directed outwards, as in N_2), and the dipole on the C atom is considerably larger than the one on oxygen. The net dipole moment for the whole molecule is much smaller as a result, and in the case of the MP2 calculation has the correct sign. However MP2 is apt to over-correct for the excessive charge separation, so the dipole moment is now too large.

The quadrupole moments are much less affected by electron correlation. The main

Table 7.3 *Distributed multipoles for HF. SCF, 5s4p2d/4s3p basis. Odd moments of rank 7 and above are not shown.*

	Total moments	Distributed multipole analysis		
Site:	c. of m.	F	centre	H
Position:	0.0	−0.087	0.779	1.645
Q_0	–	**−0.07**	**−0.54**	**0.61**
Q_1	0.757	**0.31**	**−0.15**	0.01
Q_2	1.755	**0.60**	0.04	0.06
Q_3	2.56	0.09	0.04	−0.04
Q_4	4.94	−0.007	−0.12	0.024
Q_5	8.81	−0.021	0.049	−0.012
Q_6	15.41	−0.016	−0.044	0.005
Q_8	46.14	−0.006	−0.012	−0.0011
Q_{10}	135.8	0.002	−0.003	−0.0002
Q_{12}	392.4	0.0003	−0.0006	−0.0001

feature to notice here is that in contrast to N_2, where the atomic quadrupole moments were small, we have here a large positive quadrupole on O and a large negative quadrupole on C. These features are a consequence of the concentration of the π electrons on the O atom, for electronegativity reasons, and are associated with the strong π-acceptor properties of the C atom in carbon monoxide.

Another example is the HF molecule. Distributed multipoles up to rank 12 for a 3-site model are shown in Table 7.3, together with the total moments. This table illustrates very clearly the superior convergence properties of the distributed-multipole treatment. The magnitude of the central moments, referred to the centre of mass, increases steadily with increasing rank. In fact the ratio between successive moments is close to 1.6, which is the distance from the centre of mass to the most distant atom, i.e., the hydrogen. For the 3-site distributed model, on the other hand, the moments decrease steadily with increasing rank, and the higher moments are quite negligible. Indeed, a description of the charge distribution in HF that is quite accurate as well as physically understandable is obtained from just the moments shown in bold in Table 7.3. These include a substantial positive charge on the H atom, balanced by a negative charge in the bond and a much smaller negative charge on F. These charges correspond well with conventional ideas of electronegativity and chemical bonding. Then there is a significant positive dipole moment on the F atom, which is associated with the σ lone pair, and a small negative dipole in the bond describing the asymmetry of the charge distribution there. Finally there is a quadrupole moment on the F atom, which is associated with the π electrons.

For larger molecules, we can improve considerably on the central-multipole model without going to the extent of putting sites at the centre of every bond or even at every atom. For benzene, for example, a 6-site model with a site at each carbon atom is very satisfactory. In this case the divergence spheres extend to the nearest H atom, and so have a radius of about 1 Å, instead of the 2.4 Å for the central model. This is small enough to ensure convergence of the distributed multipole expansion at accessible distances. In this model, each site has a quadrupole moment representing the anisotropy arising from the p-π electrons. In conjunction with a suitable repulsion model, again using six sites, this gave good results in Monte Carlo simulations of the orthorhombic and the high-pressure monoclinic phases of crystalline benzene, and of the liquid (Yashonath *et al.* 1988).

An even simpler 6-site model, again in conjunction with a suitable repulsion and dispersion model, was able to account for the very diverse crystal structures of benzene and the azabenzenes (Price and Stone 1983, Price and Stone 1984). Here the electrostatic properties can be described very satisfactorily by a model in which there is a movement of 0.23 electrons along every CN bond, so that the C atoms become positively charged and the N atoms negatively charged. No higher multipoles are needed, except in benzene, which has no charge flow of this type, and where a quadrupole moment $Q_{20} = -0.8376\, ea_0^2$ was assigned to each C atom.

We have seen that one of the limitations of the multipole description is that it becomes invalid when the charge distributions overlap. It may still converge, and when a distributed-multipole description is used it generally does so even when the charge distributions overlap significantly. The converged interaction energy is in error, however, and the difference between the multipole result and the exact 'non-expanded' electrostatic energy is the penetration energy, discussed in §6.4.1 (p. 94). Wheatley (1993*a*) has suggested using 'gaussian multipoles' to overcome this problem. The charge density is represented, not by point charges, dipoles and so forth, but by a superposition of gaussian charge distributions of the same form as the gaussian basis functions used in *ab initio* calculations (p. 65). A spherical charge distribution, such as that of an inert-gas atom, is then represented by a sum of spherical gaussian functions, while the non-spherical distribution of an atom in a molecule is represented by adding *p*-type, *d*-type and higher-rank gaussian functions. The electrostatic interaction between the gaussian multipoles can be calculated accurately by normal *ab initio* methods, and has no singularities. The calculation is more time-consuming than for the simple multipole expansion, but very much quicker than a calculation using the ordinary *ab initio* wavefunction. Comparative results have been given by Wheatley and Mitchell (1994).

7.6 The hydrogen bond

One of the important unifying concepts in physical chemistry is the hydrogen bond. It has been recognized for many years that there is a significant attractive interaction between a hydrogen atom attached to an electronegative atom such as oxygen or a halogen, or to a lesser extent nitrogen, and another electronegative atom. Hydrogen bonds involving C–H are also now recognized, following some initial controversy. The strength of these interactions varies greatly, but is usually in the range 10–25 $\mathrm{kJ\,mol^{-1}}$.

An early treatment of the hydrogen bond due to Pauling (1928) viewed it as electrostatic in nature, assigning point charges to the atoms involved. By and large, this remains the general view, though an oversimplified electrostatic model fails in many cases, and this has led some to the conclusion that some form of 'incipient chemical bonding' is involved. The HF dimer, for instance, is incorrectly predicted to be linear if the electrostatic interaction is described simply in terms of point dipoles on each molecule (see the discussion on p. 48). Legon and Millen (1982) gave a rule for determining hydrogen bonded structures, based on their observations using microwave spectroscopy:

> 'The gas-phase geometry of B···HX can be obtained ... as follows:
> (i) The axis of the HX molecule coincides with the supposed axis of a non-bonding pair as conventionally envisaged,
> or, if B has no non-bonding pairs but has π-bonding pairs,
> (ii) the axis of the HX molecule intersects the internuclear axis of the atoms forming the π bond and is perpendicular to the [nodal] plane ... of the π orbital.'

However the Legon–Millen rule, and the general observation that over-simplified electrostatic models do not predict hydrogen-bonded structures correctly, can be interpreted in two ways. One possibility is that charge transfer occurs, electron density moving from the proton acceptor to the proton. For this to be effective, it is clearly favourable for the proton to approach a region of high electron density in the proton acceptor, in accordance with the Legon–Millen rule. The other possibility is to acknowledge that a crude electrostatic description is inadequate, and seek to improve it. We saw in Chapter 3 (p. 48) that inclusion of the HF quadrupole moment leads to a structure of HF dimer in agreement with experiment. More generally, we can see that a structure in which the proton approaches a region of high electron density in the proton acceptor is sure to have a very favourable electrostatic interaction, without any need to postulate charge transfer.

We saw in Chapter 6 that the charge-transfer energy is not a well defined quantity, and is merely a basis-dependent part of the induction energy. Nevertheless, we can contemplate, and even calculate in a reasonably well defined way, a distinction between polarization effects in which electrons are perturbed while remaining on the same molecule (classical induction) and effects involving the transfer of electrons from one molecule to the other (charge transfer).

All of these effects—charge transfer, induction and electrostatics—evidently have a part to play in forming the hydrogen bond, as does the dispersion energy. Their relative importance can be estimated by exploring a model in which only one of them is included.

7.6.1 *The Buckingham–Fowler model*

Buckingham and Fowler (1983) formulated a simple but very successful model for predicting the structures of small molecular complexes. This model has two elements:

- An accurate description of the electrostatic interaction, using a distributed multipole expansion as described above;
- A crude description of the repulsion, consisting of a hard-sphere repulsion for each heavy (non-hydrogen) atom, with standard Van der Waals radii as listed by Pauling (1960) or Bondi (1964).

The hydrogen atom forming the hydrogen bond is not assigned a radius; it is contained within the Van der Waals sphere of the heavy atom to which it is attached.

Optimization of the energy of a complex then gives predictions of the structure. The model evidently has no predictive power as far as interatomic separations are concerned, but it can predict the orientational aspects of the geometry, and does so with very good accuracy. Errors in angles are typically of the order of a few degrees—typically within

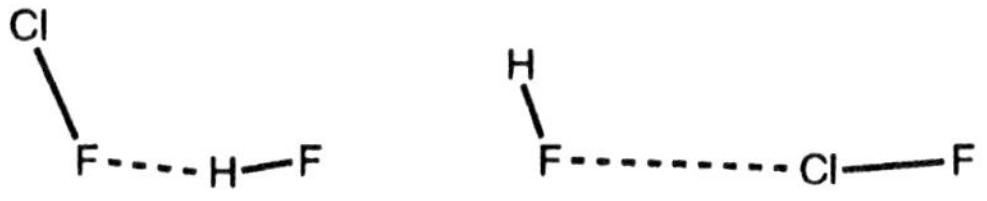

FIG. 7.2. Possible structures for HF/ClF: (left) the hydrogen-bonded structure ClF···HF, and (right) the 'anti-hydrogen-bonded' structure HF···ClF.

experimental error, since these complexes are usually rather floppy and the equilibrium geometry may be different from the average geometry determined by the experiment.

Many successful predictions were made by Buckingham and Fowler (1985), and the method has been used to study many complexes involving polar molecules, not just hydrogen-bonded ones. The model is also the basis for the prediction of the azabenzene crystal structures cited above, and has been used to predict the equilibrium structures of complexes involving aromatic molecules (Price and Stone 1987).

The success of the model emphasizes the importance of the electrostatic interaction in determining structures. The repulsion and the other components of the interaction, neglected in this model though implicit in the values used for the hard-sphere radii, are much less sensitive to orientation. One example of the failure of the model does however show that in some cases the other components can be decisive in selecting the best structure. The example is the complex formed between HF and ClF, for which two structures can be considered. One is the hydrogen-bonded structure ClF···HF, and the other is an 'anti-hydrogen-bonded' structure HF···ClF (Fig. 7.2). The model predicts both of these structures to be minima on the potential energy surface, with electrostatic energies of $-13.5\,\mathrm{kJ\,mol^{-1}}$ and $-7.5\,\mathrm{kJ\,mol^{-1}}$ respectively. However the observed structure (Novick *et al.* 1976) is the 'anti-hydrogen-bonded' one.

The reason for this incorrect prediction has to do with the atomic separations. The sum of Van der Waals radii for F and Cl is 3.15 Å, but the observed separation in HF···ClF is only 2.76 Å. We shall see in Chapter 11 that the standard Van der Waals radius for chlorine is too large for end-on approach, as in HF···ClF, where the heavy atoms are approximately collinear. *Ab initio* perturbation theory (described in Chapter 6) predicts a separation of 2.80 Å, and an electrostatic energy of $-11.5\,\mathrm{kJ\,mol^{-1}}$. Conversely, the simple hard-sphere model gives too short a separation in the hydrogen-bonded structure (2.70 Å compared with 2.98 Å predicted by perturbation theory) and at the larger separation the electrostatic energy is $-9.6\,\mathrm{kJ\,mol^{-1}}$.

We must remember, too, that this simple model ignores the effects of induction and dispersion as well as the anisotropy of the repulsion. We should therefore be cautious in using the predictions of the Buckingham–Fowler model when there are two or more possible structures. Usually one of them will be the correct one, but the relative energies may be incorrect as in the HF/ClF case.

7.6.2 *Other contributions to the energy of the hydrogen bond*

The success of the Buckingham–Fowler model, in which the charge-transfer and induction effects are omitted, suggests that they are less important in determining structure. However all three effects are expected to act in concert, favouring similar structures, so it could be argued that charge-transfer alone, or induction alone, might be equally success-

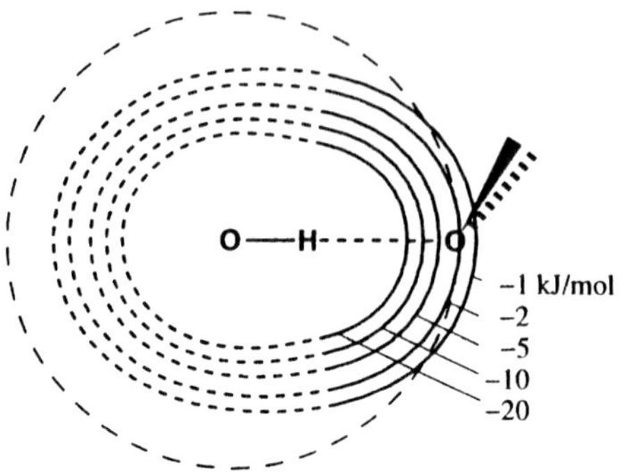

FIG. 7.3. Contours of the charge-transfer energy for the hydrogen bond in the water dimer.

ful. Because of the difficulty of calculating these terms reliably this conjecture has not been tested extensively. Against it there is the argument that the electrostatic interaction is much more strongly orientation-dependent, being attractive in some geometries and repulsive in others, while the charge-transfer and induction terms are always attractive, less sensitive to geometry and therefore less likely to be structure-determining. Calculations on the water dimer (Millot and Stone 1992, Stone 1993) show that the charge-transfer term is in fact quite sensitive to geometry, but is much smaller in magnitude than the electrostatic term. At the equilibrium structure the contributions to the interaction energy, in $\mathrm{kJ\,mol^{-1}}$, are calculated to be as follows:

Electrostatic	−25.8
Repulsion	21.3
Dispersion	−9.2
Induction	−4.5
Charge Transfer	−3.7
Total	−21.9

The form of the charge-transfer term is quite interesting. It can be fitted accurately to an attractive term that is exponential in form:

$$U_{\mathrm{ct}} = \sum_{\mathrm{H-bonds}} -K\exp[-\alpha(R_{\mathrm{OH}} - \rho(\Omega))],$$

where K is a constant and ρ is a distance that depends on the relative orientation of the molecules. (See Chapter 11 for a fuller discussion of this type of model, which is also used to describe repulsion.) The contours of this function are shown in Fig. 7.3 as a function of the position of the acceptor oxygen, the orientation of the acceptor molecule being relatively unimportant. It is apparent that in the region of the hydrogen-bonded geometry the contours are nearly spherical and centred on the H atom. (The dashed parts of the contours lie within the repulsive region, where few calculations were done, and are less well determined.) In conjunction with the repulsion, for which the contours are also approximately spherical but centred on the oxygen (as shown by the dashed circle at the O···O Van der Waals distance) this acts to favour a linear hydrogen bond. It is apparent from the table above, however, that this is a relatively minor effect.

This view has to be modified when considering larger clusters of hydrogen-bonded molecules and condensed phases. The induction energy is strongly non-additive, and it acts cooperatively in hydrogen-bonded networks of molecules. Its contribution must certainly be taken into account in such systems.

The full model of the hydrogen bond, including all the effects listed above, can account qualitatively and often quantitatively for the characteristic properties that are observed experimentally. For a hydrogen bond X–H···Y these include: (i) a lengthening of the X–H bond, (ii) a very large red-shift of the X–H infrared absorption accompanied by (iii) substantial broadening in the case of liquids and (iv) an increase in intensity by an order of magnitude. The other main effect is (v) deshielding of the proton NMR resonance. The lengthening of the bond can be attributed to electrostatic attraction between the proton and the negatively charged Y atom, together with electrostatic and exchange repulsion between the X and Y atoms. Because of the anharmonicity of the X–H bond, this stretching also has the effect of reducing the force constant, thus causing a red shift of the vibration frequency (Buckingham 1960). It is likely that charge transfer effects contribute by weakening the bond. The broadening also is attributed to anharmonicity (Sandorfy 1976), together with the disorder that occurs in the liquid. The deshielding of the proton NMR resonance can also be understood in fairly simple terms (Pople *et al.* 1959): the electric field at the proton due to the neighbouring Y atom causes the bond to stretch, as we have seen, and this inhibits the diamagnetic circulation of the electrons around the proton, especially in a magnetic field perpendicular to the bond. Since it is the diamagnetic current that is responsible for the shielding, this effect reduces it.

The large change in intensity of the X–H infrared absorption is a more subtle phenomenon (McDowell *et al.* 1992). When the X–H bond stretches in the course of vibration, the dipole moment of the X–H bond changes. This is the effect that is responsible for the normal intensity, which is proportional to the square of the dipole derivative. However, in the complex there is also an induced dipole moment in the proton acceptor Y, and this depends strongly on the position of the proton. Thus the dipole derivative for the whole complex with respect to the X–H bondlength is greatly enhanced, and the intensity is increased. McDowell *et al.* found that a calculation of the intensity enhancement for HF dimer using distributed multipoles and polarizabilities gave excellent agreement with an SCF supermolecule calculation.

7.7 Point-charge models

Although distributed multipole analysis describes the charge distribution very accurately, it has the disadvantage, a serious one in the view of some workers, of using dipoles, quadrupoles and so forth as an essential feature of the description. The disadvantage here is that the interactions involving these moments are anisotropic. This means that not only does the energy depend on orientation, but as a result there are torques on the atoms. In an energy optimization or dynamical simulation it then becomes necessary to keep track of the orientation as well as the position of each atom. It will be argued in Chapter 11 that this level of complication is inevitable if we are to attempt accurate descriptions of molecules, and further that there are advantages to offset the complications—in particular, the possibility of using fewer sites.

Nevertheless, the scientific community has generally preferred to use point-charge models, avoiding the use of higher moments. We examine here some of the ways of obtaining such models. Storer *et al.* (1995) propose to classify the methods as follows:

I Methods that extract the charges directly from experiment;

II Methods that extract them from a wavefunction by analysing the wavefunction itself;

III Methods that extract them from a quantum-mechanical calculation by analysing properties predicted from the wavefunction.

Few methods of type I exist. One is the use of X-ray diffraction data, which in principle yields directly a three-dimensional map of the charge density which can be integrated to provide atom charges. In practice the route from the measured diffraction intensities to the charge density involves a number of assumptions and approximations which reduce the reliability of the results (Spackman 1992). Another method uses infrared intensity measurements: if the atoms of a molecule are viewed as charged particles, then the intensity of a vibrational transition can be calculated in terms of the charges, since it is proportional to the derivative of the molecular dipole moment with respect to the vibrational normal coordinate. Conversely the charges can be determined from the observed intensities (Gussoni *et al.* 1986). This does not however give a complete picture of the charge distribution; the atoms may carry dipoles and higher moments as well as charges, but their movement leaves the molecular dipole unchanged and so does not contribute to the infrared intensity. A point-charge model that is to be useful in describing the interactions between molecules must represent these higher atomic moments in some way.

In fact it will be clear from the previous discussion on N_2 that point-charge models may need to be quite elaborate to describe small molecules accurately. The simplest point-charge model that reproduces the quadrupole moment of N_2 not only fails to reproduce the hexadecapole but is wholly unacceptable physically, because it assigns a positive charge to the bond. The simplest adequate point-charge description of N_2 is a 5-charge model due to Murthy *et al.* (1983). This uses charges of q and q' at $\pm z$ and $\pm z'$ respectively, with a balancing charge of $-2(q+q')$ at the centre. The four parameters—two charges and two positions—were chosen to reproduce the lowest four calculated central multipole moments. This model is successful in accounting for the lattice vibrations in the solid.

For larger molecules, especially biological molecules such as proteins and nucleic acids, point-charge models have been used for a long time with reasonable success, and there are several commercially available programs that use them. However the quality of model varies greatly (Roterman, Gibson and Scheraga 1989, Roterman, Lambert, Gibson and Scheraga 1989), and it is worth examining the criteria for a reliable model.

If we start notionally from the viewpoint of a distributed-multipole model with sites on all the atoms, the simplest route to a point-charge model is to throw away the higher moments. It is not difficult to see that this will not work. Such a truncation of the multipole series will be satisfactory only if the discarded moments are small, but the distributed-multipole picture includes atomic dipole moments, describing lone-pair electrons and other atomic distortions, and atomic quadrupole moments, arising usually from π electrons. The construction of the valence orbitals from s and p atomic orbitals guarantees that such atomic dipoles and quadrupoles will arise except in very symmetrical environments. They will be particularly important in the polar groups that are common in proteins, nucleic acids and other biological systems.

Historically, one of the popular sources of point-charge models has been the Mulliken population analysis provided by virtually all semi-empirical and *ab initio* wave-function programs. This provides a count of the number of electrons formally assigned to each atom, and can be used, following Mulliken's original suggestion, to provide some information about the nature of the bonding. In distributed-multipole terms, the Mulliken charges are just the charges that arise when the Mulliken allocation algorithm is used for the calculation of distributed multipoles, this algorithm being in fact just a generalization of Mulliken population analysis. We have already seen that this algorithm provides a poorly convergent description, so it is a particularly bad approximation to throw away the dipoles and higher moments in this case. It should be noted that Mulliken suggested his method of population analysis for a quite different purpose, and would probably have deplored the use of Mulliken charges in point charge models.

However, the use of only the charges from any distributed-multipole model will be subject to errors, if to a lesser degree, because the dipoles and quadrupoles (though not usually the higher moments) are intrinsic to the description. The examples of HF and CO discussed above (p. 113) illustrate this very clearly. The MP2 DMA charges for CO give a molecular dipole moment of −0.2 a.u., compared with the overall MP2 dipole moment of +0.17 a.u.; while the charges alone for HF give a dipole moment of 0.59 a.u. compared with the overall dipole moment of 0.76 a.u. The error in the case of CO is 0.4 a.u., or 1 D, even though this is not a particularly polar molecule.

It is therefore necessary, if a point-charge model is to be used, to modify the charges so as to take into account as far as possible the effects of these omitted moments. There are two methods that attempt to do this by fitting atomic charges so as to reproduce the correct molecular potential. The older method, introduced by Momany (1978) and refined by Cox and Williams (1981) and Singh and Kollman (1984), and now widely used, obtains the potential directly from an *ab initio* molecular charge density, using eqn (4.1.4). The potential is evaluated on a grid of points surrounding the molecule, and a least-squares fitting procedure is used to obtain a set of atom charges which reproduce the potential at the grid points as closely as possible. The charges are constrained so that the total charge on the molecule is correct, and sometimes so that the molecular dipole moment is also reproduced. It is important to choose the grid points so that they lie outside the molecular charge distribution; otherwise the potential contains penetration effects (see Chapter 6) which cannot be represented by any multipolar description. Usually the points are chosen so that they are all at least 1 Å outside the Van der Waals surface of the molecule. Unfortunately the higher moments—octopoles, hexadecapoles, and so on—make only a small contribution to the potential at such distances, and their effects on the potential at short range may be poorly reproduced by the point charges.

The other method, recently proposed by Ferenczy (1991), uses an accurate distributed multipole analysis as the source of the potential to be fitted. This has two advantages: the potential obtained this way is completely free from penetration effects, and the generation of the potential from the DMA is much faster than from the charge density. In fact, it is possible to use an analytical procedure which bypasses the intermediate step of generating the potential, but only by treating the multipole potential of each atom individually, and the fit to the total potential seems to be less accurate if this procedure is adopted (Chipot *et al.* 1994).

An alternative procedure which is based on semi-empirical calculations and is itself semi-empirical in spirit, has been proposed very recently by Storer *et al.* (1995). It takes the Mulliken charge for each atom and generates a modified charge, using a mapping procedure that depends on the nature of the neighbouring atoms and the order of the bonds to them. This reduces the error in the dipole moment calculated from the point charges from around 1 D for Mulliken charges to around 0.2–0.3 D for the modified charges. However this does not compare very well with potential-fitted point charges, which usually reproduce the dipole moment for the *ab initio* charge distribution within about 0.05 D.

All of these methods suffer from the disadvantage that atomic dipoles and quadrupoles have to be represented by point charges, some of which are necessarily on other atoms. This leads to inaccuracies in the potential close to polar and quadrupolar atoms, which often occur in just the regions where an accurate description is needed, for example near the active sites of enzymes. Wiberg and Rablen (1993) investigated a number of schemes for assigning atom charges, and concluded that "the charge distribution in a molecule is much too anisotropic to be successfully modelled by *any* single set of atom-centred charges unless only long-distance interactions are of interest. Otherwise it is necessary to include at least atomic dipole terms, and possibly higher terms as well."

One way to overcome this problem within the framework of a point-charge model is to introduce additional sites near such atoms (Williams 1993), but this increases the number of sites to be handled, and the choice of additional sites is often somewhat arbitrary. Alemán *et al.* (1994) added charges in lone-pair and bond positions. The charges are then fitted to the electrostatic potential in the usual way. They found that the r.m.s. error in the point-charge potential can be reduced by a factor of 5 or so, but there does not seem to be any systematic procedure for choosing the positions of the additional charges.

A more elaborate form of point-charge model, derived from CAMM (Cumulative Atomic Multipole Moments) has been used by Sokalski and Sawaryn (1992) who use an array of point charges near the nucleus of each atom. By choosing appropriate values for the charges, it is possible to reproduce the atomic charge, dipole and quadrupole. If the point charges are too far from the nucleus, spurious higher moments are introduced, while if they are too close, the charges have to become very large, and numerical problems can arise in calculating the interactions, which become small differences of very large quantities. Displacements of 0.25 a.u., as used by Sokalski and Sawaryn, are probably a reasonable compromise, but the charges can be as large as 200 a.u. However the main disadvantage of the method is that because there are nine sites for each atom, the number of site-site interactions is increased by a factor of 81 over a simple point-charge model. Distributed-multipole descriptions are computationally much more efficient than this. A further inefficiency is that for linear molecules, where three charges on the axis suffice to describe the only non-zero moments Q_{00}, Q_{10} and Q_{20}, the method assigns off-axis charges, so increasing the number of charges unnecessarily and also destroying the cylindrical symmetry.

Nevertheless models using off-atom charges can be useful. In aromatic systems, each C atom carries a quadrupole moment of about $-1.1\,ea_0^2$. This cannot usually be represented satisfactorily by point charges on neighbouring atoms, because there are no atoms above and below the plane. Hunter and Sanders (1990) represented it by negative charges above and below the aromatic plane and a positive charge at the C atom. This simple picture provides a very effective treatment of π–π interactions in aromatic systems (Hunter 1994); it has been used for example to account for the geometry of interactions between phenylalanine residues in proteins, where the pattern of observed orientations of pairs of neighbouring phenyl rings matches almost exactly the pattern of electrostatically attractive orientations (Hunter *et al.* 1991), and for the way in which the structure of DNA depends on the base-pair sequence (Hunter 1993).

8

DISTRIBUTED POLARIZABILITIES

Polarizabilities, like multipole moments, are commonly ascribed to the molecule as a whole, but they describe movements of charge density within the molecule, and are often associated with particular regions. The polarizabilities describe changes in the charge distribution arising from external fields, and a single-centre multipole description of these changes is subject to the same convergence problems as the ordinary multipole expansion. A distributed treatment can be expected to give better results here too.

A further consideration is that we are often concerned with the response of the molecular charge distribution to external fields from other molecules. Such fields are usually very non-uniform, so that the strength of the field varies considerably from one part of the molecule to another. We can allow for this by using higher-rank polarizabilities to describe the response to the field gradient and the higher derivatives of the field, but the Taylor series describing the variation of the field across the molecule converges poorly or not at all when the molecule is large—the sphere of convergence extends to the nearest multipole source of the field, which may be closer to the origin than the most distant part of the molecule. Here too a distributed treatment automatically takes account of variations in the strength of the field, since we use the value of the field at each site rather than the value at some arbitrary origin, and the sphere of convergence around each site only has to extend far enough to enclose the region belonging to the site.

It has been known for many years that molecular polarizabilities are additive, in the sense that the mean polarizability of a molecule can be expressed as a sum of contributions from its individual atoms or functional groups. Table 8.1, for instance, shows that the mean polarizabilities of the alkanes increase steadily with the length of the chain. The values were obtained from a study of dipole oscillator strength distributions (Thomas and Meath 1977, Jhanwar *et al.* 1981) (see §12.1.2 on p. 187); similar results have been obtained for polarizabilities derived from depolarization ratios in rotational Raman spectroscopy (Havekort *et al.* 1983). Clearly there is a sense in which we can assign a polarizability of 12 a.u. to the CH_2 unit. This can be viewed either as a sum of atomic contributions, from a carbon and two hydrogens (atomic polarizabilities) or as a sum of

Table 8.1 *Mean polarizabilities for the normal alkanes, C_nH_n (Thomas and Meath 1977, Jhanwar et al. 1981).*

Chain length, n	1		2		3		4		5		6		7		8
$\bar{\alpha}$/a.u.	17.27		29.61		42.09		54.07		66.07		78.04		90.02		102.00
Difference		12.34		12.48		11.98		12.00		11.97		11.98		11.98	

contributions from two C–H bonds and one C–C bond (bond polarizabilities). Using polarizabilities for a large number of molecules it is possible to assign atom and bond polarizabilities for a wide range of atoms and bonds. In doing this, one naturally distinguishes between different kinds of atom or bond involving the same element. For instance, the bond polarizabilities for C–C, C=C and C≡C bonds will clearly be different. Le Fèvre (1965) was probably the most thorough compiler of such data, and in addition to determining mean polarizabilities he also obtained polarizability anisotropies. More recently Miller (1990*a*) has produced an extensive tabulation of molecular polarizabilities and expressed them in terms of bond polarizabilities.

While this work clearly established that it is possible to attach a meaning to atom or bond polarizabilities, it is necessary to understand that it does not, except in a very limited way, meet our present needs. Let us suppose for the moment that we can apply an electric field to just one of the atoms in a molecule. Such a field would polarize the atom in question, distorting its charge distribution; it would induce a dipole moment, but also higher moments. These induced moments are the sources of further fields, which extend over the whole molecule and cause the charge distributions of the other atoms to distort also. These in turn cause further distortions. So the end result is that the multipole moments on all the atoms in the molecule are affected by a field applied to just one atom, though naturally the effects are smaller for atoms far away.

It is not physically possible to apply a field to a single atom of a molecule, but the response to a field that extends over the whole molecule is the sum of the responses to the fields at individual atoms. Le Fèvre's data refer to the total molecular dipole induced by a *uniform* field—either a static field or the field of a light wave at an optical frequency, which is essentially uniform on the molecular scale. The contribution to the overall induced molecular dipole from the field at a particular CH_2 group will be a sum of the induced dipoles at that CH_2 group and its neighbouring atoms. If the environment of each CH_2 group is the same, the response will be the same. This is why each additional CH_2 group adds a constant amount to the overall polarizability. It does not mean that each responds as if it were an isolated entity.

As soon as we deal with non-uniform fields that vary significantly from one part of a molecule to another, the simple additive picture breaks down. It is no longer possible to speak of 'the' electric field at the molecule, because it is different for different parts of the molecule. It becomes necessary to divide the molecule into regions over which the field is reasonably uniform, and to determine the response of each region and its neighbours to its own local field.

The induction energy, $-\frac{1}{2}\alpha_{\alpha\beta}F_\alpha F_\beta$, can be written as $-\frac{1}{2}\Delta\mu_\alpha F_\alpha$. In a uniform external field, F_α is constant, so the position of the induced dipole is immaterial. If F_α varies with position, however, the magnitude of the induction energy depends on the position of the induced dipole. Thus we cannot suppose that the induced dipoles of a CH_2 group and its neighbours can be treated as if they were all at the CH_2 group; we must take account of their positions explicitly.

A final difference between a single-site treatment of polarizability and a distributed treatment is that if we divide a molecule into regions, we have to contemplate the possibility that electronic charge may flow from one region to another under the influence of an external field. That is, we have changes in charge arising in response to differences in electrostatic potential. Such 'charge-flow' polarizabilities are not provided for in most distributed-polarizability treatments, but they can be incorporated quite naturally and consistently, as we shall see later.

8.1 The Applequist model

Silberstein (1917) was the first to address the description of molecular polarizabilities in terms of atomic polarizabilities, as we saw in §4.2, and Applequist *et al.* (1972) generalized his approach. In their treatment, each atom in a molecule is assigned a polarizability $\alpha^a_{\alpha\beta}$. (They used isotropic atom polarizabilities, but the general case is no more difficult to derive; see Birge (1980).) In an external field, each atom becomes polarized, and develops an induced dipole moment $\Delta\mu^a_\alpha$, in addition to any static dipole that may exist in the absence of a field. These induced dipoles are caused not only by the external field but by the fields arising from induced dipoles on the other atoms. The field at a due to the induced dipole on b is $T^{ab}_{\alpha\beta}\Delta\mu^b_\beta$, and the total field at atom a is the sum of the external field F^a_α and the fields due to the other atoms, so that the induced moment on atom a is

$$\Delta\mu^a_\gamma = \alpha^a_{\gamma\alpha}\Big(F^a_\alpha + \sum_{b\neq a} T^{ab}_{\alpha\beta}\Delta\mu^b_\beta\Big). \tag{8.1.1}$$

Note that the external field need not have the same value at each atom. We see that the equation for each induced moment depends on the induced moments at all the other atoms. We rewrite this set of coupled equations in the form

$$(\alpha^a)^{-1}_{\alpha\beta}\Delta\mu^a_\beta - \sum_{b\neq a} T^{ab}_{\alpha\beta}\Delta\mu^b_\beta = F^a_\alpha, \tag{8.1.2}$$

or

$$\sum_{b\beta} B^{ab}_{\alpha\beta}\Delta\mu^b_\beta = F^a_\alpha, \tag{8.1.3}$$

where the elements of the matrix B are

$$B^{ab}_{\alpha\beta} = \begin{cases} (\alpha^a)^{-1}{}_{\alpha\beta} & \text{if } a = b, \\ -T^{ab}_{\alpha\beta} & \text{if } a \neq b. \end{cases} \tag{8.1.4}$$

B here is a $3N \times 3N$ matrix if there are N atoms in the molecule.* Its rows are labelled by a and α and its columns by b and β. The solution of (8.1.3) is just

$$\Delta\mu^a_\alpha = \sum_{b\beta}(B^{-1})^{ab}_{\alpha\beta}F^b_\beta, \tag{8.1.5}$$

or

$$\Delta\mu^a_\alpha = \sum_{b\beta} A^{ab}_{\alpha\beta}F^b_\beta, \tag{8.1.6}$$

*I have interchanged the notation used by Applequist for the matrices **A** and **B** since it seems more satisfactory to use **A** for the generalized polarizability rather than for its inverse.

where the matrix $\mathbf{A} = \mathbf{B}^{-1}$ is a generalized polarizability matrix, sometimes called the *relay matrix*. The total induced moment is

$$\Delta\mu_\alpha^{\text{tot}} = \sum_a \Delta\mu_\alpha^a = \sum_{ab\beta} A_{\alpha\beta}^{ab} F_\beta^b, \tag{8.1.7}$$

and in a *uniform* external field, with $F_\beta^b = F_\beta$, the same for every atom, the polarizability for the whole molecule is

$$\alpha_{\alpha\beta}^{\text{tot}} = \sum_{ab} A_{\alpha\beta}^{ab}. \tag{8.1.8}$$

Notice that $\mathbf{B}$ depends only on the atomic polarizabilities and on their arrangement in space, as described by the interaction tensors, and not on the external field. Consequently it can be evaluated once and for all for a given molecule. As an example, let us consider a diatomic molecule with bond length R. Taking the molecular axis to be the z axis, labelling the atoms as a and b, and abbreviating $1/4\pi\varepsilon_0 R^3$ as p, we find without difficulty that

$$\mathbf{B} = \begin{array}{c} \\ \\ a\,x \\ y \\ z \\ b\,x \\ y \\ z \end{array} \begin{array}{c} \begin{array}{cccccc} a & & & b & & \\ x & y & z & x & y & z \end{array} \\ \begin{pmatrix} (\alpha_\perp^a)^{-1} & 0 & 0 & p & 0 & 0 \\ 0 & (\alpha_\perp^a)^{-1} & 0 & 0 & p & 0 \\ 0 & 0 & (\alpha_\parallel^a)^{-1} & 0 & 0 & -2p \\ p & 0 & 0 & (\alpha_\perp^b)^{-1} & 0 & 0 \\ 0 & p & 0 & 0 & (\alpha_\perp^b)^{-1} & 0 \\ 0 & 0 & -2p & 0 & 0 & (\alpha_\parallel^b)^{-1} \end{pmatrix} \end{array}. \tag{8.1.9}$$

Inversion of this matrix is straightforward (it consists of three interlaced 2×2 matrices) and the result is the matrix $\mathbf{A}$:

$$\begin{pmatrix} \frac{\alpha_\perp^a}{1-X} & 0 & 0 & -\frac{p\alpha_\perp^a\alpha_\perp^b}{1-X} & 0 & 0 \\ 0 & \frac{\alpha_\perp^a}{1-X} & 0 & 0 & -\frac{p\alpha_\perp^a\alpha_\perp^b}{1-X} & 0 \\ 0 & 0 & \frac{\alpha_\parallel^a}{1-4Z} & 0 & 0 & \frac{2p\alpha_\parallel^a\alpha_\parallel^b}{1-4Z} \\ -\frac{p\alpha_\perp^a\alpha_\perp^b}{1-X} & 0 & 0 & \frac{\alpha_\perp^b}{1-X} & 0 & 0 \\ 0 & -\frac{p\alpha_\perp^a\alpha_\perp^b}{1-X} & 0 & 0 & \frac{\alpha_\perp^b}{1-X} & 0 \\ 0 & 0 & \frac{2p\alpha_\parallel^a\alpha_\parallel^b}{1-4Z} & 0 & 0 & \frac{\alpha_\parallel^b}{1-4Z} \end{pmatrix}, \tag{8.1.10}$$

where

$$X = p^2\alpha_\perp^a\alpha_\perp^b, \quad Z = p^2\alpha_\parallel^a\alpha_\parallel^b. \tag{8.1.11}$$

We can now see that the response to a uniform field in the z direction is

$$\Delta\mu_z = \Delta\mu_z^1 + \Delta\mu_z^2 = \frac{\alpha_\parallel^a + \alpha_\parallel^b + 4p\alpha_\parallel^a\alpha_\parallel^b}{1-4Z}F_z, \tag{8.1.12}$$

while a uniform field in the x direction gives

$$\Delta\mu_x = \Delta\mu_x^1 + \Delta\mu_x^2 = \frac{\alpha_\perp^a + \alpha_\perp^b - 2p\alpha_\perp^a\alpha_\perp^b}{1-X}F_x. \tag{8.1.13}$$

These results may be compared with (4.2.4) and (4.2.6). The molecular polarizabilities are

$$\begin{aligned}\alpha_\parallel^{\text{mol}} &= \frac{\alpha_\parallel^a + \alpha_\parallel^b + 4p\alpha_\parallel^a\alpha_\parallel^b}{1-4p^2\alpha_\parallel^a\alpha_\parallel^b},\\ \alpha_\perp^{\text{mol}} &= \frac{\alpha_\perp^a + \alpha_\perp^b - 2p\alpha_\perp^a\alpha_\perp^b}{1-p^2\alpha_\perp^a\alpha_\perp^b}.\end{aligned} \tag{8.1.14}$$

This procedure gives expressions for the parallel and perpendicular polarizabilities of the diatomic in terms of the values $\alpha_\parallel$ and $\alpha_\perp$ for the individual atoms. Given the experimental polarizabilities for the molecule, we can work back to the atom polarizabilities. For a heteronuclear diatomic, the two components of the molecular polarizability would not provide enough information to characterize an $\alpha_\parallel$ and $\alpha_\perp$ for each atom, but from experimental data for many molecules, together with assumptions about transferability of the atom polarizabilities, it is possible to assign polarizabilities for a variety of atoms. Alternatively, the number of parameters to be found can be reduced by assuming that the atom polarizabilities are isotropic, as Applequist *et al.* (1972) originally did.

This treatment can very easily be extended to handle higher-rank fields and polarizabilities. Applequist (1983, 1985) did this in cartesian notation, which necessitates an elaborate scheme to cope with the redundancies resulting from the symmetry of the higher moments with respect to permutation of subscripts, and the tracelessness property. The spherical-tensor approach gives a much more compact and straightforward treatment. All we have to do is to replace $\Delta\mu_\alpha^a$ by the general induced moment ΔQ_t^a, where t as usual runs over the sequence 00, 10, 11c, 11s, etc, of multipole suffixes, and the field F_α^a by $-V_t^a$. The interaction tensor $T_{\alpha\beta}^{ab}$ becomes T_{tu}^{ab}, and we get

$$\Delta Q_{t'}^a = -\alpha_{t't}^a\Big(V_t^a - \sum_{b\neq a} T_{tu}^{ab}\Delta Q_u^b\Big), \tag{8.1.15}$$

(with an implied summation over the repeated suffixes t and u) or

$$(\alpha^a)^{-1}{}_{tu}\Delta Q_u^a - \sum_{b\neq a} T_{tu}^{ab}\Delta Q_u^b = -V_t^a. \tag{8.1.16}$$

As before, we define a matrix $\mathbf{B}$ by

$$B_{tu}^{ab} = \begin{cases}(\alpha^a)^{-1}{}_{tu} & \text{if } a=b,\\ -T_{tu}^{ab} & \text{if } a\neq b,\end{cases} \tag{8.1.17}$$

and the generalized polarizability matrix $\mathbf{A}$ is the inverse of this.

This then provides a reasonably simple and straightforward formalism for treating a molecule as an assembly of polarizable atoms. In the case where the atoms are far removed from each other (an assembly of isolated atoms rather than a molecule) the value of $p = 1/4\pi\varepsilon_0 R^3$ in eqn (8.1.9) becomes zero, and $A^{ab}_{\alpha\beta}$ (eqn (8.1.10)) becomes a diagonal matrix, with diagonal elements $(\alpha^a_\perp, \alpha^a_\perp, \alpha^a_\parallel, \alpha^b_\perp, \alpha^b_\perp, \alpha^b_\parallel)$. The overall polarizability is then just the sum of the polarizabilities of the individual atoms. When the atoms approach, the overall polarizability is modified, and when they become very close, the elements of the **A** matrix diverge. The explicit expression (8.1.10) shows that this happens, so that **A** becomes undefined, if $4p^2\alpha^a_\parallel\alpha^b_\parallel$ or $p^2\alpha^a_\perp\alpha^b_\perp$ becomes equal to 1; that is, if $R^6 = 4\alpha^a_\parallel\alpha^b_\parallel/(4\pi\varepsilon_0)^2$ or $R^6 = \alpha^a_\perp\alpha^b_\perp/(4\pi\varepsilon_0)^2$. Since $\alpha/4\pi\varepsilon_0$ for an atom is typically comparable with the atomic volume, and bond-lengths are rather smaller than Van der Waals radii, this can very easily happen. Furthermore, if R^6 is less than either of these critical values, the matrix **B** ceases to be positive definite (that is, one or more of its eigenvalues becomes negative) and the matrix **A** is then also not positive definite. Thole (1981) points out that this situation is unphysical.

As with most cases of this sort, the mathematical singularity reflects a failure in the physical model. In this case the problem is that when atoms are close together it is no longer valid to use a multipole expansion to describe the interaction between them. In the terminology used at the beginning of this chapter, the convergence spheres overlap.

Two methods have been used to overcome this difficulty. One is an empirical fix; the other involves a more fundamental review of the theory, and is discussed in the next section. The empirical approach, due to Thole (1981) and Miller (1990*b*), recognises that the interaction between the charge distributions of two atoms remains finite as they approach, except for the nucleus–nucleus terms which are not involved in electronic induction effects, and incorporates a damping term to suppress the singularity in the multipole–multipole interaction. For the dipole–dipole interaction (the only case treated in this way so far) the damped interaction proposed by Thole takes the form

$$\tilde{T}^{ab}_{\alpha\beta} = \begin{cases} T^{ab}_{\alpha\beta} & \text{when } R_{ab} > S_{ab}, \\ \dfrac{(4\nu^3 - 3\nu^4)\delta_{\alpha\beta}}{4\pi\varepsilon_0 R^3} - 3\nu^4 \dfrac{R_\alpha R_\beta}{4\pi\varepsilon_0 R^5} & \text{when } R_{ab} < S_{ab}, \end{cases} \tag{8.1.18}$$

where $\nu = R_{ab}/S_{ab}$. That is, a damped form of the dipole–dipole interaction function is used for distances less than S_{ab}. Thole suggested on empirical grounds that a suitable formula for S_{ab} is $S_{ab} = 1.662(\alpha_a\alpha_b)^{1/6}$. Miller's modification of the interaction tensor is simpler in form:

$$\tilde{T}^{ab}_{\alpha\beta} = \begin{cases} 0 & \text{if atoms } a \text{ and } b \text{ are directly bonded}, \\ T^{ab}_{\alpha\beta}[1 - \exp(-(R_{ab}/S_{ab})^{10})] & \text{otherwise}. \end{cases} \tag{8.1.19}$$

In this case, $S_{ab} = 0.7(\rho_a + \rho_b)$, where ρ_a and ρ_b are the Van der Waals radii. The only change required to the theory is now to replace the interaction tensor $T^{ab}_{\alpha\beta}$ in the matrix **B** by the damped tensor $\tilde{T}^{ab}_{\alpha\beta}$. It is possible, using this approach, to assign isotropic atomic polarizabilities that lead to molecular polarizabilities in quite good agreement with *ab initio* or experimental values (Voisin and Cartier 1993).

Table 8.2 *Applequist polarizabilities for CO.*

α^{ab}_{tu} b	C	O
u	z	z
a t		
C z	8.64	6.00
O z	6.00	6.08

α^{ab}_{tu} b	C	O
u	x	x
a t		
C x	4.78	−1.44
O x	−1.44	3.37

Table 8.3 *Damped Applequist polarizabilities for CO.*

α^{ab}_{tu} b	C	O
u	z	z
a t		
C z	9.48	0.09
O z	0.09	5.82

α^{ab}_{tu} b	C	O
u	x	x
a t		
C x	11.78	−4.07
O x	−4.07	7.23

As an example of the Applequist dipole–induced-dipole scheme and the damped version of it, some results for CO are given in Tables 8.2 and 8.3. The left-hand table of each pair gives longitudinal polarizabilities, and the right-hand table gives transverse ones. The local z axis is along the internuclear axis from C to O. Table 8.2 gives the results for the original Applequist model, based on isotropic atomic polarizabilities of 4.16 a.u. for C and 2.93 a.u. for O. Notice that the dipole–induced-dipole formalism leads to a value of 6.00 a.u. for $\alpha^{\mathrm{CO}}_{zz}$, describing the dipole induced on O by a field at C or *vice versa*, that is almost as large as the value of 8.64 a.u. for $\alpha^{\mathrm{CC}}_{zz}$, which describes the dipole induced at C by a field at C. $\alpha^{\mathrm{CO}}_{zz}$ has the same sign as $\alpha^{\mathrm{CC}}_{zz}$, because in the dipole–induced-dipole picture an electric field in the z direction at the C atom induces a dipole moment in the same direction, and the field from that induced dipole at O is also in the same direction. Similarly, $\alpha^{\mathrm{CO}}_{xx} = -1.44$ a.u. is quite comparable in magnitude with $\alpha^{\mathrm{CC}}_{xx} = 4.78$ a.u., but is negative, because, with a transverse applied field at C, the field at O due to the induced dipole on C is in the opposite direction to the applied field.

The overall polarizabilities for the molecule are $\alpha_{\parallel} = \sum_{ab} \alpha^{ab}_{zz} = 26.7$ a.u. and $\alpha_{\perp} = \sum_{ab} \alpha^{ab}_{xx} = 5.3$ a.u. These should be compared with the experimental values, which are $\alpha_{\parallel} = 15.75$ a.u. and $\alpha_{\perp} = 12.16$ a.u.; characteristically the Applequist model exaggerates the anisotropy in the molecular polarizability, even though it starts from an isotropic atomic polarizability.

The damped dipole–induced-dipole model of Thole (1981) performs rather better (Table 8.2). Here isotropic atomic polarizabilities of 9.48 a.u. for C and 5.82 a.u. for O lead to the values shown, giving $\alpha_{\parallel} = 15.5$ a.u., $\alpha_{\perp} = 10.86$ a.u. for the molecule, in quite good agreement with the experimental values. The value of $\alpha^{\mathrm{CO}}_{zz}$ in this scheme is a good deal smaller, but the transverse polarizabilities are much larger in magnitude, and the values are not justified by the accurate treatment that we shall now explore.

8.2 Distributed polarizabilities

The Applequist approach, either in its original form or as modified, treats the atoms of a molecule as distinct isolated entities, interacting only through multipole interactions. In reality, they are much more intimately connected, and a multipole description of their interactions is inadequate. In particular, it is possible for electrons to move from one atom to another in response to the external field. We can derive a more complete treatment using perturbation theory (Stone 1985).

We use a distributed-multipole description of the interaction between the molecule and the external field, so the perturbation is

$$\mathcal{H}' = \sum_{at} \hat{Q}_t^a V_t^a, \tag{8.2.1}$$

where $\hat{Q}_t^a$ is the operator for moment t of region a, and V_t^a is the t derivative of the potential at the origin of region a (i.e., at site a). Standard Rayleigh–Schrödinger perturbation theory gives the second-order energy:

$$\begin{aligned} W'' &= -\sum_n{}' \frac{\langle 0|\sum_{at} \hat{Q}_t^a V_t^a|n\rangle\langle n|\sum_{bu} \hat{Q}_u^b V_u^b|0\rangle}{W_n - W_0} \\ &= -\tfrac{1}{2}\sum_{abtu} V_t^a \alpha_{tu}^{ab} V_u^b, \end{aligned} \tag{8.2.2}$$

where

$$\alpha_{tu}^{ab} = \sum_n{}' \frac{\langle 0|\hat{Q}_t^a|n\rangle\langle n|\hat{Q}_u^b|0\rangle + \langle 0|\hat{Q}_u^b|n\rangle\langle n|\hat{Q}_t^a|0\rangle}{W_n - W_0}. \tag{8.2.3}$$

The induced moment in the external field is the derivative of the second-order energy with respect to the appropriate field component (eqn (2.3.10)):

$$\Delta Q_t^a = \frac{\partial W''}{\partial V_t^a} = -\alpha_{tu}^{ab} V_u^b, \tag{8.2.4}$$

so α_{tu}^{ab} describes the response of the multipole component Q_t^a at site a to the component V_u^b of the external field at site b. In this formulation the interactions between the different regions of the molecule are treated exactly, and we use the multipole approximation only to describe the interaction between each region of the molecule and the external field, where it is very accurate.

One important difference between distributed polarizabilities and conventional molecular polarizabilities is that the operator $\hat{Q}_{00}^a$ describing the charge of region a is not just a constant, unlike the total charge of the molecule. When we wrote down the second-order energy of a molecule in an external field in eqn (2.3.1), we were able to drop the charge terms, because the matrix elements of a constant between different eigenstates are zero because of orthogonality. The matrix elements of $\hat{Q}_{00}^a$ do not vanish in this way, and there are polarizabilities describing changes in atom charges induced by the external

Table 8.4 *Distributed polarizabilities for CO.*

α_{tu}^{ab} b / u / a t	C 00	C z	O 00	O z
C 00	1.40	−0.20	−1.40	−0.44
C z	−0.20	3.42	0.20	−0.04
O 00	−1.40	0.20	1.40	0.44
O z	−0.44	−0.04	0.44	2.01

α_{tu}^{ab} b / u / a t	C x	O x
C x	7.02	0.14
O x	0.14	3.82

field, for example by the potential difference between atoms. The total charge on the molecule must be conserved, however, and this implies that

$$\sum_b \alpha_{t00}^{ab} = 0 \quad \text{for all } a, t. \tag{8.2.5}$$

We use the term 'charge-flow polarizabilities' to emphasize that these new polarizabilities describe the movement of charge between atoms in response to the field.

Let us compare the distributed polarizabilities for CO, given in Table 8.4 with the Applequist results in Tables 8.2 and 8.3. The distributed polarizabilities are very different from either of these models. Notice first that the transverse polarizability $\alpha_{x\,x}^{\mathrm{C\,O}}$ is not only much smaller in magnitude, but *positive*. This emphasizes the limitations of the dipole–induced-dipole model, which even if damped is certain to give a negative value here; in that model, as we have seen, the secondary field at O arising from the dipole induced at C by a transverse applied field must be in the opposite direction to the applied field. In reality, the behaviour of the electron distribution cannot be described by such an over-simplified model when the atoms are so close together. The distributed polarizability method treats the response of the electrons properly, without using the multipole approximation.

The longitudinal polarizabilities are more complicated, because of the charge-flow polarizabilities. Notice however that the value of $\alpha_{z\,z}^{\mathrm{C\,O}}$ is very small and *negative*, again contrary to the simple dipole–induced-dipole model. It is helpful in understanding the remaining terms in the table to calculate the response of the molecule to a uniform electric field in the z direction. Suppose that the field has magnitude F; then the electrostatic potential is $-Fz$. If atom C is at the origin, and O at $z = b$, they experience potentials 0 and $-Fb$ respectively. The potential $V^{\mathrm{O}} = -Fb$ at O causes the charge there to change by $-\alpha_{00\,00}^{\mathrm{O\,O}}V^{\mathrm{O}} = +1.40Fb$, while the charge on C changes by $-\alpha_{00\,00}^{\mathrm{C\,O}}V^{\mathrm{O}} = -1.40Fb$. In addition to this, the potential V^{C} at C causes the charge at C to change by $-\alpha_{00\,00}^{\mathrm{C\,C}}V^{\mathrm{C}} = 0$ in this case, and the charge at O to change by $-\alpha_{00\,00}^{\mathrm{O\,C}}V^{\mathrm{C}} = 0$ also. The overall effect is a transfer of $1.40Fb$ units of charge from C to O; and this contributes $1.40Fb^2$ to the induced molecular dipole and $1.40b^2 = 6.37$ a.u. to the molecular polarizability. (The bond-length b for CO is 2.132 bohr.) The reader may check that the result does not depend on the choice of origin; in fact the charge-flow depends only on the potential difference between the atoms.

The other charge-flow terms describe the change in charge at a particular atom arising from an electric field at either atom; and the change in dipole on a particular atom arising from potential differences between the atoms. In all we have

$$\begin{aligned}
\Delta Q_{00}^{\mathrm{C}} &= -\alpha_{00\,00}^{\mathrm{C\,C}} V^{\mathrm{C}} - \alpha_{00\,z}^{\mathrm{C\,C}} V_z^{\mathrm{C}} - \alpha_{00\,00}^{\mathrm{C\,O}} V^{\mathrm{O}} - \alpha_{00\,z}^{\mathrm{C\,O}} V_z^{\mathrm{O}} \\
&= -1.40 \times 0 + (-0.20) \times F - (-1.40) \times (-Fb) + (-0.44) \times F, \\
\Delta Q_z^{\mathrm{C}} &= -\alpha_{z\,00}^{\mathrm{CC}} V^{\mathrm{C}} - \alpha_{z\,z}^{\mathrm{CC}} V_z^{\mathrm{C}} - \alpha_{z\,00}^{\mathrm{CO}} V^{\mathrm{O}} - \alpha_{z\,z}^{\mathrm{CO}} V_z^{\mathrm{O}} \\
&= (-0.20) \times 0 + 3.42 \times F - 0.20 \times (-Fb) + (-0.08) \times F, \\
\Delta Q_{00}^{\mathrm{O}} &= -\alpha_{00\,00}^{\mathrm{O\,C}} V^{\mathrm{C}} - \alpha_{00\,z}^{\mathrm{O\,C}} V_z^{\mathrm{C}} - \alpha_{00\,00}^{\mathrm{O\,O}} V^{\mathrm{O}} - \alpha_{00\,z}^{\mathrm{O\,O}} V_z^{\mathrm{O}} \\
&= (-1.40) \times 0 + 0.20 \times F - 1.40 \times (-Fb) + 0.44 \times F, \\
\Delta Q_z^{\mathrm{O}} &= -\alpha_{z\,00}^{\mathrm{OC}} V^{\mathrm{C}} - \alpha_{z\,z}^{\mathrm{OC}} V_z^{\mathrm{C}} - \alpha_{z\,00}^{\mathrm{OO}} V^{\mathrm{O}} - \alpha_{z\,z}^{\mathrm{OO}} V_z^{\mathrm{O}} \\
&= -(-0.44) \times 0 + (-0.08) \times F - 0.44 \times (-Fb) + 2.01 \times F.
\end{aligned}$$

The overall induced molecular dipole that results is

$$\begin{aligned}
b\Delta Q_{00}^{\mathrm{O}} + \Delta Q_z^{\mathrm{C}} + \Delta Q_z^{\mathrm{O}} &= F(1.40b^2 + 2(0.20 + 0.44)b + 3.42 + 2.01) \\
&= F(6.37 + 2.73 + 5.43) = 14.53F,
\end{aligned}$$

so that the molecular polarizability is 14.53 a.u. This overall polarizability is identically equal to the value that would be obtained from a conventional coupled Hartree–Fock calculation with the same basis. It does not agree particularly well with the experimental value of 15.75 a.u., but the error of about 8% is typical of SCF polarizability calculations.

This detailed description of the response of the charge distribution to an applied field may seem unnecessarily elaborate. Even for such a small molecule, however, it does make a difference. The induced dipole moments in complexes of HCl with CO (Altman *et al.* 1983*a*) and N_2 (Altman *et al.* 1983*b*) could not be understood using a conventional treatment of the polarizability (Altman *et al.* 1982). These complexes are linear: ClH$\cdots$CO and ClH$\cdots$NN. The electric field of the HCl decreases sharply with distance, so the response to a polarizability sited at the C atom of CO is much larger than to the same polarizability sited at the O atom, say, or at the centre of mass. The fact that the polarizability is formally independent of origin does not relieve us of the need to choose the origin carefully. In the case of CO, we can see that the C atom, with a longitudinal polarizability of 3.4 a.u., is 70% more polarizable than the O atom; and if we want to calculate the response accurately we have to assign the polarizability to the correct origin. The situation is analogous to a change of origin for a point charge (§2.7): we can use the 'wrong' origin only at the expense of including higher-rank terms which could be omitted if a better origin were chosen.

The induced dipole moments of ClH$\cdots$CO and ClH$\cdots N_2$ are obtained in good agreement with experiment when the distributed polarizabilities are used (Buckingham *et al.* 1986); it turns out that the dipole moment induced in the HCl by the other molecule makes a significant contribution, larger in the case of CO. Good results have been obtained for the induced moments of a number of other complexes, including some complexes involving BF_3 and the HF complexes $H_2CO\cdots HF$ and NNO$\cdots$HF (Fowler and

Stone 1987), and several acetylene complexes (Le Sueur *et al.* 1991). In the case of the NNO···HF complex, it was possible to determine the sign of the small dipole moment of nitrous oxide, then unknown, and it has since been confirmed experimentally (Jalink *et al.* 1987).

The distributed polarizability picture, then, gives more insight into the response of the charge distribution to an external field than can be obtained from an overall molecular polarizability, and it does so without invoking the multipole approximation to describe the interactions between regions of the molecule.

8.3 The induction energy in a distributed polarizability description

First we develop the classical electrostatic energy of an assembly of polarizable molecules. In this we follow to some extent the treatment of Barker (1953) and Böttcher *et al.* (1972), who discuss an assembly of polarizable dipoles, but do not treat higher-rank moments and polarizabilities.

In an external field, the perturbation experienced by molecule A takes the form

$$\sum_{a\in A}\sum_{t}\hat{Q}_t^a V_t^a, \tag{8.3.1}$$

where $\hat{Q}_t^a$ is the operator for multipole moment t of region a of the molecule, and the V_t^a are the potential V_{00}^a and its derivatives, evaluated at the origin of region a. The electrostatic interaction between molecule A and its neighbours B takes the following form (see eqn (3.3.13) on p. 43, which gives the corresponding operator):

$$\mathcal{H}' = \sum_{B\neq A} Q_t^a T_{tu}^{ab} Q_u^b. \tag{8.3.2}$$

We are using a repeated-index summation convention for the multipole suffixes t, u, etc, and for the site indices a, b, etc, but sums over the molecules themselves will be indicated explicitly. We see by comparing eqn (8.3.2) with eqn (8.3.1) that the V_t^a are given in this case by

$$V_t^a = \sum_{B\neq A} T_{tu}^{ab} Q_u^b. \tag{8.3.3}$$

When polarization is taken into account the moments of each molecule change under the influence of the electrostatic field due to its environment. Let us consider first the case where only molecule A is polarizable. In the field described by the V_t^a the moment Q_t^a changes by ΔQ_t^a. Now the value of ΔQ_t^a is determined by a competition between two effects. There is a positive energy required to distort the molecule's charge distribution from its equilibrium form in zero field, and this is counteracted by a lowering of the energy of interaction with the field. The latter becomes

$$E_f = (Q_t^a + \Delta Q_t^a)V_t^a, \tag{8.3.4}$$

while the internal energy of the molecule must depend bilinearly (in lowest order) on the ΔQ:

$$\Delta E_A = \tfrac{1}{2}\Delta Q_t^a \zeta_{tt'}^{aa'} \Delta Q_{t'}^{a'}. \tag{8.3.5}$$

(If there was a linear term it would be possible to lower the internal energy in zero field by choosing a nonzero ΔQ, but the ΔQ are defined to be zero in zero field.) Accordingly the total energy of molecule and field is minimized when

$$\partial(E_A + E_f)/\partial(\Delta Q_t^a) = 0,$$

i.e. when

$$\zeta_{tt'}^{aa'} \Delta Q_{t'}^{a'} + V_t^a = 0,$$

or equivalently when

$$\Delta Q_t^a = -\alpha_{tt'}^{aa'} V_{t'}^{a'}, \tag{8.3.6}$$

where α, the polarizability matrix, is the inverse of ζ. The polarizability $\alpha_{tt'}^{aa'}$ describes the change in moment t of region a that results from the t'th derivative of the potential at site a', as we saw in §8.2.

In terms of α the internal energy of the molecule is

$$\Delta E_A = \tfrac{1}{2}\alpha_{tt'}^{aa'} V_t^a V_{t'}^{a'}, \tag{8.3.7}$$

while the change in the energy of interaction between the molecule and the field as a result of polarization is (from eqns (8.3.4 and (8.3.6)

$$\Delta E_f = -\alpha_{tt'}^{aa'} V_t^a V_{t'}^{a'}. \tag{8.3.8}$$

The generalization to the case when all molecules are polarizable is not quite obvious. The electrostatic energy in the field takes the form

$$E_f = \sum_{A>B} (Q_t^a + \Delta Q_t^a) T_{tu}^{ab} (Q_u^b + \Delta Q_u^b), \tag{8.3.9}$$

while the internal energy of the molecules is

$$E_m = \tfrac{1}{2} \sum_A \Delta Q_t^a \zeta_{tt'}^{aa'} \Delta Q_{t'}^{a'}. \tag{8.3.10}$$

The ΔQ are found as before by minimizing the total energy to give

$$\frac{\partial(E_f + E_m)}{\partial(\Delta Q_t^a)} = \sum_{B \neq A} T_{tu}^{ab} (Q_u^b + \Delta Q_u^b) + \zeta_{tt'}^{aa'} \Delta Q_{t'}^{a'} = 0 \tag{8.3.11}$$

or

$$\Delta Q_t^a = -\sum_{B \neq A} \alpha_{tt'}^{aa'} T_{t'u}^{a'b} (Q_u^b + \Delta Q_u^b). \tag{8.3.12}$$

This is just the same as eqn (8.3.6) except that the fields V_t^a are given not by eqn (8.3.3) but by

$$V_t^a = \sum_{B \neq A} T_{t'u}^{a'b} (Q_u^b + \Delta Q_u^b). \tag{8.3.13}$$

That is, the fields depend on the polarized moments of the other molecules, and a solution of the problem requires the solution of the coupled equations (8.3.12), as we saw in

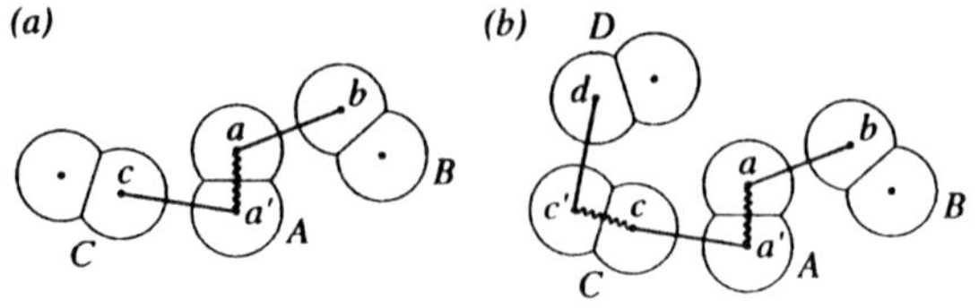

FIG. 8.1. (a) A first-order term in the induction energy of molecule A, involving the static moments of two other molecules, B and C. (b) A second-order term, in which molecule A is perturbed by the moment induced in C by the static moment of molecule D.

§ 8.1. In practical applications, it is usual to solve them iteratively, since a self-consistent solution is usually reached after only half a dozen iterations or so.

Now the energy of the system is

$$E_m + E_f = -\tfrac{1}{2}\sum_A \Delta Q_t^a \sum_{B\neq A} T_{tu}^{ab}(Q_u^b + \Delta Q_u^b) + \tfrac{1}{2}\sum_A \sum_{B\neq A}(Q_t^a + \Delta Q_t^a)T_{tu}^{ab}(Q_u^b + \Delta Q_u^b), \tag{8.3.14}$$

where eqn (8.3.11) has been used to eliminate ζ from eqn (8.3.10). This includes the electrostatic energy of the unpolarized system, which is

$$E_{\text{es}} = \tfrac{1}{2}\sum_A \sum_{B\neq A} Q_t^a T_{tu}^{ab} Q_u^b,$$

so the induction energy is

$$\begin{aligned} E_{\text{ind}} &= \tfrac{1}{2}\sum_A \sum_{B\neq A}\{2\Delta Q_t^a T_{tu}^{ab} Q_u^b + \Delta Q_t^a T_{tu}^{ab} \Delta Q_u^b - \Delta Q_t^a T_{tu}^{ab}(Q_u^b + \Delta Q_u^b)\} \\ &= \tfrac{1}{2}\sum_A \sum_{B\neq A} \Delta Q_t^a T_{tu}^{ab} Q_u^b. \end{aligned} \tag{8.3.15}$$

This separates conveniently into a sum of induction energies, one for each molecule:

$$E_{\text{ind}}^A = \tfrac{1}{2}\sum_{B\neq A} \Delta Q_t^a T_{tu}^{ab} Q_u^b. \tag{8.3.16}$$

This is only a formal separation, since the induction energy is not additive in the ordinary sense; remember that the induced moments that appear in this expression depend on the total field due to the neighbouring molecules. Nevertheless it is a convenient form for computation, if the induced moments have been found by an iterative solution of eqn (8.3.12). Notice that eqn (8.3.16) involves the *unpolarized* moments on the other molecules.

An alternative approach uses eqn (8.3.12) to develop an expansion of eqn (8.3.16) in powers of the polarizability analogous to the expansion given by Barker (1953) for the dipole-polarizability case:

$$E_{\text{ind}}^A = \tfrac{1}{2}\sum_{B\neq A} \Delta Q_t^a T_{tu}^{ab} Q_u^b$$

$$= -\tfrac{1}{2} \sum_{B \neq A} \sum_{C \neq A} (Q_v^c + \Delta Q_v^c) T_{vt'}^{ca'} \alpha_{t't}^{a'a} T_{tu}^{ab} Q_u^b$$

$$= -\tfrac{1}{2} \sum_{B \neq A} \sum_{C \neq A} Q_v^c T_{vt'}^{ca'} \alpha_{t't}^{a'a} T_{tu}^{ab} Q_u^b$$

$$+ \tfrac{1}{2} \sum_{B \neq A} \sum_{C \neq A} \sum_{D \neq C} (Q_w^d + \Delta Q_w^d) T_{wv'}^{dc} \alpha_{v'v}^{c'c} T_{vt'}^{ca'} \alpha_{t't}^{a'a} T_{tu}^{ab} Q_u^b. \qquad (8.3.17)$$

This procedure can evidently be continued to arbitrary order in the polarizabilities to give a generalization of Barker's expression. The first term in eqn (8.3.17) involves only the static multipole moments and provides a 'first-order' approximation to the induction energy, while the subsequent terms allow for the induced moments. Fig. 8.1 shows this schematically: Fig. 8.1a represents one of the contributions $Q_v^c T_{vt'}^{ca'} \alpha_{t't}^{a'a} T_{tu}^{ab} Q_u^b$ to the leading term of eqn (8.3.17), in which multipoles on sites b and c interact with the polarizabilities $\alpha_{t't}^{a'a}$, while Fig. 8.1b describes a higher-order term $Q_w^d T_{wv'}^{dc'} \alpha_{v'v}^{c'c} T_{vt'}^{ca'} \alpha_{t't}^{a'a} T_{tu}^{ab} Q_u^b$ in which the moment at site c is modified by the field from Q_w^d on site d.

8.4 Computation of distributed polarizabilities

The sum-over-states expression (8.2.3) for the distributed polarizability is formally exact, but it is not very useful for practical calculation. It is expressed in terms of the exact eigenfunctions for the unperturbed molecule, and the sum over states needs to include an integral over the continuum states. However, we can use the standard procedure for calculating second-order properties, which is coupled Hartree–Fock theory (Stevens *et al.* 1963, Gerratt and Mills 1968). This requires the matrix elements between the basis functions of the operators for the perturbation and response properties. The only difference is that instead of using the operators for the overall molecular multipole moments, as we would for the ordinary molecular polarizability, we need to use the operators for the multipole moments of the individual regions of the molecule. If these matrix elements can be provided, a standard coupled-Hartree–Fock program can be used to calculate the distributed polarizabilities.

As in the case of distributed multipoles, the definition of the distributed multipole moment operators is not unique; it is only necessary to ensure that they yield the correct matrix elements for the molecular multipole moments. The criteria for the definition are first, that the distributed polarizabilities should have good convergence properties; secondly, that they should be insensitive to the choice of basis set; and thirdly, that the computation should be reasonably tractable. The definition that we used before, for distributed multipoles, unfortunately fails to meet the first of these criteria. Recall that we express a one-electron wavefunction as a linear combination of primitive gaussian functions (eqn (7.3.1)), so a matrix element between two such wavefunctions involves a sum of integrals over products of two such functions, $\chi_s(\mathbf{r} - \mathbf{p}_s)\chi_t(\mathbf{r} - \mathbf{p}_t)$, and this product can be expressed as a linear combination of gaussian functions, all centred at the 'overlap centre'. The 'nearest-site' algorithm that was most successful for distributed multipoles assigns the contribution of this product to the site that is nearest to its overlap centre. This does not work for polarizabilities; it leads to very large charge-flow polarizabilities, counterbalanced by large dipole polarizabilities. The reason for this seems to be that the large basis sets needed for accurate polarizabilities are 'almost over-complete': some basis functions on a particular atom, especially the diffuse ones, can be expressed quite accurately in terms of the basis functions on other atoms. It is then possible to change the wavefunction coefficients significantly without changing the wavefunction itself very much, and if such a change occurs in response to an applied field, it will appear that a substantial movement of charge between atoms has occurred. This however is an artefact of the way that the matrix elements are evaluated.

For this reason it appears to be necessary to define the matrix elements of the distributed multipole operators in terms of an integral over physical space. For linear molecules, this is easy if the atom boundaries are taken to be planes perpendicular to the molecular axis; the necessary integrals, for gaussian basis functions, are incomplete gamma functions, and are easily evaluated. The distributed polarizabilities given above for CO were calculated by this method (Stone 1985). The same technique can be used for molecules like formaldehyde, if a two-site model is sufficient; in this case sites are taken at the C and O nuclei, and the regions are separated by a plane perpendicular to the CO bond. Other non-linear molecules where the sites lie on a straight line can be treated the same way. The case where the sites do not lie on a straight line is much more difficult to handle. Le Sueur and Stone (1993) have described an method in which the contributions from Gauss–Hermite integration points are partitioned between the regions of the molecule. More recently Ángyán *et al.* (1994) have used Bader partitioning (Bader 1990). This procedure is less sensitive to changes in basis set, but the integration over the atomic basins is time-consuming and the numerical precision is only of the order of 0.001 a.u. Although both of these schemes provide only an approximate partitioning between the regions, they are exact in the sense that the overall molecular polarizabilities can be correctly recovered from the distributed polarizabilities.

8.5 Karlström polarizabilities

There is an alternative approach to the calculation of polarizabilities, which may be called the Karlström approach, though a similar idea was used earlier by Amos and Crispin (1976*b*) in the calculation of dispersion interactions. This is to return to the sum-over-states expression (2.3.2) on p. 21. If the wavefunctions $|0\rangle$ and $|n\rangle$ describing the ground and excited states are Slater determinants, then each excited state $|n\rangle$ can be expressed as an excitation from some occupied orbital i to a virtual orbital v. If these are expanded in terms of atomic basis functions, as is customary, then the expression for the polarizability can be expressed as a quadruple sum over atoms (Karlström 1982), each term in the sum being a sort of 4-site polarizability:

$$\alpha_{\alpha\beta}^{abcd} = \sum_{iv}{}' \frac{\langle i^a|\hat{\mu}_\alpha|v^c\rangle\langle v^d|\hat{\mu}_\beta|i^b\rangle + \langle i^b|\hat{\mu}_\beta|v^d\rangle\langle v^c|\hat{\mu}_\alpha|i^a\rangle}{W_i^v - W_0}, \tag{8.5.1}$$

where for example $|i^a\rangle$ is that part of orbital i that is made up of basis functions from atom a. This is not very useful. Karlström (1982) suggested summing over c and d, so that only the occupied orbitals are divided between atoms. The justification for this is that the virtual orbitals are more diffuse. The result is a 2-site polarizability, in the form

$$\alpha_{\alpha\beta}^{ab} = \sum_{iv}{}' \frac{\langle i^a|\hat{\mu}_\alpha|v\rangle\langle v|\hat{\mu}_\beta|i^b\rangle + \langle i^b|\hat{\mu}_\beta|v\rangle\langle v|\hat{\mu}_\alpha|i^a\rangle}{W_n - W_0}. \tag{8.5.2}$$

This provides a rigorous partitioning of the single-site polarizability, but it does not give a very satisfactory path to distributed polarizabilities. It is roughly equivalent to eqn (8.2.3) if the operator $\hat{\mu}_\alpha^a$ is defined so that its matrix element $\langle i|\hat{\mu}_\alpha^a|v\rangle$ is $\langle i^a|\hat{\mu}_\alpha|v\rangle$, but the operator $\hat{\mu}$ remains the *molecular* operator, referred to the molecular origin, and in fact the whole idea is based on the single-site formula for the molecular polarizability, with its implicit assumption that the perturbation involves the electric field at the molecular origin. The matrix elements, and hence the partitioning between sites, were found to depend quite strongly on the choice of molecular origin. Karlström dealt with this by defining the origin for the iv matrix element as the point midway between the centres of charge of the orbitals φ_i and φ_v. The physical basis for this choice is not clear, and it is not evident that the total molecular polarizability is invariant under this change of origin. Even if this can be shown to be the case for the dipole polarizability, it is not obvious how the method should be extended to higher-rank polarizabilities, which are intrinsically origin-dependent.

8.6 Localized polarizabilities

We have seen that the polarization of one atom of a molecule by a local field will inevitably produce secondary fields at the other atoms that will polarize them too. However we have also seen that the secondary polarizations are small—much smaller than the simple Applequist dipole–induced-dipole picture would suggest. If we describe these secondary polarizations by multipole expansions about the primary atom, representing a secondary induced dipole by dipole, quadrupole, etc. components on the original atom according to eqn (2.7.6) or (2.7.7) on p. 34, we can obtain a 'local polarizability' description in which all the secondary effects have been transformed away. This possibility has not been explored very fully yet, but some work on the alkanes and alkenes (Le Sueur and Stone 1994) leads to the following conclusions:

1. The transformation procedure is in principle exact; that is, the overall molecular polarizabilities are left unchanged.
2. The resulting description has less satisfactory convergence properties, making it more important to include higher-rank polarizabilities for accurate work.
3. Care must be taken with the charge-flow polarizabilities to ensure that the sum rules for charge conservation are preserved.
4. Care must also be taken to ensure that the polarizabilities remain symmetric, in the sense that $\alpha_{tu}^{ab} = \alpha_{ut}^{ba}$.
5. Pure charge-flow polarizabilities (i.e., the $\alpha_{00\,00}^{a\,b}$, describing the charge flows induced by potential differences) are quite large, even for atoms some distance apart, and it is probably better not to attempt to localize them.

The localized picture has many attractions; it eliminates the need to take account of secondary polarizations, since they are fully described within the local polarizabilities for each site. Even if it proves necessary to retain the charge-flow polarizabilities, it provides a model that is simpler to handle than the full distributed-polarizability picture, and does not require such a large number of parameters to characterize the system.

8.7 Distributed dispersion interactions

It is natural to look for a distributed description of the dispersion energy along the lines of the distributed treatment of polarizabilities. In fact we can write it down at once: taking the distributed-multipole form of the electrostatic interaction from eqn (7.2.1) on p. 106, we can follow the working of eqns (4.3.20)–(4.3.22) (p. 61) to obtain (Stone and Tong 1989)

$$U_{\text{disp}} = -\frac{\hbar}{\pi} \sum_{a,a' \in A} \sum_{b,b' \in B} T^{ab}_{tu} T^{a'b'}_{t'u'} \int_0^\infty \alpha^{aa'}_{tt'}(\mathrm{i}\nu) \alpha^{bb'}_{uu'}(\mathrm{i}\nu)\, \mathrm{d}\nu \tag{8.7.1}$$

(with an implicit summation over the repeated suffixes t, t', u and u'). This is exact, subject as usual to convergence of the distributed multipole expansions, but unfortunately it is very cumbersome in practice, because it involves a quadruple sum over the sites. Nevertheless it is instructive to study its behaviour. Because it is expressed in terms of distributed polarizabilities, there are charge-flow terms, and if $t = t' = u = u' = 00$, the interaction functions T^{ab}_{tu} and $T^{ab}_{t'u'}$ become $1/4\pi\varepsilon_0 R_{ab}$ and $1/4\pi\varepsilon_0 R_{a'b'}$ respectively, so we have terms in the dispersion energy that behave like R^{-2} (McWeeny 1984, Stone and Tong 1989). There are also mixed charge-flow and dipole polarizability terms in R^{-3}, R^{-4} and R^{-5}. It is not difficult to show that because of the charge-conservation sum rules for the charge-flow polarizabilities, these terms nearly cancel at long range, and if they are expanded in Taylor series about the molecular origins, the R^{-n} terms with $n < 6$ all cancel, so that the leading term is in $(R_{AB})^{-6}$, in agreement with the conventional treatment.

These unfamiliar features can be transformed away by localizing the polarizabilities that occur in (8.7.1). This leads to a site-site form of the dispersion interaction:

$$\begin{aligned} U_{\text{disp}} &= -\frac{\hbar}{\pi} \sum_{a \in A} \sum_{b \in B} T^{ab}_{tu} T^{ab}_{t'u'} \int_0^\infty \alpha^{aa}_{tt'}(\mathrm{i}\nu) \alpha^{bb}_{uu'}(\mathrm{i}\nu)\, \mathrm{d}\nu \\ &= -\sum_{a \in A} \sum_{b \in B} \left(\frac{C^{ab}_6}{R^6_{ab}} + \frac{C^{ab}_7}{R^7_{ab}} + \frac{C^{ab}_8}{R^8_{ab}} + \cdots \right), \end{aligned} \tag{8.7.2}$$

and so provides a justification for the widely assumed site-site model for the dispersion interaction (see Chapter 11). It must be noted, however, that the dispersion coefficients occurring in (8.7.2) are anisotropic (Stone and Tong 1989), and that the reservations about localization listed at the end of the previous section apply here too.

9

MANY-BODY EFFECTS

In Chapter 1 we assumed that the interaction energy of an assembly of molecules can be written in the form of a series starting with pairwise interaction terms and continuing with three-body terms, four-body terms, and so on:

$$\begin{aligned} U &= W - \sum_i W_i \\ &= \sum_{i>j} U_{ij} + \sum_{i>j>k} U_{ijk} + \sum_{i>j>k>l} U_{ijkl} + \cdots \\ &= W_{2\text{body}} + W_{3\text{body}} + W_{4\text{body}} + \cdots . \end{aligned} \tag{9.0.1}$$

We have assumed further that this series converges rapidly, so that the pairwise interactions dominate and the many-body corrections are relatively small. In many cases this is true, but we have already seen in §4.2.1 that the induction energy is inherently non-additive, and in §8.3 we developed an expansion of the induction energy in which many-body contributions appear even in the leading term. In this chapter we study first a case where the non-additive effects of the induction energy dominate, and then go on to examine other non-additive effects.

9.1 Non-additivity of the induction energy

One of the most striking examples of the non-additivity of the induction energy occurs in work by Wilson and Madden (1994) on the halides (except the fluorides) of the alkaline earths, such as $MgCl_2$. These comprise small and quite highly charged cations and relatively large and polarizable anions. Simple chemical intuition leads us to expect these compounds to be purely ionic, and conventional radius-ratio considerations predict the fluorite (CaF_2) structure for radius ratios between 1 and 1.37, the rutile (TiO_2) structure between 1.37 and 2.44, and a 4-coordinated structure for larger values. While these predictions are correct for the smaller radius ratios, the structures at higher ratios are chain or layer structures in which the cations are symmetrically coordinated but the environment of the anions is very unsymmetrical.

This phenomenon has been attributed to the onset of 'covalent' bonding or to 'charge transfer'; Wells (1975) attributes it to the large polarizability of the anions, though without an explanation of how this leads to the observed structure. In fact the fundamental mechanism is very simple. Consider an anion X with polarizability α, placed symmetrically between two cations M (Fig. 9.1a). We ignore the polarizability of the cations, which is much smaller than that of the anions. (The polarizability of Cl^- in NaCl, for instance, is about 20 a.u., while the polarizability of the Na^+ ion is about 1 a.u. (Fowler

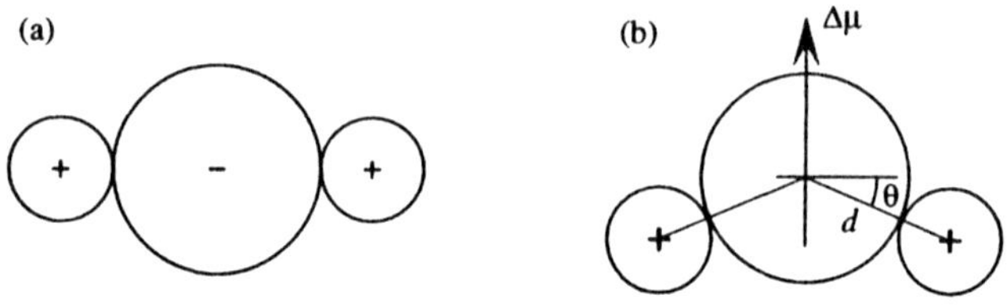

FIG. 9.1. Symmetrical and unsymmetrical arrangements of ions in a lattice.

and Madden 1984), and that of Mg^{2+} is about 0.5 a.u. (Fowler and Madden 1985).) In this configuration the induction energy is zero because the total electric field at the anion is zero. Now consider a distorted configuration (Fig. 9.1b) in which the X–M–X bond angle changes from π to $\pi - 2\theta$. The electric field at the anion is now $2(q/4\pi\varepsilon_0 d^2)\sin\theta$, where q is the charge on each cation. The induction energy is

$$U_{\text{ind}} = -\tfrac{1}{2}\alpha F^2 = -\alpha\frac{2q^2}{(4\pi\varepsilon_0)^2 d^4}\sin^2\theta,$$

while the electrostatic repulsion between the cations is

$$E_{\text{es}} = \frac{q^2}{4\pi\varepsilon_0 . 2d\cos\theta}.$$

(We need not consider the attraction between the anion and cations, which does not depend on θ.) The total energy is stationary with respect to variations of θ when

$$-\alpha\frac{2q^2}{(4\pi\varepsilon_0)^2 d^4}.2\sin\theta\cos\theta + \frac{q^2}{4\pi\varepsilon_0 . 2d}\frac{\sin\theta}{\cos^2\theta} = 0,$$

i.e., when $\sin\theta = 0$ or $\cos^3\theta = 4\pi\varepsilon_0 d^3/8\alpha$. If there are n cations arranged in a regular polygon, each M–X bond making an angle θ with the plane of the cations, the second solution becomes

$$\cos^3\theta = 4\pi\varepsilon_0 d^3/(2n^2\alpha\sin(\pi/n)). \tag{9.1.1}$$

In the $MgCl_2$ crystal, each Cl^- ion has three Mg^{2+} ions coordinated to it, with Mg–Cl–Mg angles close to 90°. (See Fig. 9.2.) This corresponds to $\theta \approx 35°$. The Cl^- ion in a crystal has a polarizability of about 20 a.u. (Fowler and Madden 1984), so a minimum other than $\theta = 0$ is possible with 3 cations if $d^3 < 300$ approximately, or $d < 6.7$ bohr. The Mg–Cl distance in $MgCl_2$ is about 2.52 Å = 4.76 bohr, and the minimum in the energy occurs for $\theta = 45°$. Consequently we can very easily attribute the unexpected structures of the alkaline-earth halides to the effects of induction, and in particular to its non-additivity. We can see from eqn (9.1.1) that the effect increases as the cation becomes smaller (reducing d) and as the anion polarizability increases. This is precisely in accordance with the observed trends.

Notice that the charge on the cation drops out of this calculation, which may therefore seem to suggest that salts involving singly charged cations are as likely to exhibit distorted structures as those involving more highly charged cations. In fact that is not the

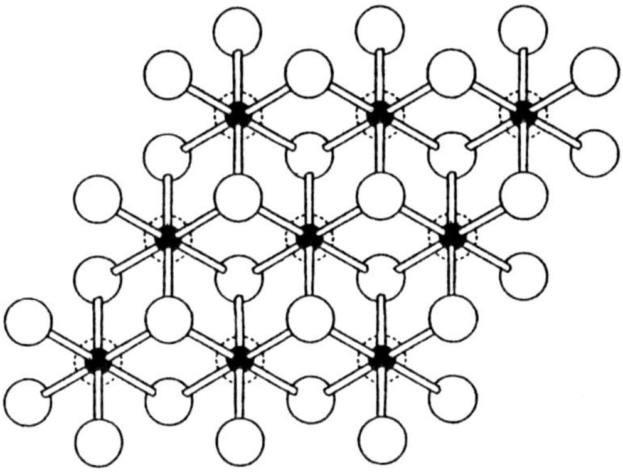

FIG. 9.2. Crystal structure of $CdCl_2$ and $MgCl_2$. One layer of the crystal is shown, comprising a plane of anions (nearest, open circles), a plane of cations (filled circles) and another plane of anions. The nearest plane of anions in the layer below is indicated by dashed open circles.

case, and the reason for this has to do with the other remarkable and apparently highly unfavourable feature of these structures. The $CdCl_2$ structure, for instance, which is also that of $MgCl_2$, can be derived from the NaCl structure by deleting alternate (111) planes of cations (Wells 1975), so it contains adjacent layers of anions, which might be expected to repel each other very strongly. However this repulsion is greatly reduced by polarization effects. There is a substantial dipole induced in each anion by the electric field of the three cations coordinated to it:

$$\Delta\mu = 3\alpha(q/4\pi\varepsilon_0 d^2)\sin\theta.$$

If $\alpha = 20$ a.u., $d = 4.76$ bohr and $\theta \approx 35°$, the magnitude of the induced dipole is about 3 a.u. or 7.5 D. The interaction between anions in two adjacent layers includes a repulsive charge–charge term $+1/4\pi\varepsilon_0 R$, an attractive dipole–charge term $-2\Delta\mu\cos\theta'/4\pi\varepsilon_0 R^2$, and a repulsive dipole–dipole term $+(\Delta\mu)^2(3\cos^2\theta' - 1)/4\pi\varepsilon_0 R^3$, where θ' is the angle between the anion–anion vector and the normal to the anion sheets. In the $MgCl_2$ structure, the anions form a cubic close-packed lattice, so an anion in one sheet fits into a hole between three adjacent anions in the adjacent sheet, the four ions forming a nearly regular tetrahedron (See Fig. 9.2.) In this geometry, $\sin\theta' = \sqrt{\frac{1}{3}}$, while R is 6.7 bohr, so the values of these three terms are $1/R = 0.15$, $-6\sqrt{\frac{2}{3}}/R^2 = -0.12$ and $9/R^3 = 0.03$. This is a grossly oversimplified calculation, but it shows how the attraction between the induced dipoles in one layer and the charges in the next compensate to a large degree for the repulsion between the ion charges, while the repulsion between the induced dipoles is relatively small because of the $3\cos^2\theta' - 1$ factor. Notice that the magnitude of the induced dipole is proportional to the cation charge, so that the compensating effect is stronger for structures containing highly charged cations.

This simplified approach does no more than suggest the way in which polarization effects can stabilize the layer structure. It is necessary to perform a sum over all the ions in the lattice to arrive at a quantitatively useful result, and the induced moments on the anions need to be determined by solving eqn (8.3.12). A better calculation also needs to take into account the damping of the induction interaction at short range, which reduces the effects just discussed, and the effect of dispersion, especially the dispersion interaction between anions in adjacent layers, which contributes greatly to the stability of the structure. Wilson and Madden (1994) discuss the contribution of these effects in detail, and describe calculations in which they are taken into account and which lead to very satisfactory predictions of the observed structures. Nevertheless the fundamental feature is the effect of induction non-additivity in stabilizing what seem at first sight to be very improbable structures.

9.2 Many-body terms in the dispersion energy

In interactions between neutral non-spherical atoms, induction cannot contribute, and the main non-additive effects are many-body repulsion and dispersion terms (Elrod and Saykally 1994). If the long-range perturbation theory is continued to third order, in a system with three or more molecules, terms arise of the form

$$\sum_{m\neq 0}\sum_{n\neq 0}\sum_{p\neq 0} \frac{\langle 000|\mathcal{H}'_{AB}|mn0\rangle\langle mn0|\mathcal{H}'_{BC}|m0p\rangle\langle m0p|\mathcal{H}'_{CA}|000\rangle}{(\Delta W^A_m + \Delta W^B_n)(\Delta W^A_m + \Delta W^C_p)},$$

where for example $\mathcal{H}'_{AB}$ is the interaction Hamiltonian between A and B. The states $|mnp\rangle$ are simple products of single-molecule states with molecule A in state m, B in state n, and C in state p. (We are using the long-range theory, so no antisymmetrization is needed, though this may not be a very good approximation at the distances where these terms are important.) Notice that molecule A is different from the others in this expression, since it is excited in both intermediate states; there are other terms in which this is true of B or C. If all terms of this sort are collected together, and the perturbation is truncated at the dipole–dipole term, the result is the Axilrod–Teller–Muto triple-dipole dispersion interaction, first given independently by Axilrod and Teller (1943) and by Muto (1943):

$$U_{3\mu} = C_9 \frac{(1 + 3\cos\widehat{A}\cos\widehat{B}\cos\widehat{C})}{R^3_{AB}R^3_{BC}R^3_{AC}}, \tag{9.2.1}$$

where R_{AB}, etc, are the lengths of the sides and $\widehat{A}$, $\widehat{B}$ and $\widehat{C}$ are the angles of the triangle formed by the three atoms or molecules. (In the case of a molecule, the reference point forming the vertex of the triangle is the origin that has been chosen for the molecule.) The coefficient C_9 can be expressed using an average-energy formula of London type:

$$C_9 = 3\overline{U}\alpha^A\alpha^B\alpha^C, \tag{9.2.2}$$

where

$$\overline{U} = \frac{U_AU_BU_C(U_A+U_B+U_C)}{(U_A+U_B)(U_B+U_C)(U_A+U_C)}, \tag{9.2.3}$$

or in terms of an integration over imaginary frequencies:

$$C_9 = \frac{3\hbar}{\pi}\int_0^\infty \alpha^A(\mathrm{i}u)\alpha^B(\mathrm{i}u)\alpha^C(\mathrm{i}u)\,\mathrm{d}u. \tag{9.2.4}$$

The latter form can be used to evaluate triple-dipole coefficients from dipole oscillator strength distributions (see §12.1.2), and Kumar and Meath (1984, 1985) have tabulated values for interactions involving a number of atoms and small molecules. (See Chapter 12.) One advantage of this approach is that it is just as easy to evaluate the coefficients for the mixed interactions, involving two or three different molecules, as for the unmixed ones. The C_9 coefficient can also be estimated from the C_6 coefficient and the polarizability (Tang 1969), using the same approach that provides a combining rule for the C_6 coefficient:

$$C_9^{ABC} \approx \frac{2S^A S^B S^C(S^A+S^B+S^C)}{(S^A+S^B)(S^B+S^C)(S^C+S^A)}, \tag{9.2.5}$$

where

$$S^A = C_6^{AA}\alpha^B(0)\alpha^C(0)/\alpha^A(0),$$

and similarly for S^B and S^C.

These expressions involve average polarizabilities. If the polarizabilities are anisotropic, (9.2.1) becomes (Stogryn 1971),

$$U_{3\mu} = \frac{\hbar}{\pi}\int_0^\infty \alpha^A_{\alpha\beta}(iu)\alpha^B_{\gamma\delta}(iu)\alpha^C_{\varepsilon\varphi}(iu)\,du\,T^{AB}_{\beta\gamma}T^{BC}_{\delta\varepsilon}T^{CA}_{\varphi\alpha}. \tag{9.2.6}$$

The principal feature that distinguishes the triple-dipole dispersion from the ordinary two-molecule dispersion terms is that it has a geometrical factor whose sign depends on the configuration of the three molecules. Also, as (9.2.2) and (9.2.4) show, C_9 is positive, whereas the ordinary dispersion interaction is negative. For three atoms in contact, forming an equilateral triangle, the geometrical factor is $(11/8)R^{-9}$, so the triple-dipole dispersion energy is repulsive in a close-packed solid, where such configurations dominate. The angular factor is negative, though much smaller, for isosceles triangles where one of the angles is 120°, which also occur in the close-packed solid, but then the distance between two of the atoms is significantly greater; while for three atoms in a line with separations of R the geometrical factor is $-\frac{1}{4}R^{-9}$. Configurations where there is only one atom-atom contact, or none, make a much smaller contribution, because of the R^{-3} factors. The upshot is that the triple-dipole dispersion makes a repulsive contribution to the total energy of the solid. In the case of argon, its contribution is estimated to be about 7% of the total binding energy.

In liquid water or ice, on the other hand, the constraints arising from hydrogen bonding ensure that one of the angles in a triangle of nearest neighbours is close to the tetrahedral angle. This is close to the angle (117°) where the angular factor changes sign, and the geometrical factor is smaller by a factor of 20 in the case when one of the angles is tetrahedral than it is for an equilateral triangle. Accordingly, the triple-dipole term is likely to be negligible in water and ice.

The Axilrod–Teller–Muto triple-dipole dispersion is only the leading term in a multipole expansion of the non-additive third-order energy:

$$U_{3\text{-body}}^{\text{long-range}} = U_{111} + U_{112} + (U_{122} + U_{113}) + (U_{222} + U_{123} + U_{114}) + \cdots, \tag{9.2.7}$$

where subscripts 1, 2, 3,... refer to dipole, quadrupole, octopole, etc. The distance dependence of U_{jkl} is $R^{-(2j+2l+2k+3)}$, so the higher terms are only significant at short range, and they are strongly anisotropic (Bell 1970, Doran and Zucker 1971), but for small clusters of inert-gas atoms they contribute around 20% of the triple-dipole term (Etters and Danilowicz 1979). There is also a fourth-order dispersion term involving the three dipole–dipole interactions between three atoms; this is attractive, and for inert gases, especially Ar and Kr, it largely cancels the higher-rank third-order terms.

9.3 Many-body terms in the repulsion energy

We have seen how the overlap of the wavefunctions of adjacent molecules, and the consequent antisymmetrization and renormalization, leads to a repulsion between the molecules. If three molecules overlap in this way, the renormalization is different in form, and the energy of the three interacting molecules is not just the sum of the three two-body interactions. We have

$$W_{ABC} = W_A + W_B + W_C + U_{AB} + U_{BC} + U_{AC} + U_{ABC}, \tag{9.3.1}$$

where U_{ABC} is a three-body correction. Wells and Wilson (1986) estimated the magnitude of this correction by *ab initio* calculation, and concluded that for three neon atoms in an equilateral triangle it is negative, i.e., attractive, but smaller in magnitude for the crystal as a whole than the repulsive contribution of the triple-dipole dispersion term. For four atoms in contact, there is a further four-body correction, which is repulsive and somewhat smaller in magnitude than the three-body term (Wells and Wilson 1989*a*). The total 4-body contribution is estimated to amount to 40% of the 3-body contribution for the He crystal (Wells and Wilson 1989*b*). The many-body corrections are largest for those configurations where the atoms are all in contact. For four Ne atoms in a tetrahedral geometry with an Ne···Ne distance of 4 bohr, the 4-body correction is 129 microhartree, and it decreases to less than 2 microhartree at a distance of 5 bohr (Wells and Wilson 1989*a*). (The Ne···Ne separation in the crystal is 5.96 bohr.) Since it is not possible to arrange more than four atoms so that each is in contact with all the others, it is reasonable to hope that the many-body expansion of the interaction energy converges. However there appears to have been no investigation of the rate of convergence after the four-body term. There has been some dispute over the relative importance of the short-range and dispersion three-body terms (Meath and Aziz 1984, Barker 1986, Meath and Koulis 1991).

9.4 Other many-body effects

It is now becoming possible to explore potential-energy surfaces for small Van der Waals clusters very accurately by fitting them to infrared spectra (see Chapter 12). Such work has demonstrated the existence of other many-body effects, small but not negligible. The spectrum of the HCl···Ar_2 complex cannot be fitted satisfactorily by a potential model unless a term is included that describes the generation of an overlap quadrupole moment when the two Ar atoms are close together. This is readily understood in terms of the qualitative description of repulsion given in Chapter 6: at short range there is a redistribution of the electron charge away from the region between the argon atoms towards the ends of the Ar_2 unit. This corresponds to a negative quadrupole moment, and contributes repulsively to the energy of the otherwise favourable structure in which the H atom of the HCl points between the two Ar atoms (Cooper and Hutson 1993, Ernesti and Hutson 1994).

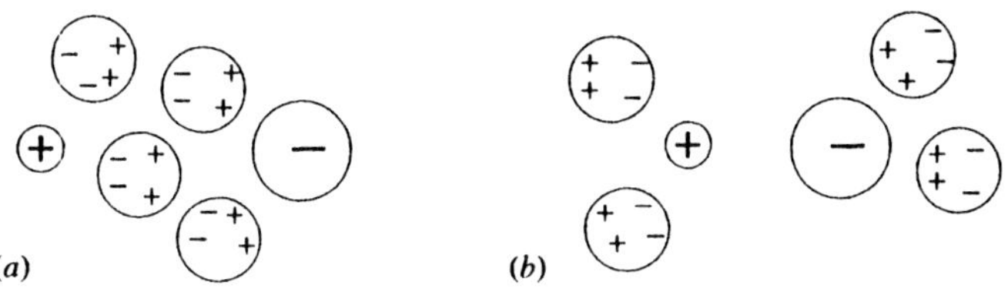

FIG. 9.3. The effect of a polarizable solvent on electrostatic interactions.

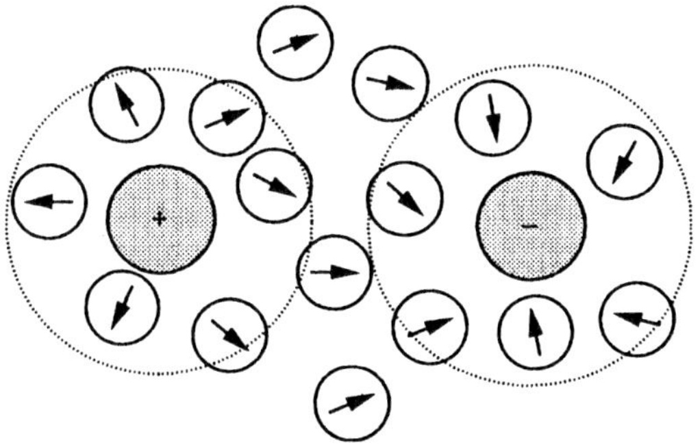

FIG. 9.4. Ions in a polar solvent.

9.5 Intermolecular forces in a medium

One important instance of many-body effects is the modification of intermolecular forces between molecules in a solvent. These modifications occur in various ways.

In the first place, the presence of a polarizable solvent changes the interactions between solute molecules. If two solute molecules carry opposite charges, for example, and there is a polarizable solvent molecule between them, its polarization has the effect of increasing the attraction between the solute ions (Fig. 9.3a). Solvent molecules in other positions, on the other hand, can cause the attraction to decrease (Fig. 9.3b). If the solvent can be treated as if it were a continuous medium, these effects can be described by assigning a dielectric constant ε to the medium, but in many cases the distance scale is too small for this to be a reasonable approximation. Indeed, a detailed study has revealed that a pair of ions with the same sign in a polarizable solvent experience an attraction, on average, due to the polarization of the solvent, while ions of opposite sign experience a repulsion (Jarque and Buckingham 1989, 1992). If the solvent polarizability is large enough, the effect can overcome the Coulomb interaction at intermediate distances.

The reorientation of solvent molecules in the field of the solute also has substantial effects on their interactions. Ions in a polar solvent are normally surrounded by a shell of solvent molecules (Fig. 9.4) and these become in effect part of the solute, so that the interaction is not between bare ions or molecules but between 'dressed' ones. The coordination shell is not rigid but constantly in motion, so that the dressed ion involves a statistical average over the positions and orientations of the solvent molecules.

In fact the interaction between molecules in solution is not properly an energy (or enthalpy) but a free energy, called the *potential of mean force*:

$$G(\mathbf{R},T) = H(\mathbf{R},T) - TS(\mathbf{R},T). \qquad (9.5.1)$$

Both the enthalpy term H and the entropy S involve a statistical average over the solvent, and so depend on the temperature, and the entropy as well as the energy depends on the positions of the solute molecules, specified by the coordinates $\mathbf{R}$. An example of the way in which the entropy of the solvent affects the interaction between solute molecules is the *hydrophobic effect*: water-repelling molecules such as hydrocarbons attract each other in aqueous solution (Ben-Naim 1980). This effect has important consequences for protein folding in aqueous media: there is a tendency for the hydrophobic side-chains to be buried inside the folded protein, and for the hydrophilic groups to take up positions on its surface. There is still some controversy about the mechanism of the hydrophobic effect. Du *et al.* (1994) studied the interfaces between water and several hydrophobic materials using sum-frequency generation spectroscopy, and found that they were characterized by a large number of dangling hydroxyl groups—approximately one for every four water molecules. Thus the insertion of a hydrophobic molecule into water requires energy to break or distort hydrogen bonds between water molecules. Nevertheless the enthalpy change on solution in water for many non-polar solutes is small but *negative*. However the entropy change is large and negative, so that the overall free energy change is large and positive (Mancera and Buckingham 1995*a*). Costas *et al.* (1994) view the solvation process as a sequence of several steps, involving the formation of a cavity, the insertion of the solute molecule, and the rearrangement of the hydrogen bonds in the water around the solute, but the steps are not very well defined and it is not clear at present that this view is helpful. The mechanism of the hydrophobic effect itself, that is, the tendency of non-polar molecules in aqueous solution to aggregate, is less controversial than the mechanism of solvation. Computer simulations show that clustering of hydrocarbon molecules in aqueous solution increases with increasing temperature, showing that the hydrophobic force between the hydrocarbon molecules is entropy-driven (Skipper 1993, Mancera and Buckingham 1995*b*). That is, the decrease in entropy when two hydrophobic molecules are inserted into water is not as great when they are close together as when they are far apart.

10

INTERACTIONS INVOLVING EXCITED STATES

10.1 Resonance interactions and excitons

When one of a pair of identical molecules is in an excited state, a new type of interaction becomes possible. Since either molecule can be excited, with the other in the ground state, there are two states involved, and they can mix under the influence of the intermolecular perturbation, one combination being stabilized and the other raised in energy. This effect operates at long range; there is no need for the molecular wavefunctions to overlap.

If the two states are $|1^A0^B\rangle$ (molecule A excited) and $|0^A1^B\rangle$ (molecule B excited), we can write down secular equations for a mixed state $c_A|1^A0^B\rangle + c_B|0^A1^B\rangle$:

$$\begin{aligned} (H_{AA} - W)c_A + H_{AB}c_B &= 0 \\ H_{BA}c_A + (H_{BB} - W)c_B &= 0 \end{aligned} \qquad (10.1.1)$$

where

$$\begin{aligned} H_{AA} &= \langle 1^A0^B|\mathcal{H}|1^A0^B\rangle = W_1^A + W_0^B, \\ H_{BB} &= \langle 0^A1^B|\mathcal{H}|0^A1^B\rangle = W_0^A + W_1^B, \\ H_{AB} &= \langle 1^A0^B|\mathcal{H}|0^A1^B\rangle. \end{aligned} \qquad (10.1.2)$$

Here $\mathcal{H} = H^0 + \mathcal{H}' = \mathcal{H}^A + \mathcal{H}^B + \mathcal{H}'$, and we have ignored the perturbation $\mathcal{H}'$ in the diagonal matrix elements. The zeroth-order Hamiltonian does not contribute to the off-diagonal matrix element H_{AB}, because, e.g., $\langle 1^A0^B|\mathcal{H}^A|0^A1^B\rangle = W_0^A\langle 1^A|0^B\rangle\langle 0^B|1^B\rangle = 0$. Using just the dipole–dipole term in the perturbation (see (3.2.2)), we have

$$\begin{aligned} H_{AB} &= -\langle 1^A0^B|T_{\alpha\beta}\hat{\mu}_\alpha^A\hat{\mu}_\beta^B|0^A1^B\rangle \\ &= -\langle 1^A|\hat{\mu}_\alpha^A|0^A\rangle T_{\alpha\beta}\langle 0^B|\hat{\mu}_\beta^B|1^B\rangle \\ &= -(\mu_\alpha^A)_{01}T_{\alpha\beta}(\mu_\beta^B)_{01}, \end{aligned} \qquad (10.1.3)$$

where $(\mu_\alpha^A)_{01}$ is the transition dipole between states 0 and 1.

If the molecules are identical, $W_1^A + W_0^B = W_0^A + W_1^B$ and we can drop the superscripts. Solving the secular equations, we find that the states and their energies are

$$\sqrt{\tfrac{1}{2}}\Big(|1^A0^B\rangle \pm |0^A1^B\rangle\Big), \qquad W = W_0 + W_1 \mp (\mu_\alpha^A)_{01}T_{\alpha\beta}(\mu_\beta^B)_{01}. \qquad (10.1.4)$$

We conclude that even if the molecules are non-polar, so that there is no dipole–dipole interaction between them when both are in their ground states, there can be an interaction of dipole–dipole form when one molecule is in an excited state, provided that there is

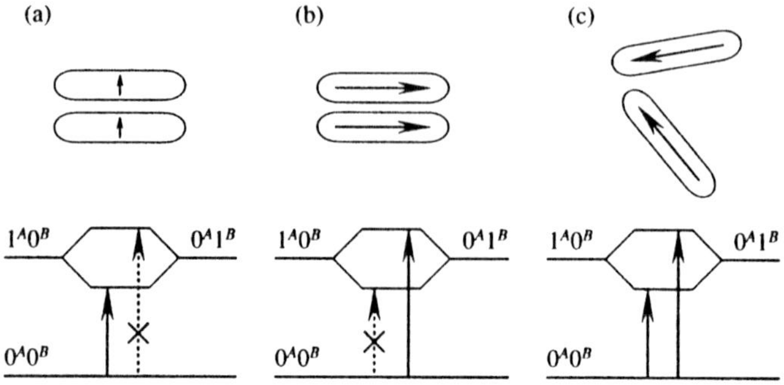

FIG. 10.1. Energy levels and allowed transitions resulting from the resonance interaction between a pair of molecules.

an allowed electric dipole transition to that state. The splitting of the excited states that occurs is called an *exciton* splitting. It is proportional to $T_{\alpha\beta}$ and hence to R^{-3}, so it tends to attract the molecules to each other if the system is in the lower of the two excited states. Notice again that this is a long-range type of interaction—there is no need for the molecules to overlap. However the molecules need to be identical; if they are not, then the energies $W_1^A + W_0^B$ and $W_0^A + W_1^B$ of the two excited states will usually be very different, and the energy change due to the interaction will be much smaller.

Consider, for example, a dipole-allowed $\sigma - \pi^*$ transition in an aromatic molecule. The transition dipole is perpendicular to the plane of the molecule, so there is an attractive interaction if the planes of the molecules are parallel (Fig. 10.1a). In this case the transition moment to the lower state is

$$\langle 0^A0^B|\mu^A + \mu^B|\sqrt{\tfrac{1}{2}}(1^A0^B + 0^A1^B)\rangle = \sqrt{\tfrac{1}{2}}((\mu^A)_{01} + (\mu^B)_{01} = \sqrt{2}\mu_{01}, \qquad (10.1.5)$$

and the intensity of this transition is twice as great as for a single molecule. For the upper state, on the other hand, the wavefunction is $|\sqrt{\tfrac{1}{2}}(1^A0^B - 0^A1^B)\rangle$, so the transition moments cancel out and the transition is forbidden.

Low-lying excited states of aromatic molecules are more commonly due to $\pi - \pi^*$ transitions, and then the transition moment lies in the plane of the molecule (Fig. 10.1b). There is still an attractive interaction, in spite of the repulsive geometry of the transition dipoles, but now it is the state $|\sqrt{\tfrac{1}{2}}(1^A0^B - 0^A1^B)\rangle$ that has the lower energy. However the transition to this state is still forbidden, and the upper state now takes all the intensity.

In fact the sign of the transition dipole is arbitrary, because it depends on the arbitrary phase assigned to the ground and excited-state wavefunctions. Consequently the geometrical features of such interactions are somewhat different from those for static moments. There are just two possibilities for the signs, since changing both of them has no effect. In example (a) above, we can have *either* transition dipoles parallel, giving the attractive lower state, *or* antiparallel, giving the repulsive upper state. In (b), transition dipoles parallel gives the repulsive upper state, while the antiparallel arrangement gives the attractive lower state. In both cases it is the state with parallel transition dipoles that has all the intensity.

Of course other orientations are possible in which the dipoles are not parallel or antiparallel, but related in some other way (Fig. 10.1c). In this case there will again be an attractive state and a repulsive one, but usually both will have non-zero intensity.

10.1.1 *Excitons*

The resonance interaction is often important in excited states of molecular crystals. Here matters become a little more complicated, because all the molecules in the crystal are involved. Davydov's (1962) classic book gives a detailed account. If there is just one molecule per unit cell, then we can label them by the cell coordinate $\mathbf{t}$ which gives the position of the cell relative to some arbitrary origin. If the state with the molecule in cell $\mathbf{t}$ excited is denoted $|1,\mathbf{t}\rangle$, then the excited states of the crystal are the Bloch functions

$$|1,\mathbf{k}\rangle = N^{-\frac{1}{2}}\sum_{\mathbf{t}}|1,\mathbf{t}\rangle\exp(i\mathbf{k}\cdot\mathbf{t}), \tag{10.1.6}$$

where $\mathbf{k}$ is the wavevector of the state and N is the number of unit cells in the crystal.* Now an optical spectrum uses light whose wavelength is long compared with the size of the unit cell—typically 100–1000 times as long—so the oscillating electric field is uniform over a macroscopic region of crystal. In this case the only Bloch function for which the intensity is non-zero is the state with $\mathbf{k}=0$, because the $\exp(i\mathbf{k}\cdot\mathbf{t})$ factors otherwise cause the transition dipole to cancel out. However the $\mathbf{k}=0$ transition acquires all the intensity of the transitions for the N molecules, just as in the two-molecule case already considered. The spectrum therefore contains a single line for the transition in question, as for an isolated molecule.

When there are two molecules in the unit cell, we are still concerned only with $\mathbf{k}=0$ states, but now there are two of them:

$$\begin{aligned}|1^A0^B\rangle &= N^{-\frac{1}{2}}\sum_{\mathbf{t}}|1^A0^B,\mathbf{t}\rangle,\\ |0^A1^B\rangle &= N^{-\frac{1}{2}}\sum_{\mathbf{t}}|0^A1^B,\mathbf{t}\rangle.\end{aligned} \tag{10.1.7}$$

We have dropped the $\mathbf{k}$ label, but we should remember that we are dealing with states with $\mathbf{k}=0$. The ket $|1^A0^B,\mathbf{t}\rangle$ denotes the state in which molecule A in cell $\mathbf{t}$ is excited and all others are in the ground state; similarly $|0^A1^B,\mathbf{t}\rangle$ has just molecule B in cell $\mathbf{t}$ excited.

Now, as before, there is an interaction between these two states, and only these, because states with different $\mathbf{k}$ cannot mix. The matrix element is

$$\begin{aligned}\langle 1^A0^B|\mathcal{H}'|0^A1^B\rangle &= N^{-1}\sum_{\mathbf{tt}'}\langle 1^A0^B,\mathbf{t}|\mathcal{H}'|0^A1^B,\mathbf{t}'\rangle\\ &= \sum_{\mathbf{t}}\langle 1^A0^B,\mathbf{t}|\mathcal{H}'|0^A1^B,0\rangle.\end{aligned}$$

We have replaced the sum over $\mathbf{t}'$ by a factor N, since the translational symmetry of the crystal ensures that all values of $\mathbf{t}'$ give the same result. The matrix elements in the remaining sum are dipole–dipole interactions between transition dipoles, as before. In each

* Some knowledge of the elementary theory of solids is being assumed here. If it is unfamiliar, this discussion may be skipped, or a suitable introduction may be found in, e.g., Kittel (1987).

of these, one of the transition dipoles is that of molecule B in cell 0, while the other is that of molecule A in the same cell or one of the neighbouring ones. Because of the R^{-3} factor in the dipole–dipole interaction, only the nearest neighbours contribute significantly. Note that all these A transition moments are parallel because $\mathbf{k} = 0$; there is no arbitrariness about their relative phases, though as before the relative phase of the A and B transition moments is arbitrary. The number of neighbouring A molecules that contribute significantly to the matrix element will depend on the details of the crystal structure, but because the two $\mathbf{k} = 0$ states are initially degenerate, the resonance interaction causes them to mix, just as for two isolated molecules; and the intensity may all attach to the lower state, or all to the upper state, or be divided between them, again just as for two isolated molecules. The delocalized excitation represented by eqns 10.1.7 is called an *exciton*, and the splitting between the states is an *exciton splitting*.

10.1.2 *Excimers*

One of the most interesting consequences of the resonance interaction is the formation of dimers in which one of the molecules is excited. Because of the resonance interaction such an excited dimer, or *excimer*, may be much more firmly bound than when both molecules are in their ground state. Indeed the ground-state molecules may not form a stable complex at all. Excimers were first observed by Forster (1955) in solutions of pyrene in benzene. Here the fluorescence spectrum at low concentration (2×10^{-4}M) shows the emission from the isolated molecule, with peaks at 3725 Å, 3840 Å and 3920 Å. As the concentration is increased, a broader emission band appears at about 4780 Å; this is due to emission from the red-shifted state of the excimer. At a concentration of (2×10^{-2}M) the monomer fluorescence virtually disappears, leaving only the excimer fluorescence.

However the most important application is in the excimer laser (Milonni and Eberly 1988). In the xenon excimer laser, for example, some of the atoms in xenon gas at high pressure are excited by a pulsed electron beam, and combine with ground-state atoms to form $Xe\cdots Xe^*$ excimers. They are quite strongly bound, and persist in the gas for some time. When a radiative transition to the ground state occurs, the repulsive form of the ground-state potential curve ensures that the dimer immediately falls apart. (See Fig. 10.2.) This means that a population inversion can easily be maintained, and laser action can occur; also, because there are no well-defined vibration–rotation states on the repulsive potential-energy surface, the radiation is broad-band and can be tuned.

10.2 Distributed transition moments

In deriving the expression for the matrix elements between the exciton states, we used the multipole expansion of the interaction and truncated it at the dipole–dipole term. We have seen enough to suggest that such a treatment may be invalid when the molecules are as close together as they are in an excimer or a crystal. It is likely that the multipole expansion will fail to converge, especially when large molecules are involved.

An example is the 'special pair' that forms a component of chlorophyll and related molecules. This essentially comprises two bacteriochlorophyll units, each of which has a conjugated system of the form shown in Fig. 10.3 (M = Mg). The precise geometry varies, as indeed does the detailed molecular structure, but usually the two units are stacked one above the other with their planes approximately parallel and approximately

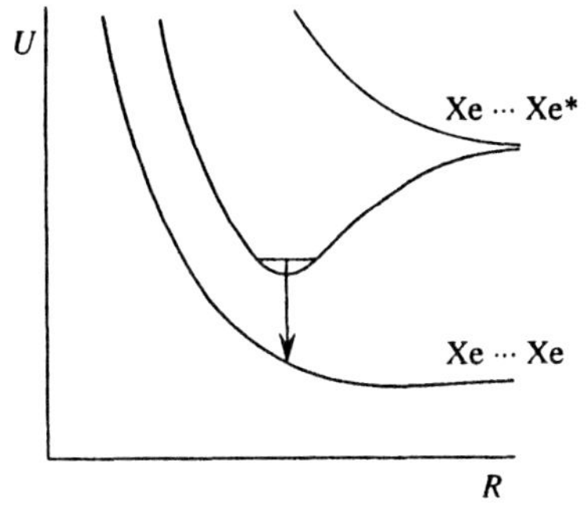

FIG. 10.2. Potential curves for Xe···Xe and Xe···Xe* (schematic).

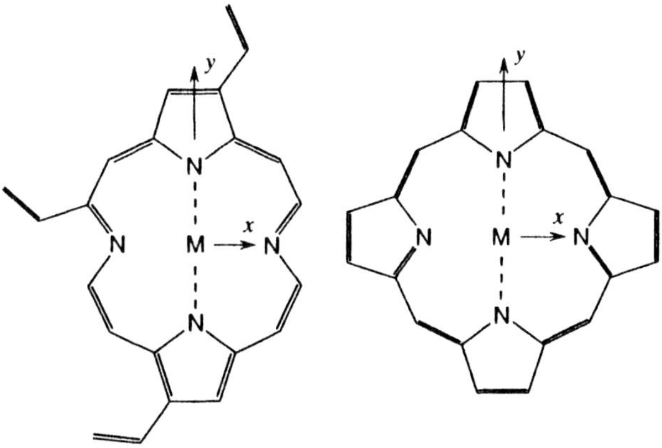

FIG. 10.3. Structure of bacteriochlorophyll (left) and porphin (right).

3.5 Å apart, and they are offset by about 7 Å in the *y* direction (see Fig. 10.3), so that the centre of one molecule is approximately above a pyrrole ring of the other (Warshel and Parson 1987). The bacteriochlorophyll molecule is some 12 Å across, compared with the separation between centres of about 8 Å, so this is a case where the multipole expansion cannot be expected to converge and the dipole–dipole model of the resonance interaction will give poor results.

A similar situation arises in some model compounds designed to mimic the special pair (Leighton *et al.* 1988). They comprise two porphin units (Fig. 10.3, either with $M =$ Zn or with $M = -\text{H}\cdots\text{H}-$, i.e., the free base) joined by flexible linking sidechains. In these compounds the porphin units are parallel, and offset in the *y* direction relative to each other by a distance of the order of 3.5 Å. (This is an example of a structure that is successfully predicted as a consequence of the electrostatic interaction, provided that a distributed-multipole model is used (Hunter 1994).) A more rigid structure can be made using the bifunctional ligand $N(CH_2CH_2)_3N$, which binds to the two Zn atoms. In this way the two porphin units are held parallel to each other, with the two Zn atoms directly over each other (Hunter *et al.* 1989) and 7 Å apart. There is a $\pi - \pi^*$ transition, the Soret band, at 415 nm in the isolated porphin molecule; in the complex it is blue-shifted to 409 nm. As the transition dipole lies in the plane of the molecule, we have case (*b*) of Fig. 10.1, and the red-shifted band is forbidden. In this case the porphin planes are 7 Å apart, but this is still small compared with the molecular size, so again we cannot use the dipole–dipole model. Instead we can use a distributed multipole model, as follows.

The transition of interest is a $\pi \leftrightarrow \pi^*$ transition. We can write the two orbitals as

$$\psi_0 = \sum c_i^0 \varphi_i, \qquad \psi_1 = \sum c_i^1 \varphi_i, \tag{10.2.1}$$

where the φ_i are p_π orbitals on the atoms of the conjugated system, and the c_i^k are molecular orbital coefficients. We need the matrix element

$$\langle \psi_1^A \psi_0^B | \mathcal{H}' | \psi_0^A \psi_1^B \rangle = \langle \psi_1^A \psi_0^B | \sum_{a\in A} \sum_{b\in B} (T^{ab} \hat{q}^a \hat{q}^b - T_\alpha^{ab} (\hat{q}^a \hat{\mu}_\alpha^b - \hat{\mu}_\alpha^a \hat{q}^b) - \cdots | \psi_0^A \psi_1^B \rangle. \tag{10.2.2}$$

Here we are using a distributed-multipole expression for the interaction operator. The summation runs, as usual, over the sites *a* of molecule *A* and the sites *b* of molecule *B*. In view of the form of the wavefunction it is natural to choose these sites to be at the atoms involved in the π system. Then the matrix element of one of the distributed charge operators is

$$\langle \psi_1^A | \hat{q}^a | \psi_0^A \rangle = \sum_{ii'} \langle c_i^1 \varphi_i | \hat{q}^a | c_{i'}^0 \varphi_{i'} \rangle = -c_a^1 c_a^0, \tag{10.2.3}$$

because only the atomic orbital on atom *a* contributes to the matrix element of $\hat{q}^a$ (if we ignore overlap) and it is normalized to 1. The minus sign takes account of the negative charge on the electron. Similarly,

$$\langle \psi_1^A | \hat{\mu}_\alpha^a | \psi_0^A \rangle = \sum_{ii'} \langle c_i^1 \varphi_i | \hat{\mu}_\alpha^a | c_{i'}^0 \varphi_{i'} \rangle = 0, \tag{10.2.4}$$

since the π-orbital distribution has zero dipole moment relative to an origin at the nucleus. There would be a non-zero quadrupole term but we neglect it. Thus (10.2.2) becomes

$$\langle \psi_1^A \psi_0^B | \mathcal{H}' | \psi_0^A \psi_1^B \rangle = \sum_{a\in A} \sum_{b\in B} c_a^1 c_a^0 c_b^1 c_b^0 / 4\pi\varepsilon_0 R_{ab}. \tag{10.2.5}$$

If we know the geometry of the complex and the wavefunctions this is easily evaluated. For most purposes a Hückel or Pariser–Parr–Pople treatment of the π orbitals is adequate, but the method is readily extended to more accurate wavefunctions.

In the case of the rigid porphin dimer complex, the point-dipole treatment, using the observed intensity to determine the magnitude of the transition dipole, predicts an exciton shift of −19 nm for the Soret band, whereas the distributed-monopole treatment, again scaled according to the observed intensity, predicts a shift of −6 nm, in exact agreement with experiment.

11

PRACTICAL MODELS FOR INTERMOLECULAR POTENTIALS

The computation of the interaction energy *ab initio* is very time-consuming, whether it is done by perturbation theory or the supermolecule method, and it is not practical in many applications. In molecular dynamics, for instance, it is necessary to calculate the forces and torques on all the molecules in an assembly of several hundred, in order to obtain the equations of motion, and this has to be repeated for each step in the simulation, which typically requires 10^5 or 10^6 steps. For this sort of problem, we usually need to use much simpler models of the interaction. For the most basic properties of condensed phases, it is sufficient to use extremely simple models—for example, the hard-sphere model provided much useful information about the behaviour of inert-gas liquids—but if we want more detailed information, we have to use better models.

Potential models are of two kinds. One kind is a simplification of the *ab initio* approach; the potential is still computed from first principles, but approximations are made in order to make the calculation more tractable. In a sense all *ab initio* calculations fall into this category, since it is virtually never possible to perform the calculation exactly; but one can distinguish between true *ab initio* calculations, where it is possible to identify a path towards better and eventually arbitrarily accurate calculations, and a model in which an approximation is imposed that offers no hope of improvement but is merely a more or less accurate prescription. One such model is the 'exchange-Coulomb' or XC model (discussed in more detail below) which makes use of the empirical observation that for inert gas atoms the repulsion (exchange) energy E_X is closely related to the electrostatic penetration (Coulomb) energy E_C: $E_X = -\gamma(1 + aR)E_C$, where a and γ are empirical constants. This relationship is quite successful in many cases, but no-one supposes that it is exact, or that it could be developed towards a relationship that would reliably yield exact values of E_X while avoiding computational effort. Its usefulness lies in the fact that E_X, which is time-consuming to calculate, can be determined reasonably accurately in terms of E_C, which can be calculated much more easily.

The other kind of model is more empirical: a suitable function, of a relatively simple form, is proposed to describe the interaction, and the parameters in the model are fitted to experimental data or to accurate calculations. In practice, a combined approach is often used. The HFD or Hartree–Fock–dispersion model, for example, uses an approximate *ab initio* treatment, namely an SCF supermolecule calculation, and corrects for the worst deficiency of such a calculation—the fact that it includes no correlation and therefore no dispersion—by adding an empirical dispersion term. This again is quite successful in some cases, but it is intrinsically limited in accuracy, however good the SCF calculation and the dispersion model may be, because its success depends crucially on the assumption, not always valid, that other correlation effects may be neglected.

Empirical models can be further subdivided. Some seek to describe a particular system, and are intended for use in calculations on that system alone. Such models can be as complicated as is necessary to describe the potential, subject to constraints imposed by the computation in which they are to be used. Indeed those constraints may determine the form of the potential; for example, calculations of the scattering of an atom with a diatomic are most easily carried out if the potential is a spherical-harmonic expansion:

$$U(R,\theta) = \sum_{k=0}^{n} V_k(R) P_k(\cos\theta)$$

and this is then a natural way to express the potential, though it does not provide any separation into contributions such as dispersion and repulsion.

Other kinds of model are intended for use with a whole set of systems, with their parameters very often determined from properties of a subset. In particular one often wishes to determine potentials for mixed ($A \cdots B$) interactions from properties of pure substances only, which in principle lead only to information about interactions between like molecules. Models of this kind need to be simpler in form, and are more usefully composed of identifiable terms such as dispersion, repulsion, and so on. The reason for this is that the derivation of potentials for mixed systems is often based on 'combining rules', and these are more successful if they can be closely based on fundamental theory. We have seen, for instance, that the geometric-mean combining rule for C_6 dispersion coefficients is derived directly from the London formula for the dispersion energy. The determination of accurate dispersion coefficients for mixed systems via dipole oscillator strength distributions (DOSDs) for the individual molecules, described in §12.1.2, can be thought of as a rather refined combination rule, though it differs from most combining rules in being exact in principle. Neither approach is useful if a potential model is used in which the dispersion energy is not clearly distinguishable from the rest of the model.

In Chapter 7 we explored models for the electrostatic interaction. Here we consider first the less tractable terms, the dispersion and repulsion, in a general way, and then consider how the various terms are commonly combined to yield a complete potential.

11.1 Potentials for atoms

11.1.1 *Hard-sphere atoms*

In this approach, each atom is treated as an impenetrable hard sphere, so that the potential is infinite if atoms overlap and zero otherwise. The radius of the sphere is usually a standard Van der Waals radius, taken from Pauling (1960) or Bondi (1964). This is a very simple treatment of the repulsion, but it is sufficient, as remarked above, to account for some of the properties of inert gas fluids, such as some of their transport properties. When applied to molecules, it is enough to account for the structures of a wide range of Van der Waals complexes when combined with an accurate electrostatic description, as we saw in §7.6.1 (Buckingham and Fowler 1983, 1985). Of course it has no predictive value whatever in terms of intermolecular separations, and it is successful in structural predictions for these complexes only because the structure is usually much more sensitive to the angular behaviour of the electrostatic interaction than to the precise interatomic separations. As a part of such models the hard-sphere repulsion still has a limited but useful contribution to make; it has the merit that it requires only one parameter, the atomic radius, that is readily and directly determinable from crystal-structure data. It has the obvious limitation that atoms in molecules are neither hard nor spheres.

11.1.2 *Lennard-Jones potentials*

The Lennard-Jones potential, first proposed in 1906 by Mie and adopted by Lennard-Jones in the early 1920s, has a repulsive term A/R^n and an attractive term $-B/R^m$, with $n > m$. Following London's work on dispersion forces it became usual to set $m = 6$, and although a variety of values of n has been used, the most common choice is $n = 12$, which is computationally convenient and reasonably successful. The Lennard-Jones potential then takes the form

$$U_{\mathrm{LJ}} = 4\varepsilon\left(\frac{\sigma^{12}}{R^{12}} - \frac{\sigma^6}{R^6}\right) = \varepsilon\left(\frac{R_m^{12}}{R^{12}} - \frac{2R_m^6}{R^6}\right), \tag{11.1.1}$$

where ε is the depth of the well and R_m the position of the minimum. $\sigma = 2^{-1/6}R_m$ is the position where the repulsive branch crosses zero.

This potential has been remarkably successful, as is shown by its widespread use 70 years after Lennard-Jones first introduced it. The attractive term has a sound theoretical justification as the leading term in the R^{-1} expansion of the dispersion energy, though the value of $C_6 = 4\varepsilon\sigma^6$ that is obtained by fitting ε and σ (or R_m) to experimental data is much too large, typically by a factor of about 2, compared with direct experimental or theoretical determination. This is in part because the Lennard-Jones potential contains no R^{-8} or R^{-10} terms, so the R^{-6} term has to be larger to compensate. The repulsive term, on the other hand, has no theoretical justification at all, beyond its steeply repulsive form.

When the atoms are different, it is common to use 'combining rules' to estimate the parameters, often using a geometric mean for ε (the 'Berthelot' rule) and an arithmetic mean for σ (the 'Lorentz' rule):

$$\varepsilon_{ab} \approx (\varepsilon_{aa}\varepsilon_{bb})^{1/2}, \qquad \sigma_{ab} \approx \tfrac{1}{2}(\sigma_{aa} + \sigma_{bb}). \tag{11.1.2}$$

The theoretical basis for these rules is very flimsy. The Berthelot rule is known to overestimate the well depth. More elaborate rules have been proposed (see Maitland *et al.* (1981) for a discussion) but still with rather dubious justification. The main reason for the continued use of these combining rules is that experimental data sufficiently good to validate or falsify them is very difficult to obtain, and meanwhile some way of estimating the parameters in mixed systems is required.

We have seen that a geometric-mean combining rule for C_6 should be successful, so an alternative possibility would be to apply a geometric-mean combining rule to the R^{-6} coefficient, i.e. to $4\varepsilon\sigma^6$, even though this is not a true C_6 but contains various other effects as well. Waldman and Hagler (1993) have shown that such a combining rule is successful for the inert gases, but only if it is used in conjunction with an arithmetic-mean combining rule for σ^6 rather than for σ itself.

11.1.3 *Born–Mayer potential*

Not long after the introduction of the Lennard-Jones potential, Born and Mayer (1932) suggested that the repulsion between atoms would have a roughly exponential dependence on distance, because of its relationship to the overlap between the wavefunctions:

$$U_{BM} = Ae^{-BR}. \tag{11.1.3}$$

If this is combined with the London formula for the dispersion, we get the exp-6 potential:

$$U_{\exp 6} = Ae^{-BR} - C/R^6. \tag{11.1.4}$$

A variant proposed by Buckingham and Corner (1947) has an R^{-8} dispersion term as well.

The exp-6 and Buckingham–Corner potentials have a deficiency which makes them unsuitable for some calculations: although the exponential term rises steeply as R decreases, it remains finite at $R = 0$, so the dispersion term dominates at very small R, and the potential reaches a maximum and then tends to $-\infty$ as $R \to 0$. This can be overcome by damping the dispersion term (see below) at the cost of complicating the form of the function. However the main disincentive to its widespread use in the days of mechanical hand-operated calculating machines, and even with early computers, was the computational inconvenience of the exponential. For this reason the Lennard-Jones potential won an early popularity which it has still not lost in spite of its obvious deficiencies.

11.1.4 *Accurate potentials for atoms*

A great variety of empirical potential functions have been proposed, especially for the inert gases. Many of these have no particular basis in theory, but use mathematical devices such as spline functions to reproduce the experimental data as accurately as possible.

For the inert gases, especially argon, the experimental data are sufficiently diverse and reliable to characterize the potentials quite well. There is a wide range of potentials fitted to such data, such as the Barker–Fisher–Watts potential for argon (Barker *et al.* 1971):

$$U(R) = \exp[\alpha(1-\bar{R})] \sum_{k=0}^{5} A_k(\bar{R}-1)^k - \sum_{n=6,8,10} C_n/(\delta + \bar{R}^n), \tag{11.1.5}$$

where $\bar{R} = R/R_m$, R_m being the separation at the minimum, and δ is a small non-physical parameter introduced to suppress the spurious singularity that would otherwise arise at $R = 0$. A detailed account of such potentials has been given by Maitland *et al.* (1981), and since our interest is primarily in molecular potentials we shall not review them fully here. However there are some general issues that are relevant to molecules as well as atoms.

Eqn (11.1.5) shares with many atom–atom potential models a separation of the interaction into a term of the form $\exp(-\alpha R)$ that can be loosely identified with the repulsion, and a series in R^{-1} that describes the dispersion. The theory gives good grounds for separating the potential in this way. However, there is no clean separation between the terms when they are derived from experimental data—the distinction between repulsion and dispersion is a theoretical one, and they are not separately observable—and any deficiencies in one tend to be taken up by the other. Notice that the 'dispersion' part of eqn (11.1.5) is undamped (the parameter δ is too small to provide any real damping) so although the potential must include dispersion damping in order to agree satisfactorily with the experimental data, it must here be contained in the 'repulsion' term. In fact, closer investigation reveals that the 'repulsion' part, with the parameters given by Barker *et al.*, has a small attractive region; since the damping as well as the repulsion itself is a repulsive effect this means that the 'repulsion' term contains part of the dispersion. Whether or not this is viewed as a satisfactory way to treat the Ar···Ar interaction, it does not provide a useful starting-point for constructing potentials for other systems. Other potentials that have been proposed for Ar···Ar are even less useful in this respect. A more robust strategy in the long term is to use functional forms for the terms in the potential that are as closely based on theory as practical constraints will allow.

11.1.5 *Dispersion*

We have seen that the dispersion term in a model potential must include a damping function to suppress the singularity as $R \to 0$. One function that has been proposed for this purpose is part of the 'HFD' or Hartree–Fock–dispersion model, proposed originally by Hepburn *et al.* (1975) and modified slightly by by Ahlrichs *et al.* (1977). Here the interaction between two atoms is described by two terms: a Hartree–Fock part, obtained from a supermolecule SCF calculation, and a dispersion part, represented by a term of the form $-F(R)(C_6/R^6 + C_8/R^8 + C^{10}/R^{10})$. The damping function $F(R)$ is

$$F(R) = \begin{cases} \exp[-(1.28(R_m/R) - 1)^2], & R < 1.28R_m, \\ 1, & R > 1.28R_m. \end{cases} \qquad (11.1.6)$$

This function was obtained by fitting to the dispersion interaction in the $b\,^3\Sigma_u^+$ state of H_2, i.e., to the dispersion interaction between two H atoms with parallel spins, which can be calculated accurately (Kołos and Wolniewicz 1974). (We refer to this for brevity as the H···H interaction in the rest of this section.) While it fits the calculated points quite well, its behaviour at $R = 0$ is unsatisfactory (it is non-analytic) and, more seriously, the same function is used for all the dispersion terms. The calculations of Kreek and Meath (1969) had already shown that different damping functions are needed for each term in the R^{-n} expansion.

A later version of the HFD model (Douketis *et al.* 1982) corrected this deficiency, writing the dispersion energy in the form

$$U_{\text{disp}} = - \sum_{n=6,8,10} F_n(R) \frac{C_n}{R^n}, \qquad (11.1.7)$$

with

$$F_n(R) = g(\rho R) f_n(\rho R), \qquad (11.1.8)$$

where f_n is supposed to be a universal damping function correcting for charge-overlap effects (penetration) in the R^{-n} term, g is a universal function correcting for exchange effects in all the dispersion terms, and ρ is a distance scaling factor. These functions, like the earlier one, were found by an empirical fit to the damping functions calculated by Kreek and Meath (1969) for the H···H dispersion interaction, where $\rho = 1$ by definition, and are

$$\begin{aligned} g(R) &= 1 - R^{1.68} \exp(-0.78R), \\ f_n(R) &= [1 - \exp(-2.1R/n - 0.109R^2/n^{1/2})]^n \end{aligned} \qquad (11.1.9)$$

Table 11.1 *Coefficients for Koide–Meath–Allnatt damping functions, with C_n coefficients for H···H.*

n	a_n/a_0^{-1}	$10b_n/a_0^{-2}$	$10^3d_n/a_0^{-3}$	C_n/ a.u.	$\lim_{R\to 0} f_nC_nR^{-n}$
6	0.3648	0.3360	1.651	6.499	0.0153
8	0.3073	0.2469	1.227	124.399	0.0099
10	0.2514	0.2379	0.5664	3285.8	0.0033
12	0.2197	0.4168	0.4168	121486.0	0.0015

Another form of damping function was proposed by Koide *et al.* (1981), who recalculated the accurate damping functions for the H···H dispersion interaction and fitted them to the form

$$f_n^0(R) = \left[1 - \exp(-a_nR - b_nR^2 - d_nR^3)\right]^n. \tag{11.1.10}$$

The coefficients in these functions, for $n \leq 12$, are listed in Table 11.1; the paper gives coefficients for n up to 20. The functions are shown in Fig. 11.1, from which it is evident that the damping functions suppress the dispersion terms more and more strongly as n increases. The leading term in eqn (11.1.10) as a power series in R is $(a_nR)^n$, so that the damped C_6R^{-6} term tends to $a_6^6C_6 = 0.0153$ a.u. (for H···H) and so on. These limits at $R \to 0$ are not quantitatively meaningful, but they show that the damped dispersion behaves in a reasonable manner at very short distances.

Yet another type of damping function was proposed by Tang and Toennies (1984). The C_n/R^n term was replaced as usual by $f_n(R)C_n/R^n$, and f_n was required to tend to 1 for large R and to zero as $R \to 0$. Tang and Toennies further assumed that $f_n(R) - 1$ has the form of a polynomial multiplied by a decaying exponential (an assumption based on the form of the exchange correction to the dispersion for the H···H system) and were led to the conclusion that $f_n(R)$ is an incomplete gamma function of order $n+1$:

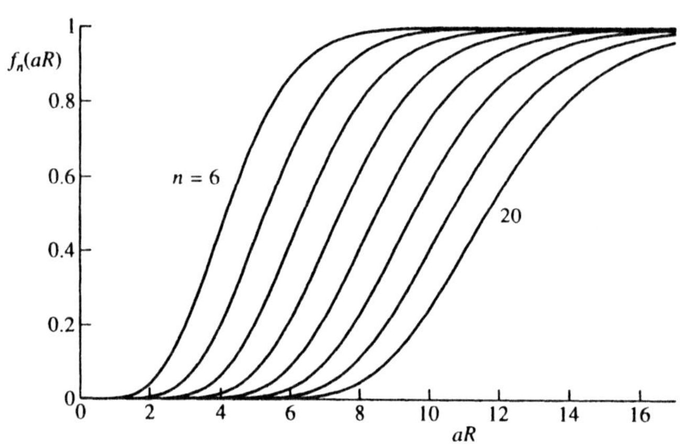

FIG. 11.1. Koide–Meath–Allnatt damping functions $f_n(aR)$ for $n = 6, 8, \ldots, 20$.

$$f_n(R) = P(n+1, bR) = 1 - \exp(-bR) \sum_{k=0}^{n} \frac{(bR)^k}{k!}. \qquad (11.1.11)$$

Here b is a distance scaling factor, usually taken to be the same as the parameter B in the Born–Mayer repulsion, on the grounds that both the repulsion and the dispersion damping are consequences of wavefunction overlap. Although f_n behaves like R^{n+1} at small R, so that it has the effect of suppressing the dispersion entirely at $R = 0$ rather than making it tend to the appropriate part of the united-atom correlation energy, this is a trivial flaw by comparison with errors in other potential terms at very small R. The Tang–Toennies functions have the merit of using no adjustable parameters, if the parameter b is taken from the repulsion term, and only one if it is fitted to the data.

These damping functions, being all derived from the H$\cdots$H potential, are all rather similar. It was assumed from the outset that they are 'universal' damping functions and can be transferred to other systems simply by scaling the distance. Douketis *et al.* (1982) acknowledged that this assumption is 'difficult, and perhaps impossible, to justify rigorously'. They remark that the accuracy of the Law of Corresponding States makes the assumption seem plausible, and indeed the reduced potential curves $U(R)/U(R_m)$ *vs.* R/R_m for the H$\cdots$H interaction and the most reliable inert-gas potentials (Aziz 1984) are very similar. The damping functions also have a very similar R dependence to the *ab initio* damping functions determined by Knowles and Meath (1986*b*, 1986*a*, 1987) from the ratio of the non-expanded dispersion terms to the C_n/R^n formula. The f_n functions of Koide *et al.* (1981) shown in Fig. 11.1 may be compared with the Knowles–Meath *ab initio* damping functions for Ar$\cdots$Ar, shown in Fig. 6.3 on p. 101. Recently Wheatley and Meath (1993*a*) have found that damping functions for other systems can be obtained by scaling the H$\cdots$H functions of Koide *et al.* (1981):

$$f_n^{\mathrm{AB}}(R) = f_n^{\mathrm{HH}}(S^{AB}R),$$

where S^{AB} is a suitable scaling factor for the $A\cdots B$ interaction. The *ab initio* calculations use Time-Dependent Coupled Hartree–Fock (TDCHF) perturbation theory in the polarization approximation, so they include penetration (charge-overlap) effects but not exchange. The original H$\cdots$H calculations, on the other hand, are essentially exact within the non-relativistic Born–Oppenheimer model.

The various HFD-type models that have been proposed for inert-gas atoms all represent the interaction as the sum of a repulsive part and an attractive (dispersion) part, but they differ in the description of the repulsive part of the potential, so that the attractive part necessarily differs also. For example, the factor $g(\rho R)$ that appears in eqn (11.1.8) describes exchange-correlation effects, but these may in other models be regarded as part of the repulsive term. Such differences between the models have been discussed in detail by Meath and Koulis (1991), who remark that it is consequently inappropriate to use a damping function designed for one model in a different context.

11.1.6 *Repulsion*

The linearity of the semi-log plots of Fig. 6.2 on p. 99 show that the simple exponential (Born–Mayer) form provides a remarkably accurate description of the repulsion at moderate energies. The HFD (Hartree–Fock–dispersion) model in its simplest form uses a

Born–Mayer repulsion fitted to the SCF energy, together with a damped dispersion term of the form discussed in the previous section. A variant uses the form $A\exp[-\alpha R-\beta R^2]$ to describe the repulsion; the parameter β is quite small. See, e.g., Dham *et al.* (1989).

The XC (exchange-Coulomb) repulsion model for inert-gas dimers is based on the idea that the repulsion (exchange) energy and the electrostatic (Coulomb) energy are closely related in such systems. The long-range electrostatic energy (the multipole expansion) is zero in such cases, and the first-order electrostatic energy E_C arises purely from penetration (see §6.4.1). The XC model uses the semi-empirical relationship

$$E_X = -\gamma(1+aR)E_C, \tag{11.1.12}$$

where a and γ are parameters. The Coulomb energy is evaluated as an integral over the charge densities: $E_C = \int \rho^A(\mathbf{r}_1)r_{12}^{-1}\rho^B(\mathbf{r}_2)\,dr_1\,dr_2$, using accurate *ab initio* (SCF) charge densities for the atoms, and fitted to a function of the form

$$E_C = -\exp[a_0R+a_1+a_1R^{-1}+a_2R^{-2}]. \tag{11.1.13}$$

This then gives a quite elaborate expression for the repulsion energy, which is quite satisfactory for the case to which it is fitted, and which describes the short-range, high-energy part of the potential curve where the simple Born–Mayer form is inadequate (Dham *et al.* 1989). It can be extended to other systems quite easily, because it starts from the monomer charge density, which is relatively easy to calculate accurately, and the calculation of the Coulomb integral is not very time-consuming.

A further model of this type is the Tang–Toennies one, which is simpler in form, using a plain Born–Mayer repulsion together with a damped dispersion with damping functions that are incomplete gamma functions, as described above:

$$U_{\mathrm{TT}} = A\exp(-bR) - \sum_{k=3}^{K} f_{2k}(bR)C_{2k}/R^{2k}. \tag{11.1.14}$$

As originally formulated, the repulsion term was fitted to SCF *ab initio* data, but if A and b are determined by fitting to the accurate potentials available for systems such as Ar_2, somewhat larger values of A are found, though the values of b do not change much. This is consistent with the observation from *ab initio* perturbation theory that the inclusion of correlation effects increases the magnitude of the repulsion by a ratio that is more or less independent of distance.

11.2 Induction

Induction is often ignored altogether in simple treatments, partly because it usually makes a relatively small contribution to the energy, and partly because its non-additive characteristics make it awkward to handle. Even in strongly polar systems like water, where the effects of induction are quite significant, it is common to omit any explicit treatment of induction and instead to use an electrostatic model with an enhanced dipole moment to describe its effects in an average way. The dipole moment of water in such cases may be increased from the isolated-molecule value of 1.8 D to as much as 2.3 D.

Just as point-charge models have long been used to give a simple description of the charge distribution, simplified models of polarization have also been used. Probably the simplest of such treatments is the 'shell model' of Dick and Overhauser (1958) and Cochran (1959). In this approach, widely used to describe ions in solids, the valence electrons are viewed as a charged massless spherical shell (equivalent for the purposes of electrostatic interactions to a point charge at the centre of the shell) and the core of the atom or ion is treated as another point charge. The two charges, the shell and the core, are taken to be attached to each other by a harmonic 'spring'. In a uniform electric field, the negatively charged shell moves in one direction and the positively charged core in the other, producing a dipole whose magnitude depends on the strength of the spring. It is a simple matter to choose the spring constant so as to reproduce the dipole polarizability of the ion. The mass of the ion is attributed to the positive core, and the shell carries the repulsive interactions.

This model has been extensively used for modelling ionic solids, where it is reasonably successful. For example, the SYMLAT program (Leslie 1983) uses an ionic model that includes the shell–model description of polarization. Freyriafava *et al.* (1993) found that it agreed well with Hartree–Fock calculations on defects in MgO. The shell model accounts only for dipole polarization, and the Hartree–Fock calculations also showed evidence of quadrupolar polarization and spherical 'breathing' distortions of the oxide anions, but their effects appeared to be unimportant in this case. Whether this was fortuitous or not could not be determined.

The shell model has not been found useful in calculations on molecules, and where induction is too important to ignore it is usual to include the dipole–dipole polarizability explicitly. In this case it is necessary to solve eqn (8.3.12) iteratively, which is time-consuming. A rather different approach has recently been used, however, by Sprik and Klein (1988) in a polarizable model of water. The polarizability was represented using a tetrahedral array of four sites, with the centre of the tetrahedron on the bisector of the HOH angle, 0.6 Å from the O atom. Two of the sites were in the plane of the molecule, displaced in the general direction of the H atoms, and the other two were out of plane, displaced in the 'lone-pair' directions. In this case, the magnitude of the charges, rather than their positions, change in response to an electric field, so this is a distributed-polarizability model, limited to charge flows. The merit of this formulation is that it is easy to describe the induction by adding a suitable term to the Lagrangian for the system, which in turn makes it easy to write down the classical equations of motion of a system containing a number of such water molecules. Wilson and Madden (1994) used a similar approach in their work on the alkaline-earth halides (§ 9.1). By treating the charges as extended gaussian charge distributions rather than point charges, Sprik and Klein also introduced damping into the model.

11.3 Model potentials for small molecules

11.3.1 *Molecule–molecule potentials*

As in the treatment of purely electrostatic interactions, it is possible to view each molecule as a unit, with a single interaction centre, or to regard the interactions as distributed over the molecule. For brevity we call the former treatment a 'molecule–molecule' description, and the latter an 'atom–atom' or 'site–site' treatment. We may expect that, as for electrostatic interactions, a molecule–molecule description will suffice for small molecules but will become unsatisfactory for larger ones.

A natural way to obtain a molecule–molecule description is to allow the parameters in a potential for atoms to become functions of the relative molecular orientation. For instance, an extension of the hard-sphere model to molecules was proposed by Kihara (1978), who treated a molecule as a convex hard solid—an ellipsoid, spherocylinder, disc or other shape. He was able to gain a useful understanding of the packing of molecules in crystals using mechanical models of this kind, but the approach has not been much used in calculations because it is difficult to manipulate mathematically.

Corner (1948) seems to have been the first to suggest using a variant of the Lennard-Jones potential in which the well-depth ε and the range parameter σ or R_m are allowed to vary with orientation. In the Gaussian Overlap Model of Berne and Pechukas (1972), the form of the angle-dependence of ε and R_m is taken from the expression for the overlap integral of two ellipsoidal gaussian charge distributions, but the only merit of this approach is its mathematical convenience. Pack (1978) explored potentials for $Ar\cdots CO_2$ in which $\varepsilon = \varepsilon_0(1 + aP_2(\cos\theta))$ and $R_m = R_0(1 + bP_2(\cos\theta))$, with θ the angle between the $C\cdots Ar$ direction and the OCO axis (see Fig. 1.3 on p. 9), and he was able to understand some of the features of $Ar\cdots CO_2$ scattering using these potentials. A similar approach has been used to describe the interaction between inert gases and SF_6 (Pack *et al.* 1982, 1984).

The Lennard-Jones potential is not a very good model for atoms, so this is not a very promising strategy for molecules if accurate potentials are needed. The exp-6 model should be better. Moreover, molecules differ from atoms in having non-zero multipole moments, so it becomes necessary to include electrostatic interactions; and even if one of the components is an inert gas, there will be an induction energy term, which may be important if the other component is strongly polar. This line of argument leads to the type of model that has been used extensively by Hutson (e.g., 1989*a*, 1989*b*) to describe the potentials for complexes between hydrogen halides and inert gases. These potentials take the form

$$U(R,\theta) = A(\theta)\exp[-\beta(\theta)R] + U_{\text{ind}} + \sum_{n=6,7,8} f_n(R)C_n(\theta)/R^n \qquad (11.3.1)$$

where the induction energy expression is the ordinary multipole expansion up to terms in R^{-7}. Such potentials have been very successful in accounting for the spectra of small Van der Waals complexes, where the molecules are nearly spherical. We shall return to this theme in Chapter 12.

11.3.2 *Atom–atom and site–site potentials*

For larger molecules, it is usual to use an atom–atom or site–site formulation. The electrostatic interaction is easily handled using distributed multipoles, though point charge descriptions are still more usual. The induction energy is often small enough to neglect; if it is included, the polarizability is usually described by a single-site dipole–dipole polarizability only, but more elaborate descriptions are easily constructed using distributed polarizabilities of one form or another.

The more difficult terms to describe satisfactorily are repulsion and dispersion, and these are usually treated in atom–atom form.

A Lennard-Jones atom–atom potential, for example, would take the form

$$U_{\mathrm{LJ}} = \sum_{a\in A}\sum_{b\in B} 4\varepsilon_{ab}\left(\frac{\sigma_{ab}^{12}}{R_{ab}^{12}} - \frac{\sigma_{ab}^{6}}{R_{ab}^{6}}\right), \tag{11.3.2}$$

where the sum is over the atoms of each molecule, and ε_{ab} and σ_{ab} are suitable parameters for atoms of type a and b. Similarly, an exp-6 atom–atom potential takes the form

$$\begin{aligned} U_{\mathrm{exp6}} &= \sum_{a\in A}\sum_{b\in B} A^{ab}\exp(-B^{ab}R_{ab}) - \frac{C_6^{ab}}{R_{ab}^6} \\ &= \sum_{a\in A}\sum_{b\in B} K\exp(-\alpha^{ab}(R_{ab}-\rho^{ab})) - \frac{C_6^{ab}}{R_{ab}^6} \end{aligned} \tag{11.3.3}$$

The potential parameters describing the repulsive part of the site–site interaction are conventionally denoted A^{ab} and B^{ab}, and must not be confused with the labels A and B that we have been using throughout to label molecules. The alternative formulation shown in eqn (11.3.3) uses $\alpha^{ab} = B^{ab}$ and $\rho^{ab} = \ln(A^{ab}/K)/\alpha^{ab}$; K is not a parameter but a convenient energy unit, usually 10^{-3} hartree. This corresponds to a temperature of about 316 K, so is a suitable unit for potentials that are to be used to describe interactions occurring at ambient temperatures. ρ^{ab} is then the distance at which the repulsion energy of atoms a and b has a value equal to the constant K, which gives it a direct physical interpretation.

Williams (1965, 1967) derived atom–atom Born–Mayer parameters for H···H, C···H and C···C interactions from data on crystalline hydrocarbons, and a set covering a wider range of atom types has been provided by Mirsky (1978). Although now quite old, these parameters are still used, and they are given in Table 11.2. A more recent set has been assembled by Filippini and Gavezzotti (1993). This is intended for use in predicting crystal structures, but is designed for use on its own, i.e., without any electrostatic terms in the potential, and is inappropriate for use when the electrostatic effects are included explicitly. The same is true of the exp-6 potentials proposed to describe hydrogen bonding (Gavezzotti and Filippini 1994).

Such a model treats the atoms as spherical, and there is now extensive evidence that the isotropic-atom approximation is inadequate in some cases, and can give qualitatively incorrect results. Probably the most direct evidence is the paper by Nyburg and Faerman (1985), which was based on distances of closest approach between atoms in a very large number of crystal structures. These showed, for instance, that iodine atoms in molecules containing C–I bonds were about 0.7 Å closer in the crystal if they were in end-on contact (C–I···I–C) than if they approached side-on. More evidence comes from the crystal structure of Cl_2. Many attempts to account for this structure using simple potentials with spherical atoms failed, predicting a cubic *Pa*3 structure instead of the observed *Cmca* structure, and the only simple model to account satisfactorily for the observed structure is one in which the Cl atoms are assumed to be slightly non-spherical, being some 5% larger in the direction perpendicular to the Cl–Cl bond than parallel to it (Price and

Table 11.2 *Exp-6 parameters for atom–atom potentials*

Atoms		C_6 a.u.	α a_0^{-1}	ρ a_0
Mirsky				
C	C	30.55	1.947	5.98
H	H	2.10	2.270	3.95
O	O	18.83	2.212	5.30
N	N	18.80	2.000	5.55
Cl	Cl	216.27	1.197	7.43
F	F	10.74	2.196	5.06
S	S	170.26	1.847	6.95
C	H	8.56	2.085	4.94
O	C	24.63	2.069	5.65
O	H	6.39	2.238	4.62
N	C	24.05	1.974	5.77
N	H	6.60	2.117	4.74
Cl	C	76.56	1.556	6.55
Cl	H	23.37	1.625	5.47
O	N	19.08	2.101	5.44
S	C	61.47	1.873	6.39
Williams				
H	H	1.98	1.979	4.22
H	C	9.07	1.942	4.91
C	C	41.22	1.905	6.19

Stone 1982). This model was also very successful in accounting for the properties of liquid Cl_2, and a similar model gave a good account of liquid and solid Br_2 and I_2 (Rodger *et al.* 1988*a*, 1988*b*).

The anisotropy of the repulsion can be determined by *ab initio* calculation, using one of the perturbation methods described in Chapter 6, and fitted to a suitable analytic function for use in other applications. One possibility is to make the parameters of the Born–Mayer potential depend on orientation:

$$U(R,\Omega) = A(\Omega)\exp(-B(\Omega)R). \tag{11.3.4}$$

(Here and subsequently we use Ω to stand for the relative orientation of the molecules, however we choose to describe it. See the discussion of coordinate systems in Chapter 1.) Consider the behaviour of B. It appears to describe the range of the potential, but if we evaluate the radial force we find that $-\partial U/\partial R = BA\exp(-BR) = BU$. Thus for a particular value of the repulsion energy U, the repulsive force is proportional to B, which therefore describes the steepness or hardness of the potential. This means that the molecular shape has to be described by $A(\Omega)$. The usual way to do this would be to expand $A(\Omega)$ in terms of spherical harmonic functions of the angular coordinates, but this leads to a very slowly convergent series, because the repulsion energy varies very sharply as a function of orientation and is not at all well described by a few terms of a spherical harmonic series.

These considerations led to the suggestion (Stone 1979, Price and Stone 1980) that the repulsive potential should take the form

$$\begin{aligned} U_{\text{rep}}^{AB} &= \sum_{a\in A}\sum_{b\in B} U_{\text{rep}}^{ab} \\ &= K\sum_{a\in A}\sum_{b\in B}\exp\bigl[-\alpha_{ab}(\Omega_{ab})\bigl(R_{ab}-\rho_{ab}(\Omega_{ab})\bigr)\bigr]. \end{aligned} \tag{11.3.5}$$

Here α_{ab} is the hardness parameter for the interaction between atoms a and b, and ρ_{ab} describes the shape. As before, K is not a parameter but a convenient energy unit, usually 10^{-3} hartree. Since $U_{ab}=K$ when $R_{ab}=\rho_{ab}(\Omega)_{ab}$, the parameter ρ_{ab} describes the shape of the contour on which the repulsion energy between atoms a and b has a value equal to this energy unit.

The variation of atomic radius and hardness with orientation is quite small, even though it can be very significant in determining structures, and spherical-harmonic expansions of α and ρ can be expected to converge very quickly. The appropriate expansion functions are again the $\bar{S}$ functions $\bar{S}_{l_a l_b j}^{k_a k_b}$:

$$\begin{aligned} \rho_{ab}(\Omega) &= \sum_{l_a l_b j k_a k_b} \rho_{l_a l_b j}^{k_a k_b}\bar{S}_{l_a l_b j}^{k_a k_b}, \\ \alpha_{ab}(\Omega) &= \sum_{l_a l_b j k_a k_b} \alpha_{l_a l_b j}^{k_a k_b}\bar{S}_{l_a l_b j}^{k_a k_b}. \end{aligned} \tag{11.3.6}$$

Often α_{ab} can be taken to be constant, so that all the anisotropy is contained in ρ_{ab}.

The $\bar{S}$ functions that can appear in this expression are restricted by the symmetry of the system, and the expansion converges rapidly, so that only a few terms are normally needed (Rodger *et al.* 1988*a*, 1988*b*, Price and Stone 1982, Wheatley and Price 1990*b*). However, the index j is not restricted to the value l_a+l_b, as it is for electrostatic interactions. One of the largest sets of $\rho_{l_a l_b j}^{k_a k_b}$ parameters has been fitted to the water dimer potential of Millot and Stone (1992). In that work a satisfactory fit was obtained using only $\bar{S}_{l0l}^{k0}$ or $\bar{S}_{0ll}^{0k}$ functions up to $l=4$. Such functions have a particularly simple form. Consider $\bar{S}_{l0l}^{k0}$, and adopt a coordinate system that coincides with the local axes for site a. Then $\Omega_a=(0,0,0)$, so

$$\begin{aligned} \bar{S}_{l0l}^{k0} &= \sum_{m_a m}\left[\begin{pmatrix} l & 0 & l \\ m_a & 0 & m\end{pmatrix}\Big/\begin{pmatrix} l & 0 & l \\ 0 & 0 & 0\end{pmatrix}\right][D_{m_a k_a}^{l_a}(0,0,0)]^*[D_{00}^{0}(\Omega_2)]^* C_{lm}(\theta,\varphi) \\ &= \sum_{m_a m}(-1)^m\delta_{m_a,-m}\delta_{m_a,k_a}C_{jm}(\theta,\varphi) \\ &= (-1)^k C_{l,-k}(\theta,\varphi) \\ &= C_{l,k}(\theta,\varphi)^*. \end{aligned} \tag{11.3.7}$$

If we express this in terms of the real components, using eqn (3.3.11) (p. 43), we find that

$$\bar{S}^{\kappa 0}_{l0l} = C_{l,\kappa}(\theta_a, \varphi_a),$$

where we have labelled the arguments of the spherical harmonic to emphasize that they are the polar coordinates describing the direction of the site–site vector in the local axis system of site a. In the same way,

$$\bar{S}^{0\kappa}_{0ll} = C_{l,\kappa}(\theta_b, \varphi_b),$$

where θ_b and φ_b here describe the direction of the vector to site a from site b in the local coordinate system of site b.

If ρ can be expressed purely in terms of these functions, then, it takes the form

$$\rho^{ab}(\Omega) = \rho^{ab}_{00} + \sum_{l>0}\sum_{\kappa} \rho^a_{l\kappa} C_{l,\kappa}(\theta_a, \varphi_a) + \sum_{l>0}\sum_{\kappa} \rho^b_{l\kappa} C_{l,\kappa}(\theta_b, \varphi_b),$$

and it is natural to express this as

$$\rho^{ab}(\Omega) = \rho^a(\theta_a, \varphi_a) + \rho^b(\theta_b, \varphi_b), \tag{11.3.8}$$

where

$$\rho^a = \sum_{l\kappa} \rho^a_{l\kappa} C_{l,\kappa}(\theta_a, \varphi_a),$$
$$\rho^b = \sum_{l\kappa} \rho^b_{l\kappa} C_{l,\kappa}(\theta_b, \varphi_b),$$

and $\rho^{ab}_{00} = \rho^a_{00} + \rho^b_{00}$. That is, there is a function ρ for each atom (or site) that describes its radius as a function of direction, and the parameter $\rho(\Omega)$ that appears in eqn (11.3.5) is then the sum of the radii of each site in the direction of the other. This is a very simple and physically appealing description, and offers obvious possibilities for transferable potentials, but it is important to remember that it is based on the approximation that only the $\bar{S}^{k0}_{l0l}$ or $\bar{S}^{0k}_{0ll}$ functions are needed in the expansion of $\rho(\Omega)$.

11.3.3 *Approximate methods for determining repulsive potentials*

One of the disadvantages of the anisotropic atom–atom model is that the parameters describing the orientation dependence of α and ρ have to be determined, usually by fitting to *ab initio* or experimental data. *Ab initio* calculations become expensive for large molecules at a reasonable level of accuracy, and have to be repeated at many different relative orientations to give an adequate coverage of the 6-dimensional coordinate space. Methods that require less computational effort can therefore be very useful, even if less accurate. Three such methods are the density overlap model, the exchange-Coulomb (XC) model, and the test particle method. The XC model has been discussed above, in the context of atoms. When it is applied to molecules, it is important to appreciate that the 'Coulomb' term is the electrostatic penetration energy, which depends on the overlap of the wavefunctions, and does not include the long-range part of the electrostatic energy, described by the multipole expansion. In fact, the only applications of the XC model to molecules (Wheatley and Meath 1993*b*) deal with cases where one component is an inert gas atom, so that the multipolar contribution to the electrostatic energy is zero. Even here, however, it is necessary to remove from the electrostatic interaction the repulsion between the nuclei, which becomes singular as their separation tends to zero. The uncorrected relationship (11.1.12) must break down in this limit because the exchange energy remains finite even when two nuclei are superimposed.

The test particle model is based on the observation that in the exponential expression eqn (11.3.5) for the repulsive potential, the effective diameter ρ can be expressed to a good approximation in the additive form (11.3.8), that is, $\rho^{ab} = \rho^a + \rho^b$. Since ρ^a describes the shape of atom a, it is natural to hope that it is a transferable quantity. If so, we can determine ρ_p for a spherical probe or test particle, such as the He atom, by studying the $p \cdots p$ interaction, which depends only on the distance between the atoms. If the additive model is valid, then $\rho^{pp} = 2\rho^p$. Then an investigation of the $a \cdots p$ interaction provides $\rho^{ap} = \rho^a + \rho^p$, and hence ρ^a. This is much more efficient computationally than exploring the $a \cdots b$ interaction directly: in the general case that would require calculations over a 6-dimensional space (one distance and 5 angles) whereas the probe calculation requires only three variables (a distance and 2 angles).

If this procedure is to work, we need to use a similar combining rule for the hardness parameter α. There are several possibilities, but the choice is not very critical because values of α fall in quite a narrow range—typically from 1.8 to 2.3 bohr^{-1}. Böhm and Ahlrichs (1982), using a nitrogen atom in its average-of-terms state as the probe, and Stone and Tong (1994), using a helium atom, found that the so-called 'energy-dependent hard-core model' was slightly more successful than other possibilities. This model is a simplification of the 'atomic distortion model' developed by Sikora (1970), in which the atoms are treated as classical deformable solids, and leads to additivity for $\eta = 1/\alpha$: $\eta^{ab} = \eta^a + \eta^b$.

The choice of test particle is rather limited. Inert-gas atoms are obvious possibilities, together with other spherical atoms such as the N atom used by Böhm and Ahlrichs (1982), which is a spherical average of the terms arising from the ground $(2s)^2(2p)^3$ configuration. The merit of the He atom is that at the SCF level it has only one occupied molecular orbital, and since the repulsion is a first-order perturbation term, the virtual orbitals are irrelevant. Consequently the $1s$ orbital can be thoroughly optimized and the calculation carried out with a contracted basis comprising just this one basis function, greatly reducing the number of two-electron integrals needed in the calculation.

Calculations for N_2, Cl_2, acetylene and H_2S give the repulsion parameters shown in Table 11.3 (Stone and Tong 1994). For N_2, the repulsion energies derived from these agree very well with direct calculations of the dimer repulsion energy at mildly repulsive geometries (up to about 20 kJ mol^{-1}), with an r.m.s. error in the energy of about 0.1 kJ mol^{-1}. For the other molecules the agreement is not so good. The r.m.s. error in the distance at which the model matches the dimer calculation at a given orientation leads to similar conclusions: this error is about 0.02 bohr for N_2, but several times larger for the other molecules.

The reasons for this are interesting, as they highlight a fundamental limitation of the simple additive model, eqn (11.3.8). On this model, the parameters in Table 11.3 can be regarded as describing the shape of each site. Fig. 11.2 shows these shapes, as well as

Table 11.3 *Repulsion parameters* $\eta = 1/\alpha$ *and* ρ *for* He, N_2, Cl_2, *acetylene and* H_2S. *Values in atomic units (bohr).*

Molecule:	He	N_2	Cl_2	HCCH	H_2S
Site:	He	N	Cl	C	S
η_{00}	0.21086	0.2433	0.2711	0.2748	0.3309
ρ_{00}	2.0817	2.955	3.427	3.415	3.998
ρ_{10}		−0.075	0.102	0.098	0.389
ρ_{20}		0.107	−0.227	0.185	−0.003
ρ_{22c}					0.012
ρ_{30}			−0.106	0.270	−0.184
ρ_{32c}					−0.469

the contours of the repulsion between each molecule and a helium atom. We see that the N atom is nearly spherical according to this picture, but the Cl atom is distinctly non-spherical and the CH group in acetylene even more so. Now, if we allow two acetylene molecules to approach in a tilted geometry, the parameter ρ in the additive model depends only on the effective radius of each site in the direction of the other, and is the same for the orientation in Fig 11.3a ($\varphi = 0°$) as for that in Fig 11.3b ($\varphi = 180°$). The Figure shows that this is not correct; there should be a difference of about 0.2 bohr. We have to conclude that although eqn (11.3.8) is a useful approximation, it will ultimately prove inadequate for accurate work.

In the density overlap model, some guidance on the form of the anisotropy in the repulsive potential is gained by a rather empirical route. It has been observed that the repulsion between pairs of inert-gas atoms is closely related to the overlap S_ρ between their charge densities. (Notice that this is a different quantity from the usual overlap between wavefunctions.)

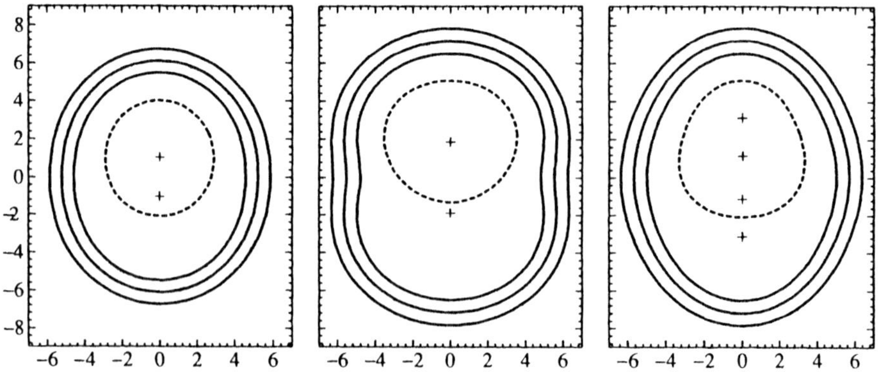

FIG. 11.2. Atom shapes. The continuous lines are contours of the $M \cdots$He repulsion energy for (left to right) $M = N_2$, Cl_2 and HCCH. The dashed lines show the site shape functions ρ^a for N, Cl and CH respectively. The scale is in bohr.

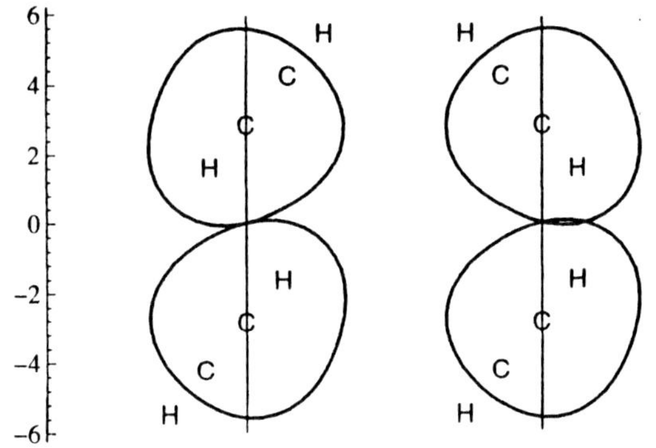

FIG. 11.3. When the site functions ρ^a and ρ^b are anisotropic, as illustrated here for acetylene, the point of contact may not be along the line from site to site, and a correction is needed to the formula $\rho^{ab} = \rho^a + \rho^b$. The scale is in bohr.

$$U_{\rm er} \approx k S_\rho^n, \qquad S_\rho = \int \rho^A(\mathbf{r})\rho^B(\mathbf{r})\,\mathrm{d}^3\mathbf{r}.$$

The value of n is close to 1—typically between 0.96 and 0.99 (Kita *et al.* 1976, Kim *et al.* 1981). Wheatley and Price (1990*a*) found that this was true for small molecules too. In this case the charge density of each molecule can be expressed to a good approximation as a sum of densities for regions of the molecule, in the same spirit as Distributed Multipole Analysis:

$$\rho^A(\mathbf{r}) = \sum_{a\in A} \rho^a(\mathbf{r}).$$

In this case the densities are anisotropic. Wheatley and Price were able to express the density overlap between two molecules as a sum of site–site terms, in the form

$$S_\rho = \sum_{a\in A}\sum_{b\in B} \exp[-\alpha^{ab}R_{ab}] \sum_{l_a l_b j} C_{l_a l_b j}\bar{S}_{l_a l_b j}, \tag{11.3.9}$$

where the $\bar{S}$ functions describe the orientation dependence in the usual way. The coefficients $C_{l_a l_b j}$ are treated as constant, though in fact they vary slightly with separation.

This method gives results comparable in accuracy with the test-particle model for N_2, and is much better for Cl_2, where the test-particle model is relatively unsuccessful. Indeed, the density overlap method gives better results for Cl_2 than a potential of the form of eqn (11.3.5) with the same number of parameters fitted directly to the calculated repulsion energies.

11.3.4 *Dispersion*

It is usual to express the dispersion interaction for molecules in atom–atom form:

$$U_{\rm disp} = -\sum_{a\in A}\sum_{b\in B}\left\{ f_6(R_{ab})\frac{C_6^{ab}}{R_{ab}^6} + f_8(R_{ab})\frac{C_8^{ab}}{R_{ab}^8} + f_{10}(R_{ab})\frac{C_{10}^{ab}}{R_{ab}^{10}} \right\}. \tag{11.3.10}$$

In principle, there will be C_7 and C_9 terms for atoms whose local symmetry is not centrosymmetric, but their coefficients are smaller than those for the even terms and they are often omitted. The $f_n(R_{ab})$ are damping functions (see §11.1.5).

The use of a damped dispersion formula removes the negative singularity of the exp-6 potential, and makes it possible to use potentials of this type in any application, and in particular for interactions between molecules as well as atoms. It is evident, however, that for molecules the potential must depend on orientation as well as on distance; that is, it must be anisotropic. The anisotropy of the dispersion was discussed in §4.3, and is satisfactorily understood at long range, but the way that the damping should be treated in the anisotropic case is not at all well understood.

11.4 Dependence on internal coordinates

All of the models discussed so far have been of 'rigid-body' type: that is, the molecules have been treated as rigid bodies with no internal degrees of freedom. For many purposes, this is adequate; the frequencies associated with internal vibrational motions are often very different from those of the intermolecular degrees of freedom, so that the two kinds of motion are coupled only weakly. In other cases, however, the interaction of intramolecular and intermolecular motions may be significant, and for some properties, such as the red-shift of the OH vibrational frequency caused by hydrogen bonding, the explicit treatment of the vibrational motion is essential. For large flexible molecules, such as polypeptides, the energy of even an isolated molecule includes interactions of an 'intermolecular' type between parts of the molecule that may be separated by many chemical bonds but are close to each other in space, so the description of the intramolecular potential energy surface has to include these terms as well as the conventional bond stretching, bending and twisting forces. Conversely, the interaction of flexible molecules with each other is often facilitated by distortions of their structures, which may for example allow a substrate molecule to squeeze into the active site of an enzyme.

11.5 Molecular mechanics

The simplest way to describe these effects is to suppose that the internal vibrations of the molecules are the same as in the isolated molecule, and to assume that the intermolecular interaction can be described by an atom–atom model whose parameters do not depend on internal coordinates. In this picture, the internal motion of the molecules affects the intermolecular interactions only because it modifies the positions of the atoms.

There has been a great deal of interest over the last ten or twenty years in the study of biological molecules. Such molecules are generally large and flexible, so it is essential to take the internal coordinates of the molecule into account. The method used is usually described as 'molecular mechanics': a function is chosen to describe the energy of the system, and then the coordinates may be varied to find an energy minimum, or the classical equations of motion of the system may be solved to explore the dynamics of the molecular motion. For biological systems it is usually necessary to include water in the simulation, either explicitly or by treating the system as embedded in a dielectric continuum.

The energy of the system is described by a function such as

$$U = \sum_{\text{bonds}} k_r(r-r_0)^2 + \sum_{\text{angles}} k_\theta(\theta-\theta_0)^2 + \sum_{\text{torsions}} \tfrac{1}{2}V_n\left(1+\cos(n\varphi-\gamma)\right)$$
$$+\sum_{i<j}\left(\frac{A_{ij}}{R_{ij}^{12}}-\frac{B_{ij}}{R_{ij}^{6}}+\frac{q_iq_j}{\varepsilon R_{ij}}\right)+\sum_{\text{H-bonds}}\left(\frac{C_{ij}}{R_{ij}^{12}}-\frac{D_{ij}}{R_{ij}^{10}}\right). \qquad (11.5.1)$$

This is the form of the AMBER force field. The first version (Weiner *et al.* 1984) was a united-atom description, in which the hydrogen atoms were not described explicitly but were incorporated into the neighbouring heavy atom. In the subsequent 'all-atom' version (Weiner *et al.* 1986), the hydrogen atoms were included explicitly. The first three terms describe the energies of intramolecular bond stretches, angle bends, and dihedral torsions respectively, while the remaining terms describe the through-space interactions. There is a Lennard-Jones repulsion–dispersion term, which is applied to pairs of atoms in the same molecule as well as to intermolecular interactions. However it is usual to omit intramolecular interactions between atoms that are separated by three or fewer bonds, since such interactions are included in the earlier terms.

The electrostatic interaction is described by atom–atom point-charge terms only, the charges usually being obtained by fitting to the molecular electrostatic potential (see §7.7). The final term in eqn (11.5.1) describes an additional interaction between the H atom of a hydrogen bond and the proton acceptor atom. This term was introduced to provide additional flexibility and accuracy in the description of hydrogen bonds.

This is evidently quite an elaborate potential. Although each of the terms is simple in form, and indeed has serious inadequacies, a very large number of parameters is needed to describe all the possible interactions that may occur. In the Lennard-Jones term for example, parameters A_{ij} and B_{ij} must be provided for each pair of atom types. This may not seem too serious a problem until one realises that AMBER provides for more than 20 different types of carbon atom alone. The number of parameters that is needed to characterize the force field is consequently extremely large.

The AMBER force field is only one of a number of energy functions in the literature. The earliest still in use is probably the MM*n* series (MM1 (Allinger 1976), later superseded by MM2 (Allinger 1977) and MM3 (Allinger *et al.* 1989, Lii and Allinger 1991)). Others include CHARMm (Brooks *et al.* 1983), ECEPP (Némethy *et al.* 1983, Sippl *et al.* 1984), GROMOS (Hermans *et al.* 1984) and OPLS (Jorgensen and Tirado-Rives 1988). They are all broadly similar, but differ in various details. For example, the earliest version of ECEPP used fixed bond lengths and bond angles, though this constraint was removed in the later versions. The MM*n* force fields use an exp-6 formulation of the repulsion–dispersion term, but the rest all use Lennard-Jones, though the parameters differ between models. The parameters for the interactions between unlike atoms are usually obtained from combining rules; some models use the Lorentz–Berthelot combining rules (see §11.1.2 on p. 157), while OPLS (for example) uses the geometric-mean combining rule for the dispersion coefficient B and the repulsion coefficient A of eqn (11.5.1), and ECEPP uses the Slater–Kirkwood combining rule (see §4.3.2 on p. 58) for the dispersion coefficients. The more important of the unlike-atom parameters in all force fields are directly fitted, and not obtained from combining rules.

The description of the internal degrees of freedom also varies. The MM3 force field uses anharmonic stretch and bend potentials, including terms up to quartic, and also includes quite a number of cross terms between bend and stretch, between bends involving different bond angles involving the same atom, and between bends and torsions. The other force fields include only the harmonic (quadratic) terms, with no cross terms between different coordinates.

The treatment of the electrostatic term differs in the value used for the dielectric constant ε. In some work it is taken to be equal to the distance R_{ij} in Ångstrom. This has no valid theoretical justification, but has been claimed to take some account of the effect of the aqueous solvent. A more significant reason for its inclusion at the time the early force fields were introduced was that the electrostatic interaction then became proportional to R_{ij}^{-2}, so that the energy expression included only even powers of R_{ij} and it was not necessary to evaluate any square roots. Nowadays this is no longer a serious problem, since many computers have a fast hardware square root function, and computer power has increased enormously in any case. It is now generally agreed that the solvent molecules should if possible be included explicitly in the simulation, and then the dielectric constant should be set equal to 1, or perhaps to a slightly higher value to take account of polarization effects (MM3 uses $\varepsilon = 1.5$ for gas-phase simulations).

Another difference concerns the description of hydrogen bonds: some models include explicit hydrogen-bond terms, as AMBER does, while others have no special hydrogen-bond functions and rely on the electrostatic terms to recover the special features of the hydrogen bond. CHARMm uses several fairly elaborate switching functions to limit the range of the energy terms, and combines groups of atoms together, representing their effects on distant atoms by a point-multipole description. The force fields also differ in the model that they use to describe explicitly included water molecules; CHARMm uses ST2, GROMOS uses SPC and OPLS uses TIP4P. (See §11.6 below.) These variations in detail between the force fields make it impossible to compare their parameter values directly, since the parameters are all interdependent, and also make it very difficult to compare their reliability. Hall and Pavitt (1984) compared the models available at the time, and concluded that AMBER was 'superior to many alternatives', but all the force fields then in use have since been improved in various ways, while other new ones have been developed. Roterman *et al.* (1989) compared the CHARMm, AMBER and ECEPP force fields, and concluded that their predictions were significantly different.

11.5.1 *Conformational dependence of potential parameters*

One of the limitations of all these models for large flexible molecules is that they cannot easily describe the way in which the charge distribution of the molecule depends on the molecular structure. Consider a molecule such as acrolein, $H_3C{=}CH{-}CH{=}O$. We can describe its charge distribution in terms of distributed multipoles, using a local coordinate system for each atom—for instance, the local axis system for the O atom might have the z axis perpendicular to the plane of the C–C=O unit, the y axis from C to O along the C=O bond, and the x axis forming a right-handed set. If the molecule is twisted around the central bond, the oxygen multipoles are approximately unchanged in this local coordinate system, but this is not a very good approximation, because the change in the central bond (especially the loss of conjugation as it is twisted out of plane) causes a redistribution of the charge. Consequently the multipole moments of the O atom depend on the torsion angle around the central bond. The same is true for the other atoms, par-

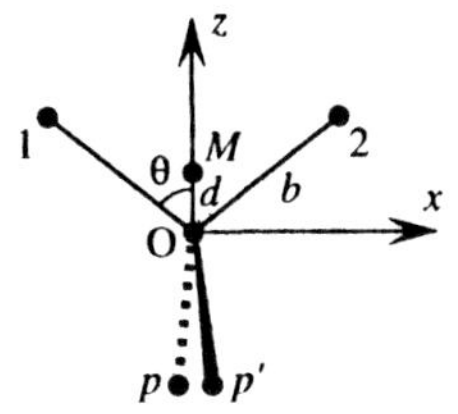

FIG. 11.4. Interaction sites for water models.

ticularly the C atoms at each end of the bond being twisted. Koch *et al.* (1995) found that the electrostatic potential on a grid of points 1.4 Å from the Van der Waals surface, calculated at one value of the torsion angle using multipole moments obtained at another, is in error by as much as $10\,\mathrm{kJ\,mol^{-1}}$ (r.m.s.). If the multipole moments, as functions of the torsional angle, are fitted to Fourier series truncated at the $\cos 3\theta$ term, the r.m.s. error is reduced to less than $1\,\mathrm{kJ\,mol^{-1}}$.

If the charge distribution is represented using point charges, the dependence on torsion angle has to include not only the intrinsic change in the charge distribution arising from the change in bonding, but the effects of moving the atoms relative to one another, since the dipoles and higher moments on each atom have to be represented in terms of point charges on other atoms. The use of charges for one conformation in calculating the potential for a different conformation leads to large errors, as one would expect. Reynolds *et al.* (1992*a*) found that by averaging the charges over conformations weighting each conformation by its Boltzmann factor, an improved point-charge description could be obtained. Nevertheless, the errors are still large, and calculations of the difference between the free energy of hydration for ethanol and propanol showed very large variations in the calculated value depending on the choice of point-charge model (Reynolds *et al.* 1992*b*). In fact these variations dominated all other sources of error. At present it seems likely that multipole models will be necessary for the calculation of such properties, and that the variation of the multipole moments with geometry will have to be described explicitly. If so, this is a rather daunting prospect.

11.6 A case study: potentials for water

Because of its obvious importance, the intermolecular potential for water has been an objective for theoreticians for many years. The first attempt was made by Bernal and Fowler (1933), who used three point charges, two at the H nuclei (points 1 and 2 in Fig. 11.4) and one on the symmetry axis (point M) together with an oxygen–oxygen Lennard-Jones term. Rowlinson (1949) used a Stockmayer potential—that is, a Lennard-Jones interaction with an embedded dipole—and in a later version (1951) he added a quadrupole. However the Bernal–Fowler type of model has been generally preferred, with variations on the position and magnitude of the charges and the parameters of the Lennard-Jones term. They include the ST2 potential (Stillinger and Rahman 1974), which has four charges in a tetrahedral array (two of them at the positions p and p' in Fig. 11.4, corresponding notionally to the lone pairs), the family of potentials known as TIPS2, TIP3P and TIP4P (Jorgensen *et al.* 1983) and the Simple Point Charge or SPC model (Berendsen *et al.* 1981). Most of these potentials have been used in molecular dynamics and Monte Carlo calculations (see Chapter 12) on liquid water, ice and hydrated proteins, with reasonable success. Such potentials cannot be expected to perform well in a variety of applications. They are very simple in form, and cannot be expected to reproduce the true potential at all accurately, so agreement with experiment depends on a cancellation of errors. In particular, they all ignore the polarization of the water molecule, which can contribute substantially to the interaction energy, and is strongly non-additive, as we saw in Chapter 9. Such non-additive effects can only be incorporated in an averaged way; thus, for example, all of these models except Rowlinson's assign a dipole moment to the water molecule that is substantially higher than the gas-phase value of 1.85 debye; typically 20–30% higher. It follows that they should not be used in circumstances for which they were not designed, and having been designed for liquid water they cannot be expected to give good results for other systems, such as an isolated pair of water molecules. Indeed, Berweger *et al.* (1995) found that even for liquid water, the optimum parameters in the SPC model depend quite strongly on the temperature and pressure. No doubt this is true of other models of this type also.

An alternative to this purely empirical approach is to use *ab initio* calculations to obtain a potential energy surface, which is then fitted with a suitable analytic function. One of the best-known potentials of this type is due to Matsuoka *et al.* (1976). Their potential, commonly known as the MCY potential, is more realistic in some respects, having an exponential repulsion term between every pair of atoms, but the electrostatic interaction is still represented by three point charges. It also has attractive exponential terms between the hydrogen and oxygen atoms, which were introduced to describe electron correlation effects. Although it was derived from calculations on the water dimer, it gives very poor results for the virial coefficient of water vapour (Reimers *et al.* 1982), which is a purely pair-potential property (see eqn (12.2.1) on p. 189). We have to conclude that either the *ab initio* calculations were not good enough or the analytic function used to represent the results was not sufficiently flexible.

In fact we know that the intermolecular electron correlation leads to the dispersion interaction, so an attractive exponential representation for it is inappropriate. Reimers *et al.* (1982) attempted to improve matters by replacing the attractive exponential terms by the damped R^{-6}, R^{-8} and R^{-10} dispersion terms of the HFD (Hartree–Fock–dispersion) model, but the calculated second virial coefficient was still very poor. Reimers *et al.* (1982) attributed this to the errors in fitting the analytic function to the *ab initio* data, rather than to errors in the calculation itself, but the calculations were not corrected for BSSE so there must be errors from this source. Matsuoka *et al.* estimated the BSSE to be about 0.8 kJ mol^{-1} at SCF level, and assumed that the correlation contribution would be about the same, as had been found for a smaller basis set. In fact it is now known that the correlation contribution to BSSE does not decrease as rapidly as the SCF contribution when the basis is improved.

Another widely used potential of this type is the RWK2 model of Reimers *et al.* (1982). This too described the electrostatics by using three point charges, chosen to reproduce the monomer dipole and quadrupole moments, but it included the accurate dispersion term from the HFD model, exponential repulsions for O···O and H···H interactions, and Morse terms for O···H interactions. This model gave a good account of the second virial coefficient of steam, though it does not agree as well with the most recent measurements of Kell *et al.* (1989) as it did with their earlier ones (Kell *et al.* 1968).

A serious limitation of all these potentials is that they omit any explicit description of induction, usually treating its effects in an average way by using an enhanced dipole moment. A number of authors have sought to rectify this omission. An early attempt was the Polarizable Electropole (PE) model of Barnes *et al.* (1979), which has an isotropic dipole–dipole polarizability, a Lennard-Jones term and point dipole and quadrupole, all at the same site. The use of an isotropic polarizability is quite a good approximation for water, for which the experimental polarizability components are (in a.u.) $\alpha_{xx} = 10.31$, $\alpha_{yy} = 9.55$ and $\alpha_{zz} = 9.91$. The PE model gave a better account of the second virial coefficient of steam and of the structure of liquid water than other models in use at the time, and Barnes *et al.* showed that the indirect effects of the induction term on the long-range interaction between pairs of molecules were considerable, and that the mean dipole of each water molecule was about 2.5 debye, even higher than the enhanced value used in most unpolarizable models.

Another potential of this type was that of Sprik and Klein (1988). The induction is here modelled by distributed charge-flow polarizabilities, as described above (p. 163), and the other features of the potential are the same as for the TIP4P potential, except that the charges are chosen to reproduce the true molecular dipole moment rather than the enhanced value that is needed when the induction is not included explicitly. A later version of this model (Sprik 1991) is similar, except that the polarizability is centred at the O atom rather than at the point M (Fig. 11.4) as was the case in the earlier version.

A more elaborate potential has been proposed by Wallqvist *et al.* (1990), who use a scheme that they call NEMO. This uses a 4-charge model for the charge distribution, derived from SCF calculations on the monomer, and isotropic dipole–dipole polarizabilities assigned to each atom so as to reproduce the induction energy calculated using Karlström polarizabilities (p. 138) calculated for the molecule. The repulsion is described by isotropic atom–atom terms fitted to the residue of the SCF interaction energy after the electrostatic and induction terms have been subtracted off. Each of these atom–atom terms is a Born–Mayer exponential together with a short-range R^{-20} term. Finally, the dispersion interaction is described by an R^{-6} term, with an anisotropic C_6 coefficient derived from the static polarizabilities using the London formula, eqn (4.3.5) on p. 59. A switching function is used to suppress the R^{-6} singularity; this is not a damping function as usually understood. This potential gives the isolated molecule a rather high dipole moment, and the mean dipole moment of the molecules in the liquid is also rather high, at about 2.9 D.

The most elaborate analytical potential so far proposed for water is that of Millot and Stone (1992). They made no attempt to simplify the potential so that it could be used in conventional molecular dynamics calculations, but sought to describe the interaction as accurately as possible. Their potential used distributed multipoles to describe the charge distribution (up to quadrupole on oxygen, charges only on hydrogen) and central polarizabilities up to quadruple–quadrupole. (Distributed polarizabilities offer no advantage for this near-spherical molecule.) The dispersion interaction was described using the anisotropic dispersion coefficients C_6 to C_{10} calculated by Rijks and Wormer (1989), though the smaller coefficients were neglected. Repulsion was described by anisotropic atom–

atom exponential terms, fitted to the repulsion–penetration energy calculated using *ab initio* intermolecular perturbation theory, as described in Chapter 6. This Anisotropic Site Potential or ASP model gave a good account of the equilibrium geometry and of the tunnelling splittings and second virial coefficient in the water dimer. Subsequently the potential was improved by including a charge transfer term, which has an attractive exponential form (Stone 1993), and by replacing the dispersion coefficients by the corrected values published later (Rijks and Wormer 1990).

These potentials are commonly used in classical calculations of properties. In simulations, for example, the equations of motion of the water molecules are solved as though they were classical particles. Since they are in fact quantum particles this is not correct, and Kuharsky and Rossky (1985) used Feynman path-integral techniques to estimate the quantum corrections. They found that the quantum liquid was less structured than the classical one, and that at ambient temperature the difference between the quantum and classical radial distribution functions was the same as the change induced by an increase in temperature of 50°.

11.7 Calculation of energy derivatives

If we wish to optimize the structure of a system, we need to evaluate not only the energy but its derivatives. For molecular dynamics calculations, in which the evolution of the system is explored by solving the classical equations of motion, the forces and torques on the molecular sites suffice. The same is true if we seek a local minimum in the energy. If we want to find saddle points on the potential energy surface, as is the case when isomerization pathways are of interest, the most effective procedure for finding them is the eigenvector-following technique of Cerjan and Miller (1981) (see also Wales (1991)), which requires second derivatives (the Hessian matrix), and this is more efficient for finding minima too than methods that use only first derivatives. When the potential is described by one of the more accurate models discussed above, in which dipoles and higher moments appear, and the repulsion and dispersion may also be orientation dependent, we require derivatives of the orientation-dependent terms. The perceived difficulty of evaluating these derivatives has been one of the reasons for the continued use of potentials that depend only on the distance between sites, but they are neither as difficult nor as time-consuming to evaluate as is often supposed.

11.7.1 *Derivatives of the electrostatic energy*

The electrostatic interaction U^{ab} between sites a and b is:

$$U^{ab} = \sum_{tu} Q_t^a Q_u^b T_{tu}^{ab}(\mathbf{q}),$$

where Q_t^a, Q_u^b are multipole moments on sites a and b, referred to the local coordinate system for each site, and so are independent of orientation. The problem therefore reduces to finding the derivatives of the interaction functions $T_{tu}^{ab}(\mathbf{q})$.

The function T_{tu}^{ab} depends on the relative orientations of the site axis systems and the inter-site vector $\mathbf{R}$. More specifically, as we saw in §3.3 (eqn (3.3.12) on p. 43) it can be

Table 11.4 *First derivatives of R and the scalar products* $\mathbf{w}^a \cdot \mathbf{R}$, $\mathbf{w}^b \cdot \mathbf{R}$ *and* $\mathbf{w}^a \cdot \mathbf{w}^b$ *(where* $\mathbf{w}^a$ *is* $\mathbf{x}^a$, $\mathbf{y}^a$ *or* $\mathbf{z}^a$, *and* $\mathbf{w}^b$ *similarly) with respect to the twelve external coordinates (translation and rotation) necessary to describe a pair of rigid interacting molecules. The differential operators are given in the left column and their arguments in the top row.*

	R	$\mathbf{w}^a \cdot \mathbf{R}$	$\mathbf{w}^b \cdot \mathbf{R}$	$\mathbf{w}^a \cdot \mathbf{w}^b$
$\dfrac{\partial}{\partial A_K}$	$-\hat{R}_K$	$-w^a_K$	$-w^b_K$	0
$\dfrac{\partial}{\partial \theta^A_K}$	$(\hat{\mathbf{R}} \times \mathbf{a})_K$	$-[(\mathbf{R}+\mathbf{a}) \times \mathbf{w}^a]_K$	$(\mathbf{w}^b \times \mathbf{a})_K$	$(\mathbf{w}^a \times \mathbf{w}^b)_K$
$\dfrac{\partial}{\partial B_K}$	$\hat{\mathbf{R}}_K$	w^a_K	w^b_K	0
$\dfrac{\partial}{\partial \theta^B_K}$	$-(\hat{\mathbf{R}} \times \mathbf{b})_K$	$-(\mathbf{w}^a \times \mathbf{b})_K$	$-[(\mathbf{R}-\mathbf{b}) \times \mathbf{w}^b]_K$	$-(\mathbf{w}^a \times \mathbf{w}^b)_K$

expressed in terms of the distance R between the sites and an $\bar{S}$ function, which in turn depends on the direction cosines

$$\begin{array}{llll} \mathbf{x}^a \cdot \mathbf{x}^b, & \mathbf{x}^a \cdot \mathbf{y}^b, & \mathbf{x}^a \cdot \mathbf{z}^b, & \mathbf{x}^a \cdot \hat{\mathbf{R}}, \\ \mathbf{y}^a \cdot \mathbf{x}^b, & \mathbf{y}^a \cdot \mathbf{y}^b, & \mathbf{y}^a \cdot \mathbf{z}^b, & \mathbf{y}^a \cdot \hat{\mathbf{R}}, \\ \mathbf{z}^a \cdot \mathbf{x}^b, & \mathbf{z}^a \cdot \mathbf{y}^b, & \mathbf{z}^a \cdot \mathbf{z}^b, & \mathbf{z}^a \cdot \hat{\mathbf{R}}, \\ \hat{\mathbf{R}} \cdot \mathbf{x}^b, & \hat{\mathbf{R}} \cdot \mathbf{y}^b, & \hat{\mathbf{R}} \cdot \mathbf{z}^b, & \end{array} \tag{11.7.1}$$

where $\hat{\mathbf{R}}$ is the unit vector in the direction of $\mathbf{R}$, while $\mathbf{x}^a$, $\mathbf{y}^a$ and $\mathbf{z}^a$ are the unit vectors of the local axis system of site a and $\mathbf{x}^b$, $\mathbf{y}^b$ and $\mathbf{z}^b$ those of site b. Following Willock *et al.* (1993, 1995), we regard these scalar products and the distance R as a set of intermediate variables $\mathbf{q} = \{q_i,\ i = 1, \ldots, 16\}$. The expression for T^{ab}_{tu} in terms of these quantities is not unique, because there are some relationships between them, such as $[\mathbf{x}^a \cdot \hat{\mathbf{R}}]^2 + [\mathbf{y}^a \cdot \hat{\mathbf{R}}]^2 + [\mathbf{z}^a \cdot \hat{\mathbf{R}}]^2 = 1$, but this is not a problem for the differentiation procedure.

For brevity we write $\mathbf{w}^a$ for any of $\mathbf{x}^a$, $\mathbf{y}^a$, or $\mathbf{z}^a$, and $\{\mathbf{w}^a\}$ for the set of all of them. Likewise $\{\mathbf{w}^b\}$ represents the set $\{\mathbf{x}^b, \mathbf{y}^b, \mathbf{z}^b\}$. Then the set of direction cosines is more compactly expressed as $\mathbf{q} = \{R, \mathbf{w}^a \cdot \hat{\mathbf{R}}, \mathbf{w}^b \cdot \hat{\mathbf{R}}, \mathbf{w}^a \cdot \mathbf{w}^b\}$.

The variables q_i depend in turn on the rigid-body translational and rotational coordinates of each molecule. This set of $6N$ external coordinates for the N molecules are the independent variables with respect to which we require derivatives of the total energy. It is convenient to introduce the alternative notation $\{A_X, A_Y, A_Z\}$ for the coordinates of the centre of mass of molecule A in the global axis system, and $\{\theta^A_X, \theta^A_Y, \theta^A_Z\}$ to describe angles of rotation of molecule A about the global axis directions.

To differentiate an interaction term, we use the chain rule. If C_K is one of the rigid-body coordinates, we have

$$\frac{\partial}{\partial C_K} T^{ab}_{tu} = \sum_i \frac{\partial q_i}{\partial C_K} \frac{\partial}{\partial q_i} T^{ab}_{tu}. \tag{11.7.2}$$

Now the form of T_{tu}^{ab} is a polynomial in the direction cosines q_i multiplied by an inverse power of R (see Appendix F), so its derivatives are very easy to evaluate. For computational purposes they can be determined, like the T functions themselves, with the help of an algebra program such as REDUCE or Maple (Hättig and Hess 1993). The derivatives of the q_i with respect to the molecular positions and orientations are more complicated, but they only have to be calculated once. To do this it is helpful to write, for instance, $\mathbf{x}_a \cdot \hat{\mathbf{R}}$ as $R^{-1}\mathbf{x}_a \cdot \mathbf{R}$, and then

$$\begin{aligned}\frac{\partial}{\partial C_K}(\mathbf{x}_a \cdot \hat{\mathbf{R}}) &= \frac{\partial}{\partial C_K}\frac{1}{R}(\mathbf{x}_a \cdot \mathbf{R}) \\ &= \frac{1}{R}\frac{\partial}{\partial C_K}\mathbf{x}_a \cdot \mathbf{R} - (\mathbf{x}_a \cdot \mathbf{R})\frac{1}{R^2}\frac{\partial R}{\partial C_K} \\ &= \frac{1}{R}\left[\frac{\partial}{\partial C_K}(\mathbf{x}_a \cdot \mathbf{R}) - (\mathbf{x}_a \cdot \hat{\mathbf{R}})\frac{\partial R}{\partial C_K}\right]. \end{aligned} \tag{11.7.3}$$

Further details have been given by Popelier and Stone (1994). The derivatives that occur in the last line of this formula are given in Table 11.4.

The K component of the force on molecule A arising from its interaction with molecule B is then

$$\begin{aligned} F_K^A &= -\frac{\partial}{\partial A_K}U^{AB} \\ &= -\sum_{ab} Q_t^a Q_u^b \frac{\partial}{\partial A_K} T_{tu}^{ab} \\ &= -\sum_{ab} Q_t^a Q_u^b \sum_i \frac{\partial q_i}{\partial A_K}\frac{\partial}{\partial q_i} T_{tu}^{ab}, \end{aligned} \tag{11.7.4}$$

while the K component of the torque is

$$\begin{aligned} G_K^A &= -\frac{\partial}{\partial \theta_K^A}U^{AB} \\ &= -\sum_{ab} Q_t^a Q_u^b \frac{\partial}{\partial \theta_K^A} T_{tu}^{ab} \\ &= -\sum_{ab} Q_t^a Q_u^b \frac{\partial q_i}{\partial \theta_K^A}\frac{\partial}{\partial q_i} T_{tu}^{ab}. \end{aligned} \tag{11.7.5}$$

11.7.2 *Second derivatives*

We can compute all the second derivatives in a similar manner by consecutive applications of the chain rule. To evaluate the Hessian matrix element H_{KL}, that is, the second derivative of the energy U with respect to two external coordinates C_K and C_L, which may refer to the same molecule or to different ones, we have

$$\frac{\partial^2 U}{\partial C_K \partial C_L} = \frac{\partial}{\partial C_K}\sum_j \frac{\partial U}{\partial q_j}\frac{\partial q_j}{\partial C_L}$$

$$= \sum_i \sum_j \frac{\partial^2 U}{\partial q_i \partial q_j} \frac{\partial q_i}{\partial C_K} \frac{\partial q_j}{\partial C_L} + \sum_j \frac{\partial U}{\partial q_j} \frac{\partial^2 q_j}{\partial C_K \partial C_L}. \tag{11.7.6}$$

Each term here is a product of a factor that is specific to the nature of the interaction and a factor that depends only on the geometry. The former set includes $\partial U/\partial q_i$, which is a first derivative and has already been calculated, and $\partial^2 U/\partial q_i \partial q_j$. The latter comprises the derivatives $\partial q_i/\partial C_K$, which have already been evaluated in the course of finding the forces and torques, and the $\partial^2 q_j/\partial C_K \partial C_L$, which like the first derivatives can be calculated once and for all. Here too it is convenient to work with the scalar products $\mathbf{w}^a \cdot \mathbf{R}$ rather than the direction cosines $\mathbf{w}^a \cdot \hat{\mathbf{R}}$. For example:

$$\begin{aligned}
\frac{\partial}{\partial C_K} \frac{\partial}{\partial C_L} (\mathbf{x}_a \cdot \hat{\mathbf{R}}) &= \frac{\partial}{\partial C_K} \frac{\partial}{\partial C_L} \frac{1}{R} \mathbf{x}_a \cdot \mathbf{R} \\
&= \frac{\partial}{\partial C_K} \left[\frac{1}{R} \frac{\partial}{\partial C_L} (\mathbf{x}_a \cdot \mathbf{R}) - (\mathbf{x}_a \cdot \mathbf{R}) \frac{1}{R^2} \frac{\partial R}{\partial C_L} \right] \\
&= \frac{1}{R} \frac{\partial^2}{\partial C_K \partial C_L} (\mathbf{x}_a \cdot \mathbf{R}) - \frac{1}{R^2} \left(\frac{\partial R}{\partial C_K} \frac{\partial (\mathbf{x}_a \cdot \mathbf{R})}{\partial C_L} + \frac{\partial R}{\partial C_K} \frac{\partial (\mathbf{x}_a \cdot \mathbf{R})}{\partial C_L} \right) \\
&\quad + (\mathbf{x}_a \cdot \hat{\mathbf{R}}) \left(\frac{2}{R^2} \frac{\partial R}{\partial C_K} \frac{\partial R}{\partial C_L} - \frac{1}{R} \frac{\partial^2 R}{\partial C_K \partial C_L} \right).
\end{aligned} \tag{11.7.7}$$

Expressions for the second derivatives are given in Table 11.5 on p. 182.

Note that these derivatives are not in general symmetric. As an example we see from Table 11.5 that $\partial^2(\mathbf{w}^b \cdot \mathbf{R})/\partial\theta_K^A \partial\theta_L^A = a_L w_K^b + \delta_{KL}(\mathbf{w}^b \cdot \mathbf{a})$, while $\partial^2(\mathbf{w}^b \cdot \mathbf{R})/\partial\theta_L^A \partial\theta_K^A = a_K w_L^b + \delta_{KL}(\mathbf{w}^b \cdot \mathbf{a})$. These expressions are clearly not the same unless $K = L$. In fact, all the rotation–rotation derivatives in Table 11.5 exhibit this asymmetry. This result, at first unexpected, is easily understood in terms of the infinitesimal rotation operators (angular momentum operators) $J_K = -\mathrm{i}\partial/\partial\theta_K$. These satisfy the familiar commutation relation $[J_K, J_L] = \mathrm{i}\varepsilon_{KLM} J_M$. It follows that, for example,

$$\left[\frac{\partial}{\partial\theta_X} \frac{\partial}{\partial\theta_Y} - \frac{\partial}{\partial\theta_Y} \frac{\partial}{\partial\theta_X} \right] U = -\frac{\partial}{\partial\theta_Z} U = \tau_Z. \tag{11.7.8}$$

When the system is at a stationary point the rotation–rotation asymmetry disappears because the torques vanish. At such a point the diagonalized Hessian yields real eigenvalues and orthogonal eigenvectors, in line with the physics of normal modes. Elsewhere the Hessian is not symmetric, but optimization methods like the Cerjan–Miller method (Cerjan and Miller 1981) can still function if a symmetrized Hessian is used. This asymmetry in the rotation–rotation block has been previously observed in the context of isotropic electrostatic and Lennard-Jones potentials (Wales 1991).

11.7.3 *Repulsion and dispersion*

As we saw on p. 167, the repulsive potential between two molecules A and B is represented to good accuracy by the empirical expression

$$U_{\text{rep}}^{AB} = \sum_{a \in A} \sum_{b \in B} \exp[-\alpha^{ab}(\Omega)(R_{ab} - \rho^{ab}(\Omega))]. \tag{11.7.9}$$

Table 11.5 *Second derivatives of* $\mathbf{R}\cdot\mathbf{R}$ *and the scalar products* $\mathbf{w}^a\cdot\mathbf{R}$, $\mathbf{w}^b\cdot\mathbf{R}$ *and* $\mathbf{w}^a\cdot\mathbf{w}^b$ *(where* $\mathbf{w}^a$ *is* $\mathbf{x}^a$, $\mathbf{y}^a$ *or* $\mathbf{z}^a$, *and* $\mathbf{w}^b$ *similarly) with respect to the twelve external coordinates (translation and rotation) necessary to describe a pair of rigid interacting molecules. The summation convention is used here.*

C \ C'	A_L	θ^A_L	B_L	θ^B_L
$\partial^2(\mathbf{R}\cdot\mathbf{R})/\partial C\partial C'$				
A_K	$2\delta_{KL}$	$2\varepsilon_{KLJ}a_J$	$-2\delta_{KL}$	$-2\varepsilon_{KLJ}b_J$
θ^A_K	$-2\varepsilon_{KLJ}a_J$	$-2(R_K+a_K)a_L + 2\delta_{KL}(\mathbf{R}+\mathbf{a})\cdot\mathbf{a}$	$2\varepsilon_{KLJ}a_J$	$2a_Lb_K - 2\delta_{KL}(\mathbf{a}\cdot\mathbf{b})$
B_K	$-2\delta_{KL}$	$-2\varepsilon_{KLJ}a_J$	$2\delta_{KL}$	$2\varepsilon_{KLJ}b_J$
θ^B_K	$2\varepsilon_{KLJ}b_J$	$2a_Kb_L - 2\delta_{KL}(\mathbf{a}\cdot\mathbf{b})$	$-2\varepsilon_{KLJ}b_J$	$2(R_K-b_K)b_L - 2\delta_{KL}(\mathbf{R}-\mathbf{b})\cdot\mathbf{b}$
$\partial^2(\mathbf{w}^a\cdot\mathbf{R})/\partial C\partial C'$				
A_K	0	$-\varepsilon_{KLJ}w^a_J$	0	0
θ^A_K	$\varepsilon_{KLJ}w^a_J$	$(R_K+a_K)w^a_L - \delta_{KL}(\mathbf{R}+\mathbf{a})\cdot\mathbf{w}^a$	$-\varepsilon_{KLJ}w^a_J$	$-b_Kw^a_L + \delta_{KL}(\mathbf{w}_a\cdot\mathbf{b})$
B_K	0	$\varepsilon_{KLJ}w^a_J$	0	0
θ^B_K	0	$-b_Lw^a_K + \delta_{KL}(\mathbf{w}_a\cdot\mathbf{b})$	0	$b_Lw^a_K - \delta_{KL}(\mathbf{w}_a\cdot\mathbf{b})$
$\partial^2(\mathbf{w}^b\cdot\mathbf{R})/\partial C\partial C'$				
A_K	0	0	0	$-\varepsilon_{KLJ}w^b_J$
θ^A_K	0	$-a_Lw^b_K + \delta_{KL}(\mathbf{w}_b\cdot\mathbf{a})$	0	$a_Lw^b_K - \delta_{KL}(\mathbf{w}_b\cdot\mathbf{a})$
B_K	0	0	0	$\varepsilon_{KLJ}w^b_J$
θ^B_K	$\varepsilon_{KLJ}w^b_J$	$a_Kw^b_L - \delta_{KL}(\mathbf{w}_b\cdot\mathbf{a})$	$-\varepsilon_{KLJ}w^b_J$	$(R_K-b_K)w^b_L - \delta_{KL}(\mathbf{R}-\mathbf{b})\cdot\mathbf{w}^b$
$\partial^2(\mathbf{w}^a\cdot\mathbf{w}_b)/\partial C\partial C'$				
A_K	0	0	0	0
θ^A_K	0	$w^a_Lw^b_K - \delta_{KL}(\mathbf{w}^a\cdot\mathbf{w}^b)$	0	$-w^a_Lw^b_K + \delta_{KL}(\mathbf{w}^a\cdot\mathbf{w}^b)$
B_K	0	0	0	0
θ^B_K	0	$-w^a_Kw^b_L + \delta_{KL}(\mathbf{w}^a\cdot\mathbf{w}^b)$	0	$w^a_Kw^b_L - \delta_{KL}(\mathbf{w}^a\cdot\mathbf{w}^b)$

Here R_{ab} is the inter-site distance, Ω represents the set of orientational coordinates describing the relative orientation of the two sites a and b, $\alpha_{ab}(\Omega)$ describes the hardness of the repulsion as a function of orientation, and $\rho_{ab}(\Omega)$ is a sum of the effective radii of the sites. They are conveniently expanded in the set of $\bar{S}$ functions (see eqn (11.3.6)), which are polynomials in the direction cosines q_i. The derivatives of α^{ab} and ρ^{ab} are therefore easily determined by the methods already described, and the derivatives of the repulsive potential then require merely another application of the chain rule.

The atom–atom dispersion can be handled in the same way. The dispersion takes the form

$$U_{\text{disp}}^{AB} = \sum_{a\in A}\sum_{b\in B}\left(f_6(R)\frac{C_6^{ab}(\Omega)}{R_{ab}^6} + f_7(R)\frac{C_7^{ab}(\Omega)}{R_{ab}^7} + \cdots \right),$$

where the $f_n(R)$ are damping functions (see §11.1.5), and the dispersion coefficients C_n can be expanded in terms of $\overline{S}$ functions. Differentiation of the dispersion coefficients follows the same procedure as for α^{ab} and ρ^{ab}, while the damping functions currently used are invariably functions only of R.

11.7.4 *Induction*

Finally, we consider the induction energy, which is a more awkward term to handle. The calculation is based on eqn (8.3.16) on p. 136, which we write in the form

$$E_{\text{ind}}^{A} = \tfrac{1}{2}\Delta Q_t^a V_t^a, \tag{11.7.10}$$

where the induced moments are given by eqn (8.3.12):

$$\Delta Q_t^a = -\sum_{B\neq A} \alpha_{tt'}^{aa'} T_{t'u}^{a'b}(Q_u^b + \Delta Q_u^b). \tag{11.7.11}$$

and V_t^a is the total field at site a due to the unpolarized moments of all the other molecules:

$$V_t^a = \sum_{B\neq A} T_{tu}^{ab} Q_u^b. \tag{11.7.12}$$

As usual, repeated suffixes imply summation. Calculation of the induction energy and its derivatives is then a two-stage process. The fields V_t^a at each site in each molecule, and their derivatives, are evaluated in the same loop over site pairs as the electrostatic terms, since they involve the same interaction functions T_{tu}^{ab}. The induced moments and their derivatives can also be evaluated in this loop, since (11.7.11) is also a sum over site pairs involving the same interaction functions. The polarizabilities $\alpha_{tt'}^{aa'}$, like the unpolarized multipole moments, are expressed in the local coordinate system for each site and are constant. The fields and induced moments are then combined at the end of the calculation to give the induction energy according to (11.7.10), and its derivatives are easily obtained by applying the formulae for differentiation of a product.

The difficulty in this procedure arises from the coupled form of (11.7.11). Each induced moment depends on the induced moments in all the other molecules, and its derivatives depend on the derivatives of the other moments. Three options are available. One is to use a first-order treatment, in which the induced moments on the r.h.s. of (11.7.11) are simply neglected. This is a good approximation when the molecules are not very polarizable or the electrostatic fields are relatively small. In this case the calculation is straightforward. The second option is to solve (11.7.11) iteratively. This is more time-consuming, though the process usually converges in half a dozen or so iterations. For geometry optimizations, a third option is available. This is to use the induced moments from the previous step of the optimization, and their derivatives, in the r.h.s. of (11.7.11) and its derivatives. If this is done, the induced moments and their derivatives are slightly in error at each step, but as the optimization procedure approaches convergence the induced moments also converge to the correct final values. This means that the iterated version of eqn (11.7.11) can be used in a geometry optimization with virtually no additional effort. Unfortunately it requires a great deal of computer memory because of the need to store the induced moments and all their derivatives. Moreover it is not suitable for a molecular dynamics calculation, because it does not conserve the energy of the system. In this case it is necessary either to use the first-order approximation throughout, or to iterate the moments to convergence at every step. Since the moments from the previous step normally provide a good starting-point, however, only a few iterations are needed to reach convergence.

12

SOURCES OF EXPERIMENTAL DATA

For the most part, this account has concentrated on methods of calculating intermolecular interactions, either analytically or from *ab initio* computations. However the purpose of such calculations is to make it possible to predict the behaviour of molecules, in small aggregates or in the bulk, and in principle a knowledge of intermolecular potentials should make it possible to predict the outcome of any experiment on molecules that does not involve actual chemical reaction. Comparison of theory with experiment in cases where the result is known allows us to validate the theory and, in some cases, to calibrate unknown parameters. In this chapter we outline the experimental techniques that are useful in the study of intermolecular potentials. This is not intended to be a detailed account; a fuller treatment may be found in the books by Gray and Gubbins (1984) and by Maitland *et al.* (1981).

12.1 Properties of individual molecules

An important source of data for intermolecular forces lies in the properties of isolated molecules. We have seen that at long range, the interactions between molecules can be described entirely in terms of such properties, so that methods of measuring them are of great importance.

12.1.1 *Multipole moments*

Dipole moments are usually obtained from measurements of the dielectric constant or from the Stark effect in rotational spectroscopy, where the absorption lines are split by the application of a static electric field, though McClellan (1963, 1974, 1989) lists a large number of other techniques in his tabulations of dipole moments. Many of these measurements give only the magnitude of the dipole moment, not its direction; in the case of the Stark effect, the experiment gives the magnitude, but not the sign, of the components of the dipole moment along the inertial axes. Often *ab initio* calculation, or even chemical intuition based on electronegativity arguments, is accurate enough to give the sign reliably. One way to determine the sign independently uses the effect of isotopic substitution on the rotational magnetic moment (Townes *et al.* 1955), though this has rather large errors and in the case of ClF gave the wrong sign (Ewing *et al.* 1972, McGurk *et al.* 1973, Janda *et al.* 1976). Another method, used to determine the sign of the dipole moment of nitrous oxide, N_2O, used an electric field to align the molecules in a molecular beam, or rather to focus only the molecules with a particular alignment, and observed the rate of reaction with barium atoms in a crossed beam. The rate increases when the N_2O molecules are aligned with the O atom leading, and from the direction of the field needed to cause an increase, it is possible to deduce the sign of the dipole moment (Jalink *et al.* 1987). It is also possible to deduce the sign from measurement of the dipole moment of a Van der Waals complex with another polar molecule if the structure of the complex and the sign of the dipole moment of the other molecule are known (Janda *et al.* 1976, Fowler and Stone 1987). A more direct approach is to use X-ray diffraction data for crystals, from which the electron density map can be determined (Spackman 1992). This gives the sign and magnitude of all the components, but not very accurately; moreover the electron density is polarized by the fields of the neighbouring molecules, so that the dipole moments obtained this way differ from those of the free molecules. Nevertheless they agree closely in magnitude and direction with values obtained by other techniques.

The same technique can be used to determine quadrupole moments, and values have been reported for a number of molecules (Spackman 1992), but more accurate values are obtained from measurements of the induced birefringence (Buckingham 1967), i.e., of the anisotropy in the refractive index induced by a non-uniform external electric field. In the non-uniform field of a two-wire cell (see Fig. 2.3 on p. 20), quadrupolar molecules in a gas become partially aligned. A molecule such as CO_2, for example, positioned midway between positively charged wires, will tend to lie with its molecular axis in the plane of the wires and perpendicular to them. Since its polarizability along the molecular axis ($\alpha_{\parallel}$) is greater than the component perpendicular to it ($\alpha_{\perp}$), the refractive index of the gas for light polarized in the plane of the wires will be slightly different from the refractive index perpendicular to the plane. The effect of this birefringence on a laser beam parallel to the wires can be measured; it is proportional to the quadrupole moment and to the polarizability anisotropy $\alpha_{\parallel} - \alpha_{\perp}$, so if the anisotropy is known the quadrupole moment can be measured. For the field produced by a two-wire source, $F_{zz} = 0$ and $F_{xx} = -F_{yy}$, and the birefringence is

$$n_x - n_y = \frac{4\pi F_{xx}}{15}\left(B_{\alpha\beta,\alpha\beta} + \frac{1}{kT}\alpha_{\alpha\beta}\Theta_{\alpha\beta}\right). \qquad (12.1.1)$$

The quantity $B_{\alpha\beta,\gamma\delta}$ in the temperature-independent term is the dipole–dipole–quadrupole hyperpolarizability. The temperature-dependent term involves a contraction of the quadrupole moment with the polarizability. Because the quadrupole moment is traceless, only the anisotropy of the polarizability enters this expression; for linear molecules $\alpha_{\alpha\beta}\Theta_{\alpha\beta} = (\alpha_{\parallel} - \alpha_{\perp})\Theta_{zz}$. If $\alpha_{\parallel} - \alpha_{\perp}$ is known, this technique provides both the magnitude and the sign of Θ_{zz}. Further details of the experimental technique and the underlying theory are given by Buckingham and Disch (1963).

Methods for determining higher moments are indirect and not very accurate. One of the more reliable methods involves the study of the collision-induced infrared spectrum of the gas. In gaseous methane, for example, there is no allowed pure rotational spectrum for the isolated molecule, or for the gas at very low pressure, but at higher pressures there is an infrared spectrum whose intensity is proportional to the square of the density. This arises from interactions between pairs of molecules, each inducing fluctuating dipole moments in the other (Birnbaum and Cohen 1975, Joslin *et al.* 1985). The magnitude of the induced moments depends mainly on the molecular octopole and hexadecapole, in this case, so measurements of the intensity lead to values for these moments. They can

Table 12.1 *Values of the mean polarizability at zero frequency and at a frequency* ω *of 15,800 cm*$^{-1}$.

Molecule	$\alpha(0)$	$\alpha(\omega)$
CH_4	2.593	2.607
CF_4	3.84	2.85
C_6H_6	10.6	10.39
SF_6	6.56	4.50

be distinguished because the octopole moment is associated with rotational transitions with $\Delta J = 0, \pm 1, \pm 2$ and ± 3, while the hexadecapole also leads to $\Delta J = \pm 4$. The transitions are not resolvable because there are also translational contributions to the spectrum which smear out the lines, but from the frequency dependence of the spectrum the octopole and hexadecapole can be determined separately. However the induced moments also depend on the polarizability and on the intermolecular potential—the latter because the magnitude of the induced moment is affected by the distance to which the molecules can approach. In fact Birnbaum and Cohen (1975) point out that the quantities determined from the spectrum are Ω^2/σ^7 and Φ^2/σ^9, where σ is the size parameter in the Lennard-Jones potential that they used, and Ω and Φ are the octopole and hexadecapole moments Ω_{xyz} and Φ_{zzzz}. (Recall (p. 29) that there is only one independent non-vanishing component of each for a tetrahedral molecule.)

12.1.2 *Polarizabilities*

The dipole–dipole polarizability α can be measured by a variety of methods. The mean polarizability $\overline{\alpha}$ is related to the refractive index n:

$$n - 1 = 2\pi\rho\overline{\alpha}(\omega), \tag{12.1.2}$$

where ρ is the number density. This relationship involves the polarizability $\overline{\alpha}(\omega)$ for the frequency at which the refractive index is measured. This will normally be different from the static polarizability, though not very different if there are no low-lying absorption bands. Measurements of the dielectric constant also yield values of the polarizability, but these include a contribution from the movement of the nuclei under the influence of the electric field (the vibrational or 'atomic' polarizability) and this contribution can be quite large for molecules with low-frequency vibrations. Some typical values, taken from the book by Gray and Gubbins (1984) (a useful source of numerical values of multipole moments and polarizabilities) are given in Table 12.1.

Experimental data on electric-dipole absorption spectra can be used to obtain accurate values of mean polarizabilities. The average polarizability at frequency ω is, from eqn (2.5.7),

$$\overline{\alpha}(\omega) = \frac{1}{3}\alpha_{\alpha\alpha}(\omega) = \frac{2}{3}\sum_n{}' \frac{\omega_n |\langle 0|\hat{\mu}|n\rangle|^2}{\hbar(\omega_n^2 - \omega^2)}, \tag{12.1.3}$$

and the numerator $|\langle 0|\hat{\mu}|n\rangle|^2$ of each term is related to the intensity of the corresponding electric dipole transition. It is customary to define a dimensionless number called the *oscillator strength* of a transition by

$$f_n = \frac{2}{3}\frac{\hbar\omega_n}{E_h}\frac{|\langle 0|\hat{\mu}|n\rangle|^2}{e^2a_0^2}. \tag{12.1.4}$$

From a knowledge of the dipole oscillator strengths at all frequencies (the 'dipole oscillator strength distribution' or DOSD), a number of properties can be determined. Other sources of information about oscillator strengths include molar refractivities and inelastic scattering cross sections of charged particles (Meath *et al.* 1981). It is useful to define a set of oscillator strength sums by

$$S_k = \sum_n f_n \left(\frac{\hbar\omega_n}{E_h}\right)^k. \tag{12.1.5}$$

The Thomas–Reiche–Kuhn sum rule states that S_0 is equal to the number of electrons in the system, and this can be used as an additional constraint on the data, while comparison of eqns (12.1.3) and (12.1.4) shows that $S_{-2} = \overline{\alpha}(0)/4\pi\varepsilon_0 a_0^3$, numerically equal to the mean static polarizability in atomic units.

Now the polarizability as a function of real frequency has singularities at every absorption frequency according to eqn (12.1.3), and although the singularities are moved off the real axis when the finite lifetime of the excited states is taken into account (see eqn (2.5.9) on p. 28), the polarizability is still strongly frequency-dependent near absorption frequencies. However these singularities are all near the real axis, and the polarizability as a function of imaginary frequency is very well-behaved, as we saw in §4.3.2. Consequently it can be accurately represented in terms of oscillator strengths and transition frequencies, and is much less sensitive to errors in these quantities than the polarizability at real frequencies. In fact, it is possible to represent the polarizability at imaginary frequency to good accuracy in terms of 'pseudo dipole oscillator strength distributions':

$$\overline{\alpha}(i\nu) = \frac{2}{3}\sum_{k=1}^{N}\frac{\tilde{\omega}_k \tilde{f}_k}{\tilde{\omega}_k^2 + \nu^2}, \tag{12.1.6}$$

where the parameters $\tilde{f}_k$ and $\tilde{\omega}_k$ are chosen to fit the experimental DOSD. Usually a good fit can be found with quite a small number of parameters; $N = 10$ is typical (Kumar and Meath 1992).

The importance of this approach to the determination of polarizabilities at imaginary frequency is that it provides a route to accurate values of isotropic C_6 dispersion coefficients, via eqn (4.3.13). Moreover it is just as easily applied to interactions between unlike molecules; most experimental methods for determining interactions between unlike systems are difficult and inaccurate. The method can also be used to calculate C_9 triple-dipole dispersion coefficients, using eqn (9.2.4). The method was first used by Dalgarno and Lynn (1957*b*), and applied to the inert gases, to some spherical ions, and to a number of small molecules to obtain values of the C_6 dispersion coefficients, both for like and unlike pairs (Dalgarno 1967). More recently, Meath and his collaborators (Jhanwar and Meath 1982, Kumar and Meath 1992) have used the method extensively to determine dipole oscillator strength distributions for a number of atoms and small molecules, and have given values of polarizabilities, C_6 dispersion coefficients, and C_9 triple-dipole dispersion coefficients.

It is possible to obtain the anisotropic terms in the same way if the polarization associated with each oscillator strength can be determined. For a linear molecule, separate oscillator-strength distributions are needed for transitions polarized parallel and perpendicular to the molecular axis. Meath and Kumar (1990) have determined these distributions for N_2 and H_2, and have derived anisotropic C_6 coefficients for interactions involving these molecules and the inert gases. CO has been treated similarly (Kumar and Meath 1994).

12.2 Experimental determination of intermolecular potentials from bulk properties

The determination of the intermolecular potential, as distinct from the constants that enter the long-range part of the potential, is a much more difficult matter. The experiments are all to a greater or lesser degree indirect; that is, they do not measure the potential itself, but some property that in turn depends on the potential. This dependence invariably involves some sort of configurational average, so that it is very difficult to work back from the measurement to the potential. Instead it is usual to assume some more or less plausible functional form for the potential, and to fit the parameters in that to the experimental data. If this is to be successful, the functional form that is chosen must be a realistic one, but it cannot contain too many adjustable parameters if numerical instabilities are to be avoided. The traditional source of data for this type of approach consists of measurements on the physical properties of gases, but in recent years spectroscopy of small molecular clusters has proved very fruitful, especially for the region near the minimum of the potential surface.

12.2.1 *The virial coefficient*

One of the classical sources of data about intermolecular potentials is the second (pressure) virial coefficient. The quantity PV/RT is 1 for an ideal gas and for real gases at the limit of zero density, but at finite densities it deviates from this value, and we can use an expansion in powers of the density, or inverse powers of the molar volume. The coefficients in this expansion depend on the temperature:

$$\frac{PV}{RT} = 1 + \frac{B(T)}{V} + \frac{C(T)}{V^2} + \cdots. \tag{12.2.1}$$

$B(T)$ is the second virial coefficient, and an expression for it can be obtained from statistical mechanics (McQuarrie 1976). It depends rigorously on the pair potential only, even if many-body terms occur in the total energy. For atoms:

$$B(T) = -2\pi N_A \int_0^\infty \left[\exp(-U(R)/kT) - 1\right] R^2 \,\mathrm{d}R, \tag{12.2.2}$$

The quantity in square brackets here is the *Mayer f-function* $f(R)$. Its radial dependence is illustrated for a Lennard-Jones potential in Fig. 12.1 for two temperatures, $kT = 1.5\varepsilon$

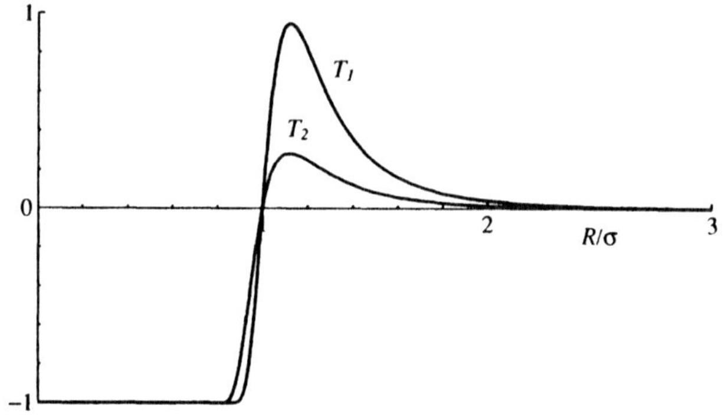

FIG. 12.1. The Mayer f-function for a Lennard-Jones potential at temperatures $T_1 = 1.5\varepsilon/k$ and $T_2 = 4\varepsilon/k$, as a function of $R^* = R/\sigma$.

and $kT = 4\varepsilon$. At all temperatures, $f(R) = -1$ when $R = 0$, passes through zero when $R = \sigma$ to a maximum at $R = R_m$ and then decays to zero. The virial coefficient is an integral of this over all space, i.e., over all positions of a second molecule. It is negative at low temperature, where the positive tail of $f(R)$ dominates, and positive at high temperatures, where the small-R region wins. The temperature T_B at which the virial coefficient is zero is called the Boyle temperature.

For atoms, the virial coefficient, as a function of temperature, is an integral transform of the Mayer function, and it is possible in principle to invert this transform to recover the Mayer function, and hence the potential, from measurements of the virial coefficient over a range of temperatures. In practice $B(T)$ cannot be measured accurately enough or over a wide enough temperature range. More detailed investigation has revealed (Keller and Zumino 1959, Frisch and Helfand 1960) that inversion of the virial coefficient yields only the width of the well as a function of depth, though the region where the potential is positive can be determined uniquely. Moreover the formula is a classical one which requires correction for quantum effects. Maitland *et al.* (1981) give a detailed discussion. For molecules, the virial coefficient is again an integral of the Mayer function over all positions of the second molecule, but now it takes six coordinates (a distance and five angles) to describe the position of the second molecule relative to the first, if both are non-linear. Consequently the virial coefficient is a 6-dimensional integral, and there is no hope of inverting virial coefficient data to obtain the potential.

Nevertheless, virial coefficient data provide an important test of a potential, as well as a source of data for fitting parameters. The virial coefficient depends rigorously on the pair potential only, and contains no contributions from three-body corrections. These appear in the third virial coefficient C, which involves a similar integration over all configurations of three molecules, while the four-body terms first appear in the fourth virial coefficient, and so on. Accurate measurement of C is even more difficult than B, and because the pair-potential terms appear in the expression for C as well as the three-body terms, it has not so far proved possible to extract useful three-body data from this source.

12.2.2 *Transport coefficients*

The viscosity, thermal conductivity and diffusion coefficient of a gas at low pressure also depend only on the pair potential, but like the second virial coefficient they can usually be used only to gauge the validity of a given potential and not to determine it in the absence of other information. They involve a more complicated sequence of integrations than the virial coefficient. For example, the thermal conductivity λ of a monatomic gas may be expressed in the form

$$\lambda = \frac{75}{64}\left(\frac{k_B^3 T\pi}{m}\right)^{1/2} \frac{1}{\overline{\Omega}^{(2,2)}(T)} f_\lambda, \tag{12.2.3}$$

where m is the molecular mass, $\overline{\Omega}^{(2,2)}(T)$ is a 'collision integral' and f_λ is a correction factor that normally differs from unity by only 1 or 2 per cent. The collision integrals are defined by

$$\overline{\Omega}^{(2,2)}(T) = \frac{1}{6(k_B T)^4} \int_0^\infty Q^{(2)}(E) \mathrm{e}^{-E/k_B T} E^3 \, \mathrm{d}E, \tag{12.2.4}$$

where E is the relative kinetic energy of a collision between two atoms and $Q^{(2)}(E)$ is the 'transport cross-section', given by

$$Q^{(2)}(E) = 3\pi \int_0^\infty (1 - \cos^2\chi) b \, db. \tag{12.2.5}$$

Here χ is the classical deflection angle for a collision with energy E and impact parameter b:

$$\chi(E,b) = \pi - 2b \int_{R_0}^\infty \frac{dR}{R^2\sqrt{1 - b^2/R^2 - U(R)/E}}, \tag{12.2.6}$$

where $U(R)$ is the interatomic potential and R_0 is the distance of closest approach in the collision.

The viscosity can be expressed in terms of the same collision integrals:

$$\eta = \frac{5}{16} \frac{(m\pi k_B T)^{1/2}}{\overline{\Omega}^{(2,2)}(T)} f_\eta, \tag{12.2.7}$$

where f_η, like f_λ, is a correction factor close to unity. The self-diffusion coefficient D can be expressed in a similar way but involves a different collision integral. For full details the reader is referred to Hirschfelder *et al.* (1954) or Maitland *et al.* (1981).

In spite of the complexity of this formulation, it is possible to use it in reverse to determine the potential for monoatomic gases from viscosity or thermal conductivity data. The procedure is iterative, but converges rapidly if the first guess at the potential function has the correct well depth (Maitland *et al.* 1981).

Methods of this sort have been important in the past in determining accurate potentials for inert gases, where the potential depends on one coordinate only, but they are much less useful for studying interactions between molecules. The difficulty in applying them to molecules is two-fold: the calculation of the physical property from the potential is very much more difficult, because of the orientation dependence of the interaction potential, and even if this can be achieved reliably, direct inversion procedures leading from the experimental data to the potential do not appear to be feasible, because the experimental data do not provide enough information to characterize the many-dimensional potential surface.

12.3 Spectroscopic methods

The most important source of accurate data on intermolecular potentials is undoubtedly spectroscopy. The Faraday Discussion on the Structure and Dynamics of Van Der Waals Complexes (Faraday 1994) and the November 1994 issue of Chemical Reviews (Chemical Reviews 1994) both illustrate the wide range of techniques that are now being applied to this problem.

One long-standing technique is molecular beam microwave spectroscopy. The use of a molecular beam offers many advantages: the effects of collisions with other molecules are virtually eliminated, and the very low rotational-vibrational temperature (typically a few K) means that only a few rotational levels are occupied. This technique was pioneered by Klemperer and refined by Flygare, and has been used by them and many others to determine the microwave spectra of a wide variety of weakly-bound molecular complexes. From these spectra the rotational constants can be obtained, and by using Stark spectroscopy it is possible to measure the magnitude of the components of the dipole moment of the complex along one or more of the inertial axes. If the molecules contain quadrupolar nuclei such as nitrogen or chlorine, there may be hyperfine splittings in the spectra that can be used to determine further features of the structure.

The disadvantage of microwave spectroscopy is that it yields averaged information for the vibrational ground state. The potential energy surface of a weakly-bound complex is often very complex, with shallow minima separated by low barriers. There are numerous low-frequency vibrations, often anharmonic and large-amplitude, and the low barriers lead to tunnelling splittings. The structure that emerges from microwave spectroscopy is averaged over these large-amplitude vibrations, and so may give a quite misleading picture of the complex. Much more information can be obtained from far-infrared spectroscopy ('vibration–rotation–tunnelling' or FIR-VRT spectroscopy), where the energy levels depend on the details of the potential energy surface far from the minima. Until recently, this information was difficult to obtain, because there were few satisfactory light sources and detectors for the far infrared, but now that tunable far-infrared lasers and sensitive detectors have been developed, detailed information on potential energy surfaces is becoming available (Saykally and Blake 1993). Systems that have been studied in this way include Ar···DCl, for which Hutson (1992) has determined an accurate potential energy surface using microwave, FIR-VRT and mid-infrared data. This system has a global minimum about $170\,\mathrm{cm}^{-1}$ deep at the hydrogen-bonded geometry (Ar···HCl), and a secondary minimum about $145\,\mathrm{cm}^{-1}$ deep at the linear Ar···ClH configuration. Hutson's potential predicts the energy levels extremely accurately, but recent work by Elrod *et al.* (1993*a*) has revealed small inaccuracies in the description of the secondary minimum.

Other systems where detailed potential energy surfaces have been obtained from far infrared spectroscopy include Ar···H_2O (Cohen and Saykally 1993) and Ar···H_3N (Schmuttenmaer *et al.* 1994). In these systems there is no first-order electrostatic interaction, and the binding is due to a balance between induction, dispersion and repulsion.

As Saykally and Blake (1993) have pointed out, the anisotropy of the repulsion is particularly important in determining the structure.

It is also possible to use infrared spectroscopy at more conventional frequencies, by observing combination bands between the intermolecular and intramolecular vibrations, or by studying the changes in frequency and rotational structure of the intramolecular bands on complex formation. The early work involved conventional gas-phase absorption spectroscopy with very long path lengths (Watanabe and Welsh 1964), and experiments of this kind are still used (McKellar 1994), but most recent work has used molecular beam spectroscopy. This method has been used to study the intermolecular vibrational modes in $(HF)_2$ and $(DF)_2$ (Nesbitt 1994) and in Ar_nHF (McIlroy *et al.* 1991, McIlroy and Nesbitt 1992), where in conjunction with far infrared vibration–rotation–tunnelling spectroscopy (Elrod *et al.* 1991, 1993*b*) it provides accurate information about many-body terms in the potential (see Chapter 9).

12.3.1 *Rotational Rydberg–Klein–Rees method*

The determination of the potential from the spectroscopic data is a difficult problem. In a few favourable cases, the spectroscopic measurements can be used directly to determine the intermolecular potential. The method is a version of the RKR method for determining the potential function for a diatomic molecule. This is a semi-classical procedure that uses the Bohr quantization condition:

$$\frac{\sqrt{2\mu}}{\hbar}\int_{a_{vJ}}^{b_{vJ}}\left[E(v,J)-V(R;J)\right]^{1/2}\mathrm{d}R=(v+\tfrac{1}{2})\pi, \tag{12.3.1}$$

where a_{vJ} and b_{vJ} are the classical turning-points of the vibrational motion in the state with vibrational quantum number v and rotational quantum number J, and $V(R;J)$ is the effective vibrational potential in rotational state J:

$$V(R;J)=V(R)+\frac{J(J+1)}{2\mu R^2}. \tag{12.3.2}$$

The standard RKR treatment (Child 1991) leads to the formulae

$$\begin{aligned} b_{vJ}-a_{vJ}&=\frac{2\hbar}{\sqrt{2\mu}}\int_{-1/2}^{v}\frac{dv'}{\left[E(v,J)-E(v',J)\right]^{1/2}},\\ \frac{1}{a_{vJ}}-\frac{1}{b_{vJ}}&=\frac{2\sqrt{2\mu}}{\hbar}\int_{-1/2}^{v}\frac{B(v',J)dv'}{\left[E(v,J)-E(v',J)\right]^{1/2}}, \end{aligned} \tag{12.3.3}$$

where $B(v,J)$ is the generalized rotational constant:

$$B(v,J)=\frac{1}{2J+1}\left(\frac{\partial E}{\partial J}\right)_v.$$

This treatment normally relies on information about rotational levels of a series of vibrational states. Nesbitt *et al.* (1989, 1993) showed that a similar method can be used when the rotational levels are available only for a single vibrational state. It relies on centrifugal stretching data to supply the missing information, so it needs high-precision rotational energy levels. Furthermore it can be used not just for atom–atom potentials but for complexes such as Ar$\cdots$HF and $N_2\cdots$HF, which are linear and relatively stiff; in this case the method provides a radial potential that is averaged over the angular motions. For most Van der Waals molecules, unfortunately, the angular motions are very floppy and strongly coupled to the radial motion, so an average radial potential is not very helpful in understanding the details of the potential energy surface.

12.3.2 *Expansion in an angular basis*

Usually, therefore, the determination of potential energy surfaces from spectroscopic data involves the calculation of energy levels from a postulated potential energy surface, which is then modified iteratively to bring the calculated levels into agreement with experiment (Le Roy and Van Kranendonk 1974). This is a difficult calculation, but where it can be carried out it gives high-quality information about the potential energy surface. In the most favourable cases, such as Ar$\cdots$HCl, the potential surface obtained this way is believed to be very accurate (Hutson 1992). Systems like this, comprising an atom and a diatomic molecule, are the simplest Van der Waals complexes where orientation-dependent potentials arise, but they will serve to illustrate the methods used and the approximations needed to determine the energy levels from the potential energy surface.

We follow the notation of Hutson (1990*b*), which should be consulted for further details. The position of the atom A relative to the centre of mass of the diatomic molecule B is described by the vector $\mathbf{R}$, or (R, β, α) in space-fixed polar coordinates. We can define a body-fixed axis system with its z axis in the direction of $\mathbf{R}$; the x and y axes of this system are arbitrary. The vector from one atom of the diatomic to the other is $\mathbf{r}$, or (r, θ, φ) in polar coordinates referred to the body-fixed frame. Consequently r is the bond-length, and θ is the angle between $\mathbf{r}$ and $\mathbf{R}$.

The Hamiltonian for the system is now

$$\begin{aligned}\mathcal{H} &= -\frac{\hbar^2}{2\mu}\nabla_R^2 + \mathcal{H}_B + V(\mathbf{R}, \mathbf{r}) \\ &= -\frac{\hbar^2}{2\mu}R^{-1}\frac{\partial^2}{\partial R^2}R + \frac{\hbar^2 \mathbf{l}^2}{2\mu R^2} + \mathcal{H}_B + V(R, r, \theta), \end{aligned} \qquad (12.3.4)$$

where $\mathcal{H}_B$ is the Hamiltonian for the isolated diatom B, $\mu = M_A M_B/(M_A + M_B)$ is the reduced mass of the whole system, V is the intermolecular potential, and $\mathbf{l}$ is the angular momentum operator for the end-over-end rotation of the complex. The r-dependence of the potential can normally be ignored: it causes mixing between the diatom vibrational states, but not usually to a significant extent.

However V depends on the orientation of the diatom relative to the intermolecular vector $\mathbf{R}$, and the rotation of the diatom is approximately quantized along $\mathbf{R}$. That is, the appropriate wavefunctions are the spherical harmonics $Y_{jK}(\theta, \varphi)$ with component K of angular momentum along the body-fixed z axis, though they are mixed by the intermolecular interaction. Rotation of the system as a whole relative to the space-fixed frame changes the angles α and β; the third Euler angle γ is the same as φ. Accordingly the appropriate basis functions to describe the rotational motion are

$$\Phi_{jK}^{JM}(\alpha,\beta,\theta,\varphi) = \left(\frac{2J+1}{4\pi}\right)^{1/2} D_{MK}^{J}(\alpha,\beta,0)^{*} Y_{jK}(\theta,\varphi). \qquad (12.3.5)$$

Here D_{MK}^{J} is a Wigner function used as a rotational wavefunction (see Appendix B). The total angular momentum **J** is the vector sum of the angular momentum **j** of the diatom and the angular momentum **l** of the end-over-end rotation of the complex.

The wavefunction for the system is then written as

$$\psi_n^{JM}(\mathbf{R},\mathbf{r}) = \frac{1}{R}\sum_{jK} \Phi_{jK}^{JM}(\alpha,\beta,\theta,\varphi)\chi_{jKJ}^{n}(R),$$

where n is a label distinguishing different solutions with the same rotational quantum numbers. If we substitute this into the Schrödinger equation $\mathcal{H}\psi_n = E\psi_n$, multiply by $\Phi_{jK}^{JM*}(\alpha,\beta,\theta,\varphi)$ and integrate, we get

$$\begin{aligned}\left\{-\frac{\hbar^2}{2\mu}\frac{d^2}{dR^2} + (jKJ|V|jKJ) + \frac{\hbar^2}{2\mu R^2}(jKJ|(\mathbf{J}-\mathbf{j})^2|jKJ) + E_j^0 - E\right\}\chi_{jKJ}^{n}(R)\\ = -\sum_{j'}{}'(jKJ|V|j'KJ)\chi_{j'KJ}^{n}(R)\\ - \sum_{K'=K\pm1}\frac{\hbar^2}{2\mu R^2}(jKJ|(\mathbf{J}-\mathbf{j})^2|jK'J)\chi_{jK'J}^{n}(R). \qquad (12.3.6)\end{aligned}$$

Here $\mathbf{l}^2$ of eqn (12.3.4) has been replaced by $(\mathbf{J}-\mathbf{j})^2$, and the use of parentheses rather than angle brackets in the matrix elements denotes integration over the angular variables only. The prime on the summation over j' indicates that $j' = j$ is excluded. Eqn (12.3.6) is a set of coupled differential equations in R for the radial functions $\chi_{jKJ}^{n}(R)$; they are called the *close-coupling equations* or the *coupled-channel equations*. The matrix elements $(jKJ|(\mathbf{J}-\mathbf{j})^2|jK'J)$ that occur in them can be evaluated by standard techniques of angular momentum theory, not discussed here; they are

$$\begin{aligned}(jKJ|(\mathbf{J}-\mathbf{j})^2|jKJ) &= J(J+1) + j(j+1) - 2K^2,\\ (jKJ|(\mathbf{J}-\mathbf{j})^2|j,K\pm1,J) &= [J(J+1)-K(K\pm1)]^{1/2}[j(j+1)-K(K\pm1)]^{1/2}.\end{aligned} \qquad (12.3.7)$$

The matrix elements $(jKJ|V|j'KJ)$ are more troublesome in general, but for small molecules, the potential can be expanded in Legendre polynomials:

$$V(R,\theta) = \sum_{\lambda} V_\lambda(R)P_\lambda(\cos\theta), \qquad (12.3.8)$$

and then the angular integration can again be carried out by standard techniques to give

$$(jKJ|V|j'K'J) = \sum_{\lambda} V_\lambda(R)g_\lambda(jj'K)\delta_{KK'}, \qquad (12.3.9)$$

where

$$g_\lambda(jj'K) = (-)^K[(2j+1)(2j'+1)]\begin{pmatrix} j & \lambda & j' \\ 0 & 0 & 0\end{pmatrix}\begin{pmatrix} j & \lambda & j' \\ -K & 0 & K\end{pmatrix}.$$

Thus the matrix elements are independent of J and diagonal in K.

There are two interrelated aspects to the solution of eqn (12.3.6). One concerns the method to be used to find the radial functions $\chi^n_{jKJ}(R)$, and the other concerns the dimension of the problem. If all the $\chi^n_{jKJ}(R)$ are included, up to some maximum value $j_{\max}$ of j, there are $2j+1$ values of K for each j, giving nearly $(2j_{\max}+1)^2$ coupled equations in all. (Values of $j < K$ cannot occur, but K is usually small for the states of interest.) This truncation of the problem, with no other approximations made, is called the *close-coupling approximation*. It is usually impracticable to carry out the calculations without further approximations, because the time required is proportional to the cube of the number of coupled equations, i.e., to $(2j_{\max}+1)^6$, and further approximations are then needed to reduce the dimension. The most important of these is the *helicity decoupling* approximation, in which the Coriolis terms coupling K with $K \pm 1$ on the right of eqn (12.3.6) are neglected. There is then only one value of K in the equations, so that the dimension is reduced substantially, from $(2j_{\max}+1)^2$ to $j_{\max}+1-K$. This is a good approximation if the coupling term between K and K' is sufficiently small.

One way to solve the coupled equations of eqn (12.3.6) uses the formalism of scattering theory. For a given value of the energy, the differential equations are solved numerically, starting in the classically forbidden regions at large and small R and approaching the classically allowed region around the potential minimum. The large-R and small-R solutions have to match, and this is only possible for certain values of the energy, which are the eigenvalues of the problem. The strategy for finding the eigenvalues efficiently has been described by Johnson (1978) and Manolopoulos (1988), and a program is available for carrying out the whole calculation by this method (Hutson 1990*a*).

This is a laborious procedure, because of the need to iterate towards an acceptable value of the energy and because the numerical integration of the radial differential equations is itself a time-consuming process. An alternative is to assume that the radial and angular motions are approximately separable, so that a Born–Oppenheimer-like treatment can be used. The wavefunction is written in the form

$$\chi_{nJ}(R)\psi_J(\Omega;R),$$

That is, the angular equations are solved at a particular value of R, ignoring the radial kinetic energy term in the Hamiltonian. This leads to a set of secular equations whose solutions are the rotational wavefunctions ψ_J and the rotational energies, as a function of R. The rotational energy serves as the potential energy for the radial motion, and (numerical) solution of the radial problem gives the radial wavefunction $\chi_{nJ}(R)$ and the total energy. This procedure is like the conventional Born–Oppenheimer separation of electronic and nuclear motion, but it does not depend, as that approximation does, on the very different timescales of the two kinds of motion. Rather it relies on the anisotropy of the interaction being only weakly dependent on R. The method was introduced by Holmgren *et al.* (1977) and called by them the Born–Oppenheimer angular–radial separation, or BOARS, method. Hutson and Howard (1980) showed that this adiabatic approximation leads to significant errors, but that they can be effectively corrected by perturbation theory, leading to a much more accurate procedure that they called the Corrected Born–Oppenheimer, or CBO, method.

An alternative procedure that does not require this approximation is to expand the radial wavefunction in terms of some basis set. The whole problem then reduces to the solution of a set of secular equations. This is feasible if the number of rotational states that have to be included is not too great, which in turn requires that the potential is not very anisotropic. The reason for this is that a term in P_λ in eqn (12.3.8) couples state j with states from $|j-\lambda|$ to $j+\lambda$, so if the expansion of the potential converges slowly it will be necessary to take a large value of $j_{\rm max}$ to describe its effects. This increases the dimension of the problem, as we have seen. Consequently the method has been used only for near-spherical molecules like H_2 (Bačić and Light 1989).

So far we have been considering the simplest case of a complex like $Ar\cdots H_2$, formed from an atom and a diatomic molecule. Complexes comprising an atom and any linear molecule can be treated by similar methods, if the molecule can be treated as rigid, but as soon as we contemplate a complex of two diatomics, or of an atom and a non-linear molecule, the problem becomes very much less tractable. For an atom and a non-linear molecule we require two angular coordinates; for two linear molecules we need three; and for two non-linear molecules we need five. The number of angular functions needed for an adequate description increases by an order of magnitude for each new variable, and the computational effort is roughly proportional to the cube of the number of basis functions.

An additional difficulty is that the intermolecular potential becomes more complicated. Moreover, as the molecules become larger, the expansion of the potential in Legendre polynomials, or the equivalent in the many-variable case, converges more and more slowly, and eventually ceases to converge at all, as we have seen for the electrostatic interaction. It then becomes necessary to use multi-site potentials of the type discussed in Chapter 7. Some calculations have been carried out with such potentials; for example a site–site potential was used in calculations of the vibration–rotation–tunnelling spectrum of NH_3 dimer (Van Bladel *et al.* 1992, Van der Avoird *et al.* 1994), but it was necessary to expand the potential in terms of single-site angular functions in order to calculate the matrix elements, truncating the expansion at quite low rank to keep the calculation tractable. As a result of the truncation the error in the expanded potential was about 8%. Because of such difficulties, two other approaches to the problem have been used.

12.3.3 *Discrete variable representation*

The first of these uses the so-called Discrete Variable Representation or DVR (Bačić and Light 1989). In simple terms, this describes the wavefunction by its value at each of a large array of points. The relationship to conventional expansion in terms of a set of basis functions is that if such a set of (orthonormal) functions is given, a DVR can be constructed using the gaussian quadrature points and weights derived from that set of basis functions, and the two procedures—DVR and basis function expansion—are then precisely equivalent. The advantage of the DVR over the conventional procedure is that the matrix elements of the intermolecular potential energy involve no integrations, but only the evaluation of the potential energy at the DVR points. It therefore becomes possible to handle potential energy functions of arbitrary complexity. Moreover the size of the problem can be reduced by deleting points in regions of high potential energy where the wavefunction is known to be negligible—a much simpler and safer procedure than dropping expansion functions from a conventional basis.

The difficulty with the DVR is that the gaussian quadrature points are only easily obtained for functions of one variable, and no generalization to systems of more than one dimension has yet been found. A system with many degrees of freedom has to be described in a direct-product fashion; that is, if we have a set of quadrature points x_i, $1 \leq i \leq n_x$, in coordinate x, a set y_j, $1 \leq j \leq n_y$, for coordinate y, and so on, the DVR points for the many-dimensional problem must be $(x_i, y_j, \ldots)$, $1 \leq i \leq n_x$, $1 \leq j \leq n_y, \ldots$. Although points can be deleted from this set as already mentioned, the remaining points may not cover the space very efficiently. An example of this is the case of two linear molecules: their rotational wavefunctions can be described by products of spherical harmonics $Y_{l_1 m_1}(\theta_1, \varphi_1) Y_{l_2 m_2}(\theta_2, \varphi_2)$, and a DVR for this basis can be set up. However, it uses the four variables θ_1, φ_1, θ_2 and φ_2, and only three variables are of interest, the potential energy of the system being independent of $\varphi_1 + \varphi_2$. No three-dimensional DVR for this system has been found. See Althorpe and Clary (1995) for a fuller discussion.

Variants on the DVR include the collocation method, a less efficient but much simpler approach (Peet and Yang 1989, Cohen and Saykally 1990, Leforestier 1994). Here one expands the wavefunction in the usual way in terms of a set of basis functions:

$$\Psi = \sum c_j \varphi_j,$$

and then solves the eigenvalue equation

$$\mathbf{Hc} = E\mathbf{Rc},$$

where the elements of the matrices $\mathbf{H}$ and $\mathbf{R}$ are simply the functions $\mathcal{H}\varphi_j$ and φ_j evaluated at a point $\mathbf{r}_i$ in configuration space.

$$H_{ij} = (\mathcal{H}\varphi_j)|_{\mathbf{r}=\mathbf{r}_i}, \qquad R_{ij} = \varphi_j(\mathbf{r}_i).$$

If there are N basis functions it is necessary to choose N points in configuration space so that the matrices are square, but they are not symmetrical and the eigenvalues and eigenvectors may be complex. In practice the imaginary parts are small if the basis functions and sampling points are well chosen, and this provides a criterion for judging the quality of the calculation. Leforestier's method cited above involves using more points than functions; in this case there are more equations than unknowns and they are solved by a least-squares procedure. The increased number of points allows for higher accuracy but the computational effort remains modest.

12.3.4 *Diffusion Monte Carlo*

The quantum Monte Carlo method for solving the Schrödinger equation was first set out by Anderson (1975, 1976, 1980), though it had been discussed much earlier. More recent reviews have been given by Suhm and Watts (1991) and Anderson (1995), who discuss the technical details involved in making the method work efficiently and accurately.

The time-dependent Schrödinger equation for the motion of the nuclei is, in atomic units,

$$\mathrm{i}\frac{\partial \Psi}{\partial t} = -\sum_k \nabla_k^2 \Psi + V\Psi. \tag{12.3.10}$$

where V is the potential energy. If we change variables, setting $\tau = \mathrm{i}t$, this takes the form of a diffusion equation:

$$\frac{\partial \Psi}{\partial \tau} = \sum_k \nabla_k^2 \Psi - V\Psi. \tag{12.3.11}$$

Any linear combination of the exact stationary-state wavefunctions is a solution of eqn (12.3.10):

$$\Psi = \sum_n \varphi_n \exp(-\mathrm{i}t(E_n - E_R)),$$

where E_R is an arbitrary reference energy, and when expressed in terms of the imaginary time variable τ this becomes a sum of decaying functions, provided that we choose E_R so that all the $E_n - E_R$ are positive:

$$\Psi(\mathbf{r}, \tau) = \sum_n \varphi_n \exp(-(E_n - E_R)\tau).$$

The ground-state term has the smallest E_n and so decays most slowly, so we can start from an arbitrary initial guess and follow the diffusion process until all the other terms have disappeared. Ideally we wish to choose the E_R equal to E_0, and then the ground state term will not decay at all. Initially we do not know E_0; the object of the calculation is usually to find it.

The procedure that is used to solve eqn (12.3.11) is a type of random walk; hence the term 'Monte Carlo'. An initial guess at the wavefunction is made by choosing a large set of points, sometimes called psi-particles or 'psips', in configuration space; the local density of the points is proportional to the wavefunction. In the absence of the potential-energy term, the wavefunction evolves by diffusion: a delta function $\prod_k \delta(\mathbf{r}_k - \mathbf{r}_k(0))$ becomes, after (imaginary) time τ, a gaussian distribution:

$$U(\mathbf{r}, \tau) = \prod_k (4\pi D_k \tau)^{-3/2} \exp\left[-\frac{(\mathbf{r}_k - \mathbf{r}_k(0))^2}{4D_k\tau}\right], \tag{12.3.12}$$

where $D_k = 1/2m_k$ is an effective diffusion coefficient for nucleus k. This can be modelled by making each point follow a random walk: each point is displaced randomly in configuration space at each step of the process, with a probability distribution given by eqn (12.3.12). The influence of the potential energy is then taken into account for a point at $\mathbf{r}$ by either destroying or replicating it according to a random number whose probability density is proportional to $\exp[E_R - V(\mathbf{r})]$. That is, points in regions of high potential energy are likely to be destroyed and points in regions of low potential energy are likely to be replicated. The replicas then evolve independently of their parents.

The attraction of this method is that no integrations are needed; all that is required is the value of the potential energy at randomly generated points in configuration space. Consequently the intermolecular potential can be as elaborate as necessary, and need not be constrained into mathematically convenient but physically unsound forms. For example, calculations have been carried out on the water dimer by Gregory and Clary (1994) using the potential of Millot and Stone (1992) (see p. 177), and the tunnelling splittings so obtained were in good agreement with more conventional calculations using the same potential (Althorpe and Clary 1994). Because the method is statistical there are errors, which decrease only slowly ($\propto N^{-1/2}$) as the number of steps in the calculation is increased; but reliable estimates of the errors can be obtained, and there are methods of reducing the error by biasing the calculation using an approximate wavefunction obtained by some other method (Suhm and Watts 1991). It is also possible to reduce the error by eliminating the internal motions of the molecules from the calculation and using only the intermolecular coordinates; this is the rigid body diffusion Monte Carlo or RBDMC method (Buch 1992).

The main limitation is the difficulty of handling excited states. States whose symmetry is different from the ground state can be handled relatively easily by the fixed-node method; that is, by destroying any diffusing point that crosses a nodal surface. This ensures that the wavefunction goes to zero at the nodal surface. States with the same symmetry as the ground state or some other lower-energy state are much more difficult to handle, though methods have been suggested (Suhm and Watts 1991). An important variant of the fixed-node method is the method of correlated sampling for calculating small energy differences such as tunnelling splittings (Wells 1985). When a system tunnels through a symmetrical barrier, as in the donor–acceptor inversion of HF dimer, there are two wavefunctions, one symmetric and one antisymmetric with respect to the barrier, but both very similar except for sign everywhere except in the barrier region. If both states are described using the same set of diffusing points, the statistical uncertainties are strongly correlated and the energy difference can be calculated much more accurately than the absolute energy. The method has been used to calculate tunnelling splittings in water dimer and trimer (Gregory and Clary 1995*a*). It is possible to include three-body terms, both the non-additive induction energy and the triple-dipole dispersion, in the case of the trimer (Gregory and Clary 1995*b*); these terms would pose extremely difficult problems in evaluation of matrix elements if a conventional basis-set expansion were used, even if the dimension of the problem were not prohibitive.

12.4 Molecular beam scattering

In molecular-beam spectroscopy, a single beam is used; if a mixed complex is to be studied, a mixture of gases is fed to the nozzle. An alternative way to study molecular interactions is to use two crossed beams, one for each component of the complex, and to study the scattering of the molecules following collisions at the crossing-point. The scattering is most conveniently studied mathematically in the centre-of-mass coordinate system, which moves relative to the laboratory frame, so a transformation between these coordinate systems is necessary; see, for example, the book by Murrell and Bosanac (1989). In the centre-of-mass frame the position of one molecule relative to the other is described by a distance R and an angle ϑ (Fig. 12.2), and we can think of a particle of mass μ, the reduced mass of the system, being scattered from a fixed point at the origin. A classical scattering event begins with R large and $\vartheta = \pi$; as the molecules collide R reaches a minimum value R_c, and then the molecules separate again to large R along a path at

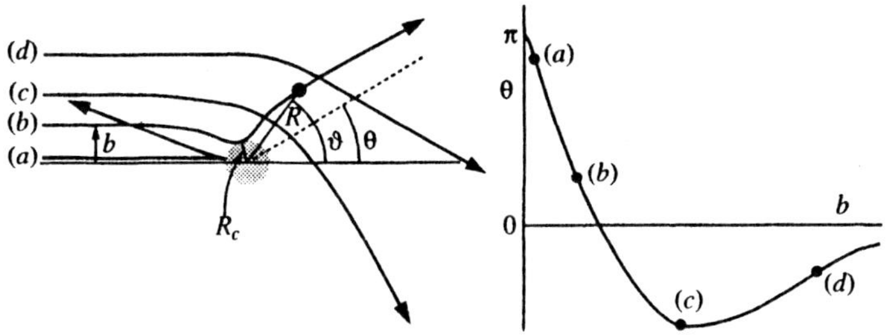

FIG. 12.2. (Left) Classical scattering trajectories for several values of the impact parameter b. (Right) Scattering angle as a function of impact parameter

$\vartheta = \theta$. The parameters describing the collision of two atoms are the kinetic energy E and the 'impact parameter' b which is the distance by which the centres of mass would miss each other if there were no interaction. The angular momentum of the system is related to E and b: its magnitude is $L = b\mu v = b\sqrt{2\mu E}$, where v is the initial relative velocity and μ the reduced mass. The collision angle θ is therefore a function of E and b. For a purely repulsive potential, $\theta = \pi$ when $b = 0$; that is, in a head-on collision the atoms rebound along their approach paths. As b increases, the scattering angle decreases from π, and is zero when b is so large that the atoms do not influence each other.

If the atom–atom potential has an attractive long-range part, as is usual, then it is possible for θ to become negative for some values of b; that is, the trajectory is pulled towards the origin instead of being repelled from it. Since the limit for large b is still $\theta = 0$, the deflection angle reaches a minimum for some value of b and then increases again to zero. The scattering intensity near the minimum value of θ is high, because a range of values of b all contribute to the scattering near this angle; this effect is called a rainbow, because it is analogous to the refraction of sunlight in rain droplets that causes an optical rainbow. It occurs when the trajectory passes through the deepest part of the attractive well without experiencing the repulsive wall, and therefore the position of the rainbow gives information about the depth of the well (Mason 1957).

The observed scattering (still in the centre-of-mass coordinate system) is described by the *cross-section* σ. A particle will be scattered into a direction between θ and $\theta + d\theta$ if its impact parameter falls in some range b to $b + db$. The area of a cross-section through the incoming beam corresponding to this range of b values is $2\pi b\,db$, so the flux of particles is $F.2\pi b\,db$, where F is the incident flux per unit area. The corresponding outgoing flux is spread over solid angle $2\pi \sin\theta\,d\theta$, so the outgoing flux per unit solid angle for unit incident flux per unit area is

$$\sigma(\theta) = 2\pi b\,db/2\pi\sin\theta\,d\theta = \left|\frac{b}{\sin\theta d\theta/db}\right|; \qquad (12.4.1)$$

the modulus sign appears because it is immaterial whether θ increases or decreases as b is increased. $\sigma(\theta)$ is called the *differential cross-section*.

In an observation we cannot distinguish between trajectories like (b) and (d) in Fig. 12.2, for which the scattering angles are θ_b and $\theta_d = -\theta_b$. Similarly, a scattering angle of

θ is indistinguishable from $\theta + 2\pi$, where the trajectory orbits the origin. The experimental scattering angle Θ is obtained from the microscopic scattering angle θ by constructing $\Theta = |\theta - 2n\pi|$, choosing n so that the result is between 0 and π. The experimental differential cross-section is a sum over all the values of b that scatter into a given experimental angle Θ:

$$\sigma(\Theta) = \sum_i \left| \frac{b_i}{\sin\theta (d\theta/db)_i} \right|. \tag{12.4.2}$$

The differential cross-section is a function of the collision energy E, so a further integration over the collision energy profile is needed to yield the observed property, though it is sometimes possible to carry out time-of-flight measurements to isolate the scattering at a particular collision energy. The integral of the differential scattering over all directions is the total scattering cross-section:

$$\sigma = 2\pi \int_0^\pi \sigma(\Theta) \sin\Theta \, d\Theta.$$

The differential cross-section (12.4.2) is singular when either $(d\theta/db) = 0$ or $\sin\theta = 0$. In practice, because of the variability of the collision energies, one just sees a maximum in the differential cross-section. When $(d\theta/db) = 0$ this is called a rainbow, as already discussed; when $\sin\theta = 0$ it is known as a *glory*, again by analogy with the atmospheric phenomenon.

It will be apparent that the scattering data do not provide information directly about the intermolecular potential. Rather it is necessary, as for other experimental properties, to postulate some form of interatomic potential and to calculate the differential cross-section from it, modifying the potential until agreement with the observations is achieved. In a few cases the classical cross-section can be calculated analytically; alternatively it can be found by constructing a large number of collision trajectories by solving the classical equations of motion.

For atoms, the potential depends only on the centre-of-mass separation R, and atomic beam scattering provides useful information about it. For molecules, as usual, the problem is complicated by the presence of rotational and vibrational degrees of freedom, and trajectory calculations have to sample a range of initial conditions for these degrees of freeedom as well as for the collision parameter and the energy.

A further complication is that classical mechanics is often inadequate; this is so especially for light atoms and molecules and for molecules with small moments of inertia. In such cases interference effects appear, for example between scattered waves at a particular Θ that arise from different values of b. The possibility of inelastic scattering also arises: the collision may transfer energy from the translational motion into rotational or vibrational degrees of freedom of the molecules. At very low collision energies, and with the molecules in their ground states, no such transfer is possible, because there is not enough energy to excite the molecules, but as the collision energy is increased, more and more excited states become accessible. Each possible internal state of the scattered molecules is called a *channel* in the language of scattering theory, and as the collision energy increases more channels open.

The quantum-mechanical solution of the scattering problem starts from the same point as the calculation of spectra, namely the coupled-channel equations—eqn (12.3.6) for the scattering of an atom from a diatomic. In the scattering problem, however, the collision energy E can take any positive value. The standard procedure is to integrate the equations outwards from small R, where the wavefunction is zero because of the large repulsion energy, to large R, where the wavefunctions become plane waves. The difficulty with this is the same as in the spectroscopic calculation: there are usually too many possible states or channels to consider. At long range only the open channels can contribute to the wavefunction, but at short range the closed channels may also contribute. It is always necessary to truncate the set of coupled-channel equations, but further approximations are usually needed. It is not necessary here to explore the details; they can be found, for instance, in the book by Murrell and Bosanac (1989). They also give an account of semi-classical methods in scattering theory, described in more detail by Child (1991); such methods can be used for simple atomic scattering problems but unfortunately they are not easily applied to systems with rotational and vibrational degrees of freedom.

An example of the use of molecular beam scattering to verify the accuracy of a calculated potential is the case of $H_2 \cdots D_2$ (Buck *et al.* 1983). Here an accurate *ab initio* potential was used in close-coupling calculations to obtain the differential cross-sections, not only for elastic scattering, but also for inelastic scattering in which one or both molecules were excited to a higher rotational state. These different types of scattering event can be distinguished by time-of-flight techniques: following inelastic scattering where the molecules become excited, they have less translational energy and so travel more slowly and reach the detector later than elastically scattered molecules. In this case the agreement between experiment and theoretical calculation was complete, so the measurements did not provide new information about the potential but merely validated the calculation.

Usually it is not possible to calculate the potential so accurately. An example of the use of molecular beam scattering to study intermolecular potentials involving a polyatomic molecule is the study by Yang and Watts (1994) of the interactions between acetylene and Ar, H_2 and N_2. In the case of $Ar \cdots C_2H_2$, a potential determined by Thornley and Hutson (1992) was used to calculate the differential cross-section, and was found to give the rainbow peak in the wrong place, at about 9° in laboratory coordinates instead of 7° as observed. The rotational spectra from which the potential was derived depend mainly on the vibrationally-averaged structure, and are not sensitive to the depth of the well, which, as we have seen, is the main factor determining the position of the rainbow angle. Yang and Watts were able to reproduce the differential cross-section using a different potential with several adjustable parameters, but this is not necessarily a better potential since it is fitted only to the scattering and takes no account of the spectroscopic data. Clearly the most satisfactory procedure for determining intermolecular potentials is to use experimental data from as many sources as possible, so as to take advantage of the sensitivity of different measurements to different features of the potential.

The conclusion is that molecular beam scattering data can provide useful information about potentials, and in principle a given potential can be used to calculate scattering cross-sections for comparison with experimental data. For practical purposes, however, the potential often needs to be described by a single-site spherical-harmonic expansion, and this faces the same convergence limitations as in calculations of spectra. Consequently it is difficult to study molecules unless they are 'nearly spherical'—that is, unless any anisotropies can be described by just a few terms of a spherical harmonic expansion.

12.5 Measurements on liquids and solids

Experiments on condensed phases can also provide information about intermolecular potentials, but they are complicated by the effects of the many-body terms. It is usual to attempt to describe the properties of liquids and solids in terms of an effective pair potential that includes the many-body effects in some averaged fashion.

12.5.1 *Crystals*

Crystal structures can provide quite direct information about some aspects of the potential. The interatomic distance in a simple solid such as that of an inert gas is close to the equilibrium separation R_m of the pair potential, though it is not quite the same because the attractions between more distant pairs of atoms tend to make the interatomic distance slightly shorter, while the repulsive effect of the triple-dipole dispersion tends to increase it. Similarly the lattice energy per atom is close to the value of $-\frac{1}{2}n\varepsilon$ obtained by counting nearest-neighbour pair interactions only. (n is the number of nearest neighbours of each atom.) However this too is modified by the same effects as the interatomic distance, and also by the zero-point energy associated with the lattice vibrations. For any given potential and crystal structure it is easy to calculate the optimum interatomic separation. The corresponding lattice energy can also be calculated, though it is rather more difficult to determine accurately because of the need to estimate the zero-point energy. The lattice vibration frequencies can however be determined from the potential, and provide a further point of comparison with experiment, both through direct spectroscopic measurement and via measurements of the heat capacity. Spectroscopy normally yields only the $\mathbf{k} = 0$ vibration frequency, in which the molecules in different unit cells all move in phase.

The structures of molecular crystals are sensitive to details of the potential and are surprisingly difficult to predict. The case of Cl_2 has already been mentioned (p. 165) as one where the anisotropy of the repulsion and electrostatic were needed for a successful prediction of the structure. Willock *et al.* (1993, 1995) have studied crystal structures of molecules such as *m*-nitroaniline, which have large values for the hyperpolarizability β, and so may be useful as media for frequency-doubling laser light. However it is not enough for the molecules to be non-centrosymmetric; the crystal structure must be non-centrosymmetric too. Consequently it is important to be able to predict whether a molecule with a high hyperpolarizability forms a centrosymmetric or non-centrosymmetric crystal. Willock *et al.* found that conventional point-charge models were inadequate for this purpose, and that accurate distributed-multipole models were needed, because the electrostatic effects of the lone pair and pi-electron density make a qualitative difference to the crystal packing.

The use of anisotropic site–site models to describe the intermolecular potentials in crystals calls for the development of new techniques, as in other applications. The optimization of the structure requires the first and second derivatives of the potential, and these can be obtained as described in §11.7; further details of the application to the crystal structure problem are given by Willock *et al.* (1995). One feature that requires attention is that the interaction with other molecules involves a summation over all the unit cells of the lattice, out to infinity. For terms with R^{-n} distance dependence this summation is only conditionally convergent when $n \leq 3$, i.e., for charge–charge, charge–dipole and dipole–dipole interactions. Here it is necessary to use the Ewald summation technique (Allen and Tildesley 1987); its original application was to charge–charge interactions (Ewald 1921), but the extension to charge–dipole and dipole–dipole interactions has been described by Smith (1982*a*, 1982*b*).

12.5.2 *Neutron and X-ray scattering*

Both X-ray and neutron scattering experiments on crystals are used to determine the structure. In liquids, scattering experiments also give information about the structure, but it is less well defined. The experiments yield the structure factor $S(\mathbf{k})$, which is related to the intensity of the scattering at wave-vector $\mathbf{k}$. (For particles of wavelength λ scattered through an angle α the wave-vector is $(4\pi/\lambda)\sin\frac{1}{2}\alpha$.) The structure factor is the Fourier transform of the radial distribution function $g(r)$, which is the probability density for finding a second atom at distance r from a given first atom, and the Fourier transform can be inverted to derive the distribution function from the structure factor. In the case of molecules, the scattering is a superposition of contributions from pairs of atoms, including scattering by pairs of atoms in the same molecule. With neutron scattering, the amplitude of the scattering is usually very different for different isotopes, so isotopic substitution can be used to separate the structure factors for different types of atom, and hence their radial distribution functions.

The radial distribution function and the structure factor involve an average over the ensemble, so as with many other properties it is not possible to derive much information about the potential directly from the measurement. The positions of the peaks in the radial distribution function correspond to favoured atom–atom distances, which gives a guide to the value of R_m, but otherwise it is necessary, as with so many properties, to postulate a potential and calculate the experimental property from it as a test. In the case of X-ray and neutron scattering this calculation is almost always carried out by a simulation technique, as discussed below.

12.6 Simulation methods

Some properties of liquids can be calculated by the methods of statistical mechanics, but it is often necessary to make approximations. The prediction of properties such as liquid structure from a given potential cannot be done by analytical methods at all. It is now usual to calculate such properties by simulation methods, in which a collection of molecules in an imaginary box is allowed to interact under the influence of a pair potential. A random configuration of the molecules is chosen initially, and then either a sequence of new configurations is generated randomly (Monte Carlo method) or the classical equations of motion of the system are integrated numerically (molecular dynamics method). In either case there is an initial equilibration, after which a sequence of configurations is collected and the appropriate average is calculated.

It is usual to apply 'periodic boundary conditions', in which the imaginary box of molecules is replicated infinitely many times in all directions, so that the box can be considered as being immersed in an infinite medium. The calculation is still only carried out with the original finite box, but molecules leaving at one side are replaced by their images at the opposite side, so the density remains constant; and molecules near a side of the box interact with images of molecules at the opposite side, so there are no surface effects. There may however be artifacts arising from the periodicity of the arrangement.

In the Monte Carlo treatment, it is usual to use the Metropolis method of sampling. Here a new configuration is obtained randomly, usually by changing some of the coordinates by small amounts, and its energy is computed. If the new energy is lower than the previous energy the configuration is retained, while if it is higher by an amount ΔE it is retained with probability $\exp(-\Delta E/kT)$ and otherwise rejected. A retained configuration becomes the reference for the next trial. It can be shown that this procedure leads to a Boltzmann distribution for temperature T if continued for long enough. The retained configurations are used as the basis for the averaging.

The Monte Carlo treatment deals only with static properties such as structure; it provides no information about molecular motion. That has to be obtained by molecular dynamics.

Simulation methods have become a very large field of study, and a detailed description is beyond the scope of this book. The book by Allen and Tildesley (1987) provides an excellent introduction to the field. It has been conventional for many years to use extremely simple potentials in both molecular dynamics and Monte Carlo simulations, because each step of the calculation calls for the calculation of the energy of the system, and in the case of molecular dynamics, of the force and torque on each molecule too. The more realistic potentials described in this book lead to more complicated expressions for the forces and torques as well as for the energy, but it is a misconception to suppose that there is an inordinate increase in computer time. The greater complication of the calculation is balanced by the fact that fewer sites are needed when anisotropic site–site potentials are used. Rodger *et al.* (1987) found that in molecular dynamics simulations of liquid chlorine, the computation time per step for an anisotropic two-site potential was almost the same as would be needed for a three-site Lennard-Jones potential. Since the anisotropic potential can provide a much better description (Rodger *et al.* 1988*a*), it is clearly to be preferred. Some discussion of the technical issues involved in using such potentials in molecular dynamics calculations may be found in a further paper (Rodger *et al.* 1992).

A recent development in simulation methods has been the introduction of the Car–Parrinello technique (Car and Parrinello 1985), in which the motion of the nuclei is handled classically, but the energy and forces at each configuration of the system are found by solving the electronic problem 'on the fly'. A good account of the method is given by Remler and Madden (1990). At the moment the method is principally used for systems in which the electronic structure is sensitive to the nuclear geometry—generally ones with loosely bound valence electrons, the time may come when computer power is sufficiently great for the method to be applied more generally. Laasonen *et al.* (1993) have already shown that it can be used in a simulation of liquid water, though as yet with only a small number of molecules. If such methods become widespread, the potential energy surface will become redundant as an intermediate step between the electronic wavefunction calculation and the molecular dynamics calculation. As a means of understanding the way that molecules behave, however, it will continue to be valuable for a long time yet.

APPENDIX A

CARTESIAN TENSORS

The essential feature of tensor algebra is that it expresses physical ideas in a form which is independent of coordinate system. It is then often unnecessary to specify a coordinate system at all, and when one is needed, it can be chosen to simplify the treatment of the problem. The description of a physical problem in terms of tensors has two further virtues:

- It is very compact, making the essential features of the mathematics more apparent;
- It is immediately obvious from the form of an expression whether it has the correct (e.g. scalar or vector) behaviour.

This is a condensed summary of the basic ideas. For a fuller account see, for example, Jeffreys (1931).

A.1 Basic definitions

A *scalar* is a quantity described by a single number, whose value is the same in all coordinate systems. Example: energy.

A *vector* is a quantity described by 3 numbers (labelled by a subscript taking values 1, 2 or 3 (or x, y, z)) whose values in one (primed) coordinate system are related to those in a rotated (unprimed) system by

$$v_{\alpha'} = \sum_{\alpha=1}^{3} a_{\alpha'\alpha} v_{\alpha}. \tag{A.1.1}$$

Examples are velocity; angular momentum; dipole moment. The quantities $a_{\alpha'\alpha}$ form an orthogonal matrix A describing the rotation. Consequently

$$\begin{aligned} \sum_{\alpha=1}^{3} a_{\alpha'\alpha} a_{\beta'\alpha} &= \delta_{\alpha'\beta'} \qquad (AA^T = I \text{ in matrix notation}), \\ \sum_{\alpha'=1}^{3} a_{\alpha'\alpha} a_{\alpha'\beta} &= \delta_{\alpha\beta} \qquad (A^T A = I). \end{aligned} \tag{A.1.2}$$

The coefficient $a_{\alpha'\alpha}$ is the cosine of the angle between the α' and α axes. In the equivalent matrix notation, T denotes the transpose.

It is customary to use the *repeated-suffix summation convention*, due originally to Einstein, according to which any subscript appearing twice in one term is automatically summed over. Thus (A.1.2) can be written

$$a_{\alpha'\alpha} a_{\beta'\alpha} = \delta_{\alpha'\beta'}, \tag{A.1.3a}$$

$$a_{\alpha'\alpha}a_{\alpha'\beta} = \delta_{\alpha\beta}. \tag{A.1.3b}$$

No subscript may occur more than twice in a term. Thus the expression $(\mathbf{r}\cdot\mathbf{s})^2$ may not be written as $r_\alpha s_\alpha r_\alpha s_\alpha$ (because of the possibility of ambiguity with $(\mathbf{r}\cdot\mathbf{r})(\mathbf{s}\cdot\mathbf{s})$) but must be written as, e.g., $r_\alpha s_\alpha r_\beta s_\beta$. α and β here are *dummy suffixes*; since they are summed over, the actual symbol used is irrelevant. α and β in (A.1.3b), on the other hand, are *free suffixes*; each such suffix must occur precisely once in each term of the equation, and the equation holds for each value (1, 2 or 3) that they can take.

A *tensor* of rank n is a quantity described by 3^n numbers (labelled by n subscripts each taking the value 1, 2 or 3). Tensors of rank 2 often describe the relationship between two vectors; e.g. the polarizability $\alpha_{\xi\eta}$ describes the dipole μ_ξ induced by an electric field F_η:

$$\mu_\xi = \alpha_{\xi\eta}F_\eta. \tag{A.1.4}$$

In order that such equations remain valid in any coordinate system, it is necessary that the values of tensor components in the primed coordinate system are related to those in the unprimed system by

$$T_{\alpha'\beta'\ldots\nu'} = a_{\alpha'\alpha}a_{\beta'\beta}\ldots a_{\nu'\nu}T_{\alpha\beta\ldots\nu}. \tag{A.1.5}$$

It follows from (A.1.2) that the converse relationship holds:

$$T_{\alpha\beta\ldots\nu} = a_{\alpha\alpha'}a_{\beta\beta'}\ldots a_{\nu\nu'}T_{\alpha'\beta'\ldots\nu'}. \tag{A.1.6}$$

[Proof: substitute (A.1.5) into the R.H.S of (A.1.6) and use (A.1.2).] We see that scalars and vectors are tensors of rank 0 and 1 respectively. Note that the order of subscripts on $a_{\alpha\alpha'}$ is immaterial: $a_{\alpha\alpha'}$ is the same as $a_{\alpha'\alpha}$, being the direction cosine between the two axes concerned. Thus $a_{1'2} = a_{21'}$, but neither of course is the same as $a_{12'} = a_{2'1}$.

However the order of the subscripts on the tensor itself is usually significant; that is, $T_{\alpha\beta\gamma\ldots\nu} \neq T_{\beta\alpha\gamma\ldots\nu}$ in general. If in a particular case $T_{\alpha\beta\gamma\ldots\nu} = T_{\beta\alpha\gamma\ldots\nu}$ for any choice of coordinate system, and for all values of the remaining suffixes $\gamma\ldots\nu$, then the tensor T is said to be *symmetric* with respect to the first two subscripts.

Addition and *subtraction* of tensors is straightforward: the quantity

$$W_{\alpha\beta\ldots\nu} = T_{\alpha\beta\ldots\nu} + U_{\alpha\beta\ldots\nu} \tag{A.1.7}$$

is a component of a tensor W which is the sum of the tensors T and U.

The *outer product* of two tensors is obtained by multiplying components without summing:

$$X_{\alpha\beta\ldots\lambda\mu\nu\pi\ldots\zeta} = T_{\alpha\beta\ldots\lambda\mu}U_{\nu\pi\ldots\zeta}, \tag{A.1.8}$$

and is readily shown to be a tensor (use (A.1.5)), with a rank which is the sum of the ranks of the factors.

Contraction involves setting two subscripts equal and performing the sum implied by the notation; e.g.

$$Y_{\gamma\delta\ldots\nu} = T_{\alpha\alpha\gamma\delta\ldots\nu} \equiv \sum_\alpha T_{\alpha\alpha\gamma\delta\ldots\nu}. \tag{A.1.9}$$

It is easy to show that if T is a tensor of rank n, then Y is a tensor of rank $n-2$. (Use (A.1.5) and (A.1.2).) Applying this procedure to an outer product, taking the contracted suffices one from each factor, yields an *inner product*, e.g.

$$Z_{\alpha\beta\ldots\lambda\pi\ldots\zeta} = X_{\alpha\beta\ldots\lambda\mu\mu\pi\ldots\zeta} = T_{\alpha\beta\ldots\lambda\mu}U_{\mu\pi\ldots\zeta}. \tag{A.1.10}$$

Clearly Z is a tensor: it is a contraction of X which is a tensor, as stated above. It is also true that if (A.1.10) holds in all coordinate systems, and Z and T are known to be tensors, then it follows that U is a tensor (*quotient rule*).

The simplest case of an inner product is the scalar product of two vectors: for example $s = u_\alpha v_\alpha = \mathbf{u}\cdot\mathbf{v}$. It is a quantity of rank 0, i.e., a scalar—it is independent of coordinate system. Another common case is the vector formed by taking the inner product of a second-rank tensor with a vector, as in eqn (A.1.4) above.

A.1.1 *Isotropic tensors*

Normally the application of (A.1.5) yields a set of numbers, describing the tensor in the new axes, which are quite different from those describing it in the old axes—e.g. $T_{1'1'\ldots1'} \neq T_{11\ldots1}$. However some tensors retain the same numerical values in all axis systems, and are called *isotropic*. Apart from scalar multiplicative factors (scalars being always isotropic by definition) the important isotropic tensors are the *Kronecker* tensor:

$$\delta_{\alpha\beta} = \begin{cases} 1 & \text{when } \alpha = \beta, \\ 0 & \text{otherwise,} \end{cases} \tag{A.1.11}$$

and the *Levi-Civita* tensor:

$$\varepsilon_{\alpha\beta\gamma} = \begin{cases} 1 & \text{when } \alpha\beta\gamma \text{ is a cyclic permutation of 123,} \\ -1 & \text{when } \alpha\beta\gamma \text{ is a cyclic permutation of 321,} \\ 0 & \text{otherwise.} \end{cases} \tag{A.1.12}$$

There are no other isotropic tensors of rank less than 4; moreover all isotropic tensors of any rank can be expressed in terms of outer products of δ's and ε's.

Notice that $\delta_{\alpha\beta}$ behaves like a subscript substitution operator: in $\delta_{\alpha\beta}t_{\beta\gamma}$, for example, there is only one term in the implied sum over β that does not vanish, namely the one for which $\beta = \alpha$. Consequently $\delta_{\alpha\beta}t_{\beta\gamma} = t_{\alpha\gamma}$. Notice also that $\delta_{\alpha\alpha} = 3$.

An important use of the Levi-Civita tensor occurs in the *vector product* of two vectors; in tensor notation $\mathbf{v} = \mathbf{r}\wedge\mathbf{s}$ is $v_\alpha = \varepsilon_{\alpha\beta\gamma}r_\beta s_\gamma$.

A.1.2 *Polar and axial tensors*

The transformation rule in eqn (A.1.1) or (A.1.5) is appropriate only for proper rotations of the axes. For improper rotations, which involve a reflection and change the handedness of the coordinate system, two possibilities arise. Either the rule in (A.1.5) may apply as it stands, in which case we have a *polar* tensor, or there may be an additional change of sign for improper rotations, in which case we have an *axial* tensor. Eqn (A.1.5) may be written in a more general form to take account of these possibilities:

$$T_{\alpha'\beta'\ldots\nu'} = \det(a)^p a_{\alpha'\alpha}a_{\beta'\beta}\ldots a_{\nu'\nu}T_{\alpha\beta\ldots\nu}, \tag{A.1.13}$$

where $\det(a)$ is the determinant of the matrix whose elements are the direction cosines $a_{\alpha'\alpha}$, and p is 0 for a polar tensor and 1 for an axial one. Since $\det(a)$ is $+1$ for a proper rotation and -1 for an improper one, this gives the extra change of sign for an axial tensor under improper rotation.

It follows that any product, outer or inner, of two polar tensors or two axial tensors is polar, while the product of a polar tensor with an axial tensor is axial. The fundamental axial tensor is the Levi-Civita tensor, which is axial by definition in order that $\varepsilon_{123} = +1$ in left-handed coordinate systems as well as right-handed ones. Therefore the vector product of two polar tensors is axial. The prime example of this is angular momentum: $\mathbf{l} = \mathbf{r} \wedge \mathbf{p}$ is the vector product of the polar vectors $\mathbf{r}$ and $\mathbf{p}$ describing position and momentum, expressed in tensor notation as $l_\alpha = \varepsilon_{\alpha\beta\gamma} r_\beta p_\gamma$.

APPENDIX B

SPHERICAL TENSORS

Here we summarize the basic concepts of the spherical tensor formalism. This is not intended as an explanatory account, but merely as a summary of the principal formulae and definitions that we need. A fuller account may be found in one of the many textbooks on angular momentum theory, such as Brink and Satchler (1993) or Zare (1988).

B.1 Spherical harmonics

Recall that the spherical harmonics, usually denoted $Y_{lm}(\theta,\varphi)$, are the functions of θ and φ that satisfy the eigenvalue equations for the angular momentum operators:

$$\hat{\mathbf{L}}^2 Y_{lm} = \hbar^2 l(l+1) Y_{lm},$$
$$\hat{L}_z Y_{lm} = \hbar m Y_{lm}.$$

They are normalized so that $\int |Y_{lm}|^2 \sin\theta\, \mathrm{d}r\mathrm{d}\theta\,\mathrm{d}\varphi = 1$, and because they are eigenfunctions of the hermitian operators $\hat{\mathbf{L}}^2$ and $\hat{L}_z$, they are orthogonal:

$$\int Y_{lm}^* Y_{l'm'} \sin\theta\, \mathrm{d}r\mathrm{d}\theta\,\mathrm{d}\varphi = \delta_{ll'}\delta_{mm'}.$$

The angular momentum operator $\hat{\mathbf{L}} = \mathbf{r}\times\hat{\mathbf{p}}$ has components defined by

$$\begin{aligned}
\hat{L}_x &= yp_z - zp_y = -\mathrm{i}\hbar\left(y\frac{\partial}{\partial z} - z\frac{\partial}{\partial y}\right),\\
\hat{L}_y &= zp_x - xp_z = -\mathrm{i}\hbar\left(z\frac{\partial}{\partial x} - x\frac{\partial}{\partial z}\right),\\
\hat{L}_z &= xp_y - yp_x = -\mathrm{i}\hbar\left(x\frac{\partial}{\partial y} - y\frac{\partial}{\partial x}\right).
\end{aligned} \qquad \text{(B.1.1)}$$

In spherical polar coordinates, $\hat{L}_z = -\mathrm{i}\hbar\partial/\partial\varphi$, while

$$\hat{\mathbf{L}}^2 = -\hbar^2\left(\frac{1}{\sin\theta}\frac{\partial}{\partial\theta}\sin\theta\frac{\partial}{\partial\theta} + \frac{1}{\sin^2\theta}\frac{\partial^2}{\partial\varphi^2}\right).$$

The operators $\hat{L}_\pm = \hat{L}_x \pm \hat{L}_y$ are important in the theory. The spherical harmonics satisfy

$$\hat{L}_\pm Y_{lm} = +\sqrt{l(l+1) - m(m\pm 1)}\,\hbar Y_{l,m\pm 1}, \qquad \text{(B.1.2)}$$

so that $\hat{L}_+$ and $\hat{L}_-$ have the effect of shifting the eigenvalue m up or down by one unit. For this reason they are called shift or ladder operators.

It is more convenient for our purposes to use renormalized spherical harmonics, first defined by Racah, which differ from the Y_{lm} by a constant factor:

$$C_{lm}(\theta,\varphi) = \sqrt{\frac{4\pi}{2l+1}}\,Y_{lm}(\theta,\varphi).$$

These evidently satisfy the same eigenvalue equations, but their normalization is different: they satisfy $C_{l0}(0,0) = 1$. The use of these functions avoids the factors of $\sqrt{4\pi}$ that otherwise clutter up the equations. The explicit form of the spherical harmonics is derived in some elementary texts on quantum mechanics, and is

$$C_{lm}(\theta,\varphi) = \varepsilon_m \left[\frac{(l-|m|)!}{(l+|m|)!}\right]^{1/2} P_l^m(\cos\theta)\mathrm{e}^{im\varphi}, \tag{B.1.3}$$

where $P_l^m(\cos\theta)$ is an associated Legendre function (Abramowitz and Stegun 1965, Zare 1988), and the phase factor ε_m, important for maintaining the phase relationships required by eqn (B.1.2), is $(-1)^m$ for $m > 0$ and 1 for $m \le 0$.

We can also define related functions called the regular and irregular spherical harmonics:

$$\begin{aligned} R_{lm}(\mathbf{r}) &= r^l C_{lm}(\theta,\varphi), \\ I_{lm}(\mathbf{r}) &= r^{-l-1} C_{lm}(\theta,\varphi), \end{aligned} \tag{B.1.4}$$

where r, θ and φ form the spherical polar representation of the vector argument $\mathbf{r}$. These functions satisfy Laplace's equation: $\nabla^2 R_{lm} = 0$ everywhere, while $\nabla^2 I_{lm} = 0$ except at the origin. This is easily demonstrated using the fact that

$$\nabla^2 = \frac{1}{r^2}\frac{\partial}{\partial r}r^2\frac{\partial}{\partial r} - \frac{\hat{\mathbf{L}}^2}{\hbar^2 r^2}.$$

The first few of the regular spherical harmonics are

$$\begin{aligned} R_{00}(\mathbf{r}) &= 1, \\ R_{10}(\mathbf{r}) &= z = r\cos\theta, \\ R_{11}(\mathbf{r}) &= -\sqrt{\tfrac{1}{2}}(x+iy) = -\sqrt{\tfrac{1}{2}}r\sin\theta\mathrm{e}^{\mathrm{i}\varphi}, \\ R_{1,-1}(\mathbf{r}) &= \sqrt{\tfrac{1}{2}}(x-iy) = \sqrt{\tfrac{1}{2}}r\sin\theta\mathrm{e}^{-\mathrm{i}\varphi}, \\ R_{20}(\mathbf{r}) &= \tfrac{1}{2}(3z^2-r^2) = \tfrac{1}{2}r^2(3\cos^2\theta - 1). \end{aligned}$$

Here x, y and z are the cartesian components of $\mathbf{r}$. As the functions with nonzero m are complex, it is helpful to define real functions R_{lmc} and R_{lms}, for $m > 0$, by

$$\left.\begin{aligned} R_{lmc} &= \sqrt{\tfrac{1}{2}}[(-1)^m R_{lm} + R_{l,-m}], \\ \mathrm{i}R_{lms} &= \sqrt{\tfrac{1}{2}}[(-1)^m R_{lm} - R_{l,-m}], \end{aligned}\right\} \quad m > 0. \tag{B.1.5}$$

This relationship between the real and complex components can be written in the form

$$R_{lm} = \sum_{\kappa} R_{l\kappa} X_{\kappa m}. \tag{B.1.6}$$

Here we use the Greek label κ generically for the real components; for a given value of l it takes the values 0, $1c$, $1s$, ..., lc, ls. The transformation coefficients $X_{\kappa m}$ are zero except for

$$\begin{aligned} X_{00} = 1, \quad X_{mc,m} &= (-1)^m \sqrt{\tfrac{1}{2}}, \quad X_{mc,-m} = \sqrt{\tfrac{1}{2}}, \\ X_{ms,m} &= (-1)^m \mathrm{i} \sqrt{\tfrac{1}{2}}, \quad X_{ms,-m} = -\mathrm{i}\sqrt{\tfrac{1}{2}}. \end{aligned} \tag{B.1.7}$$

The matrix X is unitary.

It is sometimes more convenient to express the relationship between real and complex components differently:

$$\begin{aligned} R_{lmc} &= b_m R_{lm} + b_{\overline{m}} R_{l\overline{m}}, \\ \mathrm{i} R_{lms} &= b_m R_{lm} - b_{\overline{m}} R_{l\overline{m}}, \end{aligned} \tag{B.1.8}$$

where

$$b_m = \begin{cases} (-1)^m \sqrt{\tfrac{1}{2}}, & m > 0, \\ \tfrac{1}{2}, & m = 0, \\ \sqrt{\tfrac{1}{2}}, & m < 0, \end{cases} \tag{B.1.9}$$

and $\overline{m} \equiv -m$. If this is done, there is no need to restrict m in eqn (B.1.8) to be positive, and the definitions ensure that $R_{l\overline{m}c} = R_{lmc}$ and $R_{l\overline{m}s} = -R_{lms}$. We also see that

$$R_{lm} = (R_{lmc} + \mathrm{i} R_{lms})/2b_m, \tag{B.1.10}$$

for all m, and that $R_{l0c} = R_{l0}$ and $R_{l0s} = 0$.

The labels c and s stand for 'cosine' and 'sine' respectively, since R_{lmc} and R_{lms} are proportional to $\cos m\varphi$ and $\sin m\varphi$ respectively. Some of these functions are tabulated in Table E.1. Real irregular spherical harmonics are defined similarly, and can be obtained from the corresponding regular harmonics by dividing by r^{2l+1}. We shall also use real forms C_{lmc} and C_{lms} of the ordinary spherical harmonics.*

B.2 Rotations of the coordinate system

The spherical harmonics Y_{lm} and C_{lm} and the regular and irregular solid harmonics R_{lm} and I_{lm} all depend on the angular coordinates in the same way. Accordingly they all transform under rotations in the same way. A rotation is described by the Euler angles defined in Fig. 1.2 on p. 8, and the effect of such a rotation on a spherical harmonic is

*Some authors use a different notation, writing, for example, Z_{lm} for C_{lmc} and $Z_{l\overline{m}}$ for C_{lms}. This notation has several disadvantages: it requires a new letter, Z instead of C, so that the link with the complex form is less apparent; the form Z_{lm} is not obviously a real rather than complex component; and the regular and irregular harmonics either require two new, less mnemonic, letters than R and I, or must be expressed less compactly as $r^l Z_{lm}$ and $r^{-l-1} Z_{lm}$. The only compensating advantage, if indeed it is an advantage, is the loss of the c or s suffix.

$$R(\alpha,\beta,\gamma)Y_{lm} = \sum_{m'} Y_{lm'} D^l_{m'm}(\alpha,\beta,\gamma), \tag{B.2.1}$$

where the $D^l_{m'm}(\alpha,\beta,\gamma)$ are the *Wigner rotation matrices*.That is, the rotated function $R(\alpha,\beta,\gamma)Y_{lm}$ can be expressed as a linear combination of the original set of functions, with coefficients that are elements of the Wigner rotation matrix. Here we are using the active convention for rotations; that is, the l.h.s. of eqn (B.2.1) describes the function that we obtain by rotating the spherical harmonic Y_{lm} according to the Euler angles α, β and γ, while the r.h.s. is a linear combination of the original set of unrotated functions. The alternative passive convention involves rotating the axes rather than the functions; see Brink and Satchler or Zare. The Wigner rotation matrices are defined by eqn (B.2.1); since $D^l_{m'm}(\alpha,\beta,\gamma)$ is the coefficient of $Y_{lm'}$ in the rotated function $R(\alpha,\beta,\gamma)Y_{lm}$, we can use the orthogonality of the spherical harmonics to write

$$D^l_{m'm}(\alpha,\beta,\gamma) = \int Y^*_{lm'} R(\alpha,\beta,\gamma) Y_{lm} \sin\theta \, d\theta \, d\varphi.$$

Explicit tables of the Wigner rotation matrices are given in Brink and Satchler for $l \leq 2$.

B.3 Spherical tensors

A spherical tensor of rank l is now defined as any set of $2l+1$ quantities, labelled by $m = l, l-1, \ldots, -l$ like the spherical harmonics, for which the same transformation law holds as for the spherical harmonics. That is, if

$$R(\alpha,\beta,\gamma)T_{lm} = \sum_{m'} T_{lm'} D^l_{m'm}(\alpha,\beta,\gamma),$$

where the $D^l_{m'm}$ are the Wigner rotation matrices as in eqn (B.2.1), then the set T_{lm} is a spherical tensor of rank l.

For our purposes, the main importance of this is that the multipole moments, when expressed in spherical tensor form, do satisfy this requirement. However any quantum-mechanical operator can be expressed in spherical-tensor form. For example, the spherical tensor components of a vector operator with cartesian components v_x, v_y and v_z are

$$\begin{aligned} v_{11} &= -\sqrt{\tfrac{1}{2}}(v_x + iv_y), \\ v_{10} &= v_z, \\ v_{1,-1} &= \sqrt{\tfrac{1}{2}}(v_x - iv_y). \end{aligned} \tag{B.3.1}$$

Instead of regarding the $D^l_{mk}(\alpha,\beta,\gamma)$ as matrices, defined for particular values of α, β and γ, it is possible to regard them as functions of α, β and γ for particular values of l, m and k. In this context they are called Wigner functions. In fact it turns out that $[(2l+1)/8\pi^2]^{1/2} D^l_{mk}(\alpha,\beta,\gamma)^*$ is the normalized rotational wavefunction of a symmetric top whose orientation is described by the Euler angles α, β and γ and which has angular momentum quantum number l, with component m in the global z direction and component k in the molecule-fixed z direction. From this it follows that the set of $D^l_{mk}(\alpha,\beta,\gamma)^*$, for any fixed k, is a spherical tensor of rank l.

B.4 Coupling of wavefunctions and spherical tensors

Given a set of eigenfunctions $\varphi_{l_1 m_1}$ with angular momentum eigenvalues l_1 and m_1, and another set $\psi_{l_2 m_2}$ with eigenvalues l_2 and m_2, it is possible to construct from the products $\varphi_{l_1 m_1}\psi_{l_2 m_2}$ a set of angular momentum functions Ψ_{LM} for each value of the eigenvalue L from $l_1 + l_2$ by integer steps down to $|l_1 - l_2|$ (the Clebsch–Gordan series). For example, from the 15 products $Y_{1m_1}Y_{2m_2}$ of the rank 1 and rank 2 spherical harmonics, such as arise in the wavefunction for an atom with one p electron and one d electron outside closed shells, we can construct functions with $L = 3$, 2 and 1. The formula for the new functions is

$$\Psi_{LM} = \sum_{m_1 m_2} \langle l_1 l_2 m_1 m_2 | LM \rangle \varphi_{l_1 m_1} \psi_{l_2 m_2},$$

where the coefficient $\langle l_1 l_2 m_1 m_2 | LM \rangle$ is called a *Wigner* or *Clebsch–Gordan* or *vector coupling* coefficient. It is zero unless $m = m_1 + m_2$ and L is one of the values $l_1 + l_2$, $l_1 + l_2 - 1, \ldots, |l_1 - l_2|$. The latter condition is called the triangle condition, and can be expressed symmetrically: $l_1 + l_2 + L$ must be an integer, and none of the three may exceed the sum of the other two. There is a formula for the Wigner coefficient (Brink and Satchler 1993, Zare 1988), but it is quite complicated and values are usually obtained from tables, for example in Varshalovich *et al.* (1988, pp. 271–278).

The same procedure can be applied when the factors in the products are not wavefunctions $\varphi_{l_1 m_1}$ and $\psi_{l_2 m_2}$ but spherical tensor operators $R_{l_1 m_1}$ and $S_{l_2 m_2}$. In this case it often happens that $l_1 = l_2$ and we require $L = 0$. In this case the coupled function is a scalar: it is invariant under rotations. The Wigner coefficients take a particularly simple form in this case, and the coupled operator is

$$(R \times S)_{00} = \sum_m (2l+1)^{-1/2} (-1)^{l-m} R_{lm} S_{l,-m}. \tag{B.4.1}$$

The reason for the importance of this case is that the Hamiltonian for an isolated system is invariant under rotations, so these invariant combinations of operators are the only ones that can occur in such a Hamiltonian. It occurs so often, in fact, that it is usual to define the 'scalar product' of two sets of spherical-tensor operators as

$$R \cdot S = \sum_m (-1)^m R_{lm} S_{l,-m}, \tag{B.4.2}$$

i.e., with a different phase and normalization from eqn (B.4.1). In the particular case where R and S are vectors (rank 1), this expression yields, using (B.3.1),

$$R \cdot S = -R_{11} S_{1,-1} + R_{10} S_{10} - R_{1,-1} S_{11} = R_x S_x + R_y S_y + R_z S_z,$$

in agreement with the conventional cartesian definition. Indeed, this is the main reason for the use of the definition (B.4.2), the scalar product of two vectors being then the same whether the cartesian or the spherical tensor formulation is used.

Where we have a spherical-tensor 'scalar product' of the form $\sum_m (-1)^m A_{l,-m} B_{lm}$, it can always be replaced by a scalar product in terms of the real components. For any $m > 0$ we can use eqn (B.1.5) to show that

$$\begin{aligned}A_{lmc}B_{lmc}+A_{lms}B_{lms} &= \tfrac{1}{2}\left((-1)^m A_{lm}+A_{l,-m}\right)\left((-1)^m B_{lm}+B_{l,-m}\right)\\ &\quad -\tfrac{1}{2}\left((-1)^m A_{lm}-A_{l,-m}\right)\left((-1)^m B_{lm}-B_{l,-m}\right)\\ &= (-1)^m\left(A_{lm}B_{l,-m}+A_{l,-m}B_{lm}\right)\end{aligned}$$

When we sum over m from 1 to l and add the $m=0$ term we get

$$\sum_m (-1)^m A_{l,-m}B_{lm} = \sum_\kappa A_{l\kappa}B_{l\kappa}. \tag{B.4.3}$$

B.4.1 *Wigner 3j symbols*

A similar situation sometimes arises when we have three sets of spherical-tensor quantities, $R_{l_1m_1}$, $S_{l_2m_2}$ and $T_{l_3m_3}$ and we wish to construct a scalar from them. In this case we can begin by constructing a rank l_3 tensor from R and S (this is possible only if l_1, l_2 and l_3 satisfy the triangle condition) and then form the scalar product of this with T:

$$\begin{aligned}(R\times S\times T)_{00} &= (2l_3+1)^{-1/2}\sum_{m_3}(-1)^{l_3-m_3}\sum_{m_1m_2}\langle l_1l_2m_1m_2|l_3m_3\rangle R_{l_1m_1}S_{l_2m_2}T_{l_3,-m_3}\\ &= (-1)^{l_1-l_2+l_3}\sum_{m_1m_2m_3}\begin{pmatrix} l_1 & l_2 & l_3\\ m_1 & m_2 & m_3\end{pmatrix} R_{l_1m_1}S_{l_2m_2}T_{l_3m_3},\end{aligned} \tag{B.4.4}$$

where the quantity in parentheses is a *Wigner 3j symbol*, defined by

$$\begin{pmatrix} l_1 & l_2 & l_3\\ m_1 & m_2 & m_3\end{pmatrix} = (2l_3+1)^{-1/2}(-1)^{l_1-l_2-m_3}\langle l_1l_2m_1m_2|l_3,-m_3\rangle.$$

The phase factor here is chosen to make the $3j$ symbol as symmetric as possible: it is invariant under even permutations of its columns, while odd permutations multiply it by a factor $(-1)^{l_1+l_2+l_3}$.

One example is the $\overline{S}$ function introduced in §3.3 on p. 41. It takes the form

$$\begin{aligned}\overline{S}^{k_1k_2}_{l_1l_2j} &= i^{l_1-l_2-j}\left[\begin{pmatrix} l_1 & l_2 & j\\ 0 & 0 & 0\end{pmatrix}\right]^{-1}\\ &\quad\times\sum_{m_1m_2m}\left[D^{l_1}_{m_1k_1}(\Omega_1)\right]^*\left[D^{l_2}_{m_2k_2}(\Omega_2)\right]^* C_{jm}(\theta,\varphi)\begin{pmatrix} l_1 & l_2 & j\\ m_1 & m_2 & m\end{pmatrix}.\end{aligned} \tag{B.4.5}$$

Here θ and φ are the polar angles describing the direction of the intermolecular vector **R**. We see that apart from some additional numerical constants, this has the form of eqn (B.4.4), with j and m taking the place of l_3 and m_3. (Remember that $\left[D^{l_1}_{m_1k_1}(\Omega_1)\right]^*$ is the m_1 component of a spherical tensor of rank l_1.)

APPENDIX C

INTRODUCTION TO PERTURBATION THEORY

C.1 Non-degenerate perturbation theory

We give here a summary of the principles of perturbation theory, for reference and to define the notation.

We are faced with a Hamiltonian $\mathcal{H}$ that is too complicated to handle directly. We suppose that it differs by a small 'perturbation' $\mathcal{H}'$ from a closely related 'zeroth-order' Hamiltonian $\mathcal{H}^0$ describing a problem that we can solve:

$$\mathcal{H} = \mathcal{H}^0 + \lambda\mathcal{H}'. \tag{C.1.1}$$

Here λ may be a physical quantity describing the strength of the perturbation, such as the magnitude of an electric or magnetic field, but often it is just a parameter that we can vary hypothetically from 0 for the unperturbed problem to 1 for the problem that we want to solve.

Suppose that the eigenfunctions of the unperturbed problem are $|n^0\rangle$, with eigenvalues W_n^0:

$$\mathcal{H}^0|n^0\rangle = W_n^0|n^0\rangle. \tag{C.1.2}$$

We want to find $|n\rangle$ and W_n satisfying $\mathcal{H}|n\rangle = W_n|n\rangle$. We assume at this stage that $|n\rangle$ is a non-degenerate state, well separated in energy from other states. Expand $|n\rangle$ and W_n as power series in λ:

$$|n\rangle = |n^0\rangle + \lambda|n'\rangle + \lambda^2|n''\rangle + \cdots, \tag{C.1.3}$$
$$W_n = W_n^0 + \lambda W_n' + \lambda^2 W_n'' + \cdots. \tag{C.1.4}$$

Without loss of generality, we can require that $|n^0\rangle$ is normalized and that all the corrections to the zeroth-order wavefunction are orthogonal to it:

$$\langle n^0|n'\rangle = \langle n^0|n''\rangle = \cdots = 0, \tag{C.1.5}$$

so that $\langle n|n^0\rangle = \langle n^0|n^0\rangle = 1$. This is *intermediate normalization*. Substitute the expressions (C.1.3) and (C.1.4) into $(\mathcal{H} - W_n)|n\rangle = 0$, and we get

$$\left((\mathcal{H}^0 - W_n^0) + \lambda(\mathcal{H}' - W_n') - \lambda^2 W_n'' + \cdots\right)\left(|n^0\rangle + \lambda|n'\rangle + \lambda^2|n''\rangle + \cdots\right) = 0. \tag{C.1.6}$$

For sufficiently small λ, we expect the power series to converge, and in that case we can equate coefficients of powers of λ:

$$(\mathcal{H}^0 - W_n^0)|n^0\rangle = 0, \tag{C.1.7}$$
$$(\mathcal{H}^0 - W_n^0)|n'\rangle + (\mathcal{H}' - W_n')|n^0\rangle = 0, \tag{C.1.8}$$

$$(\mathcal{H}^0 - W_n^0)|n''\rangle + (\mathcal{H}' - W_n')|n'\rangle - W_n''|n^0\rangle = 0, \tag{C.1.9}$$
$$(\mathcal{H}^0 - W_n^0)|n'''\rangle + (\mathcal{H}' - W_n')|n''\rangle - W_n''|n'\rangle - W_n'''|n^0\rangle = 0, \tag{C.1.10}$$

and so on. The first of these equations is the *zeroth-order* problem, an eigenvalue equation which we suppose solved. The rest are inhomogeneous differential equations. Multiply the first-order equation (C.1.8) by $\langle n^0|$ and integrate:

$$\langle n^0|\mathcal{H}^0 - W_n^0|n'\rangle + \langle n^0|\mathcal{H}' - W_n'|n^0\rangle = 0. \tag{C.1.11}$$

Because $\mathcal{H}^0 - W_n^0$ is hermitian, it follows that for any wavefunction ψ,

$$\langle n^0|\mathcal{H}^0 - W_n^0|\psi\rangle = \langle\psi|\mathcal{H}^0 - W_n^0|n^0\rangle^* = 0, \tag{C.1.12}$$

and since $|n^0\rangle$ is normalized we obtain

$$W_n' = \langle n^0|\mathcal{H}'|n^0\rangle. \tag{C.1.13}$$

Thus the *first-order energy* W_n' is the expectation value of the perturbation operator for the unperturbed wavefunction $|n^0\rangle$. Indeed the total energy to first order is

$$W_n^0 + \lambda W_n' = \langle n^0|\mathcal{H}^0 + \lambda\mathcal{H}'|n^0\rangle. \tag{C.1.14}$$

An alternative way to write this result is the following:

$$W_n = \langle n^0|\mathcal{H}|n^0\rangle + O(\lambda^2). \tag{C.1.15}$$

Thus the expectation value of the complete Hamiltonian over the unperturbed wavefunction gives the total energy correct to first order in λ.

Now multiply the second-order equation (C.1.9) by $\langle n^0|$, to give

$$\langle n^0|\mathcal{H}^0 - W^0|n''\rangle + \langle n^0|\mathcal{H}' - W_n'|n'\rangle - \langle n^0|W_n''|n^0\rangle = 0. \tag{C.1.16}$$

Once again, the first term is zero. Moreover W_n' disappears from the second term because of the orthogonality between the unperturbed wavefunction and the corrections, eqn (C.1.5), so (C.1.16) becomes

$$W_n'' = \langle n^0|\mathcal{H}'|n'\rangle, \tag{C.1.17}$$

and we have the *second-order energy*. To evaluate this, we need the first-order wavefunction $|n'\rangle$, and to obtain it, we have to solve the inhomogeneous differential equation (C.1.8).

C.1.1 *Rayleigh–Schrödinger perturbation theory*

The standard way to do this, used by Rayleigh for classical oscillators and adapted for quantum mechanics by Schrödinger, is to expand $|n'\rangle$ in terms of the unperturbed eigenfunctions:

$$|n'\rangle = \sum_k{}' c_k|k^0\rangle, \tag{C.1.18}$$

where the prime on the summation sign conventionally indicates that we are to omit the term with $k = n$. This is to ensure that $\langle n^0|n'\rangle = 0$, in accordance with (C.1.5). Substitute this expression into (C.1.8) and remember that $\mathcal{H}^0|k^0\rangle = W_k^0|k^0\rangle$:

$$\sum_k{}' c_k(W_k^0 - W_n^0)|k^0\rangle + (\mathcal{H}' - W_n')|n^0\rangle = 0. \tag{C.1.19}$$

Multiply by $\langle p^0|$, and remember that the eigenfunctions $|k^0\rangle$ of the unperturbed Hamiltonian are orthonormal. We obtain

$$c_p = -\frac{\langle p^0|\mathcal{H}'|n^0\rangle}{W_p^0 - W_n^0} = -\frac{H'_{pn}}{\Delta_{pn}}, \tag{C.1.20}$$

so that the first-order wavefunction is

$$\begin{aligned} |n'\rangle &= -\sum_p{}' |p^0\rangle \frac{\langle p^0|\mathcal{H}'|n^0\rangle}{W_p^0 - W_n^0} \\ &= -\sum_p{}' |p^0\rangle \frac{H'_{pn}}{\Delta_{pn}}. \end{aligned} \tag{C.1.21}$$

Here the second form introduces an abbreviated notation for the 'energy denominator' $\Delta_{pn} = W_p^0 - W_n^0$ and for the matrix element $H'_{pn} = \langle p^0|\mathcal{H}'|n^0\rangle$. We substitute this result into (C.1.17) to obtain the second-order energy:

$$\begin{aligned} W_n'' &= \langle n^0|\mathcal{H}'|n'\rangle = -\sum_p{}' \frac{\langle n^0|\mathcal{H}'|p^0\rangle\langle p^0|\mathcal{H}'|n^0\rangle}{W_p^0 - W_n^0} \\ &= -\sum_p{}' \frac{H'_{np}H'_{pn}}{\Delta_{pn}} \\ &= -\sum_p{}' \frac{|H'_{pn}|^2}{\Delta_{pn}}. \end{aligned} \tag{C.1.22}$$

Here we have a sum over the excited states of the system. It is important to realise that this extends over *all* excited states, including continuum states.

C.1.2 *Unsöld's average-energy approximation*

A crude but useful approximation to the second-order energy can be obtained by replacing all the energy denominators in the Rayleigh–Schrödinger expression (C.1.22) for the second-order energy by some 'average' value Δ:

$$\begin{aligned} W_n'' = -\sum_p{}' \frac{|H'_{pn}|^2}{\Delta_{pn}} &\approx -\sum_p{}' \frac{|H'_{pn}|^2}{\Delta} \\ &= -\frac{1}{\Delta}\Big(\sum_p H'_{np}H'_{pn} - H'_{nn}H'_{nn}\Big) \\ &= -\frac{1}{\Delta}\Big(\langle n^0|(\mathcal{H}')^2|n^0\rangle - \langle n^0|\mathcal{H}'|n^0\rangle^2\Big). \end{aligned} \tag{C.1.23}$$

If $|n\rangle$ is the ground state and Δ the first excitation energy, then $\Delta_{pn} \geq \Delta$ and the approximate expression must be larger in magnitude than the exact. Often Δ is taken to be the first ionization energy, but this does not give a lower bound to $|W_n''|$ because of the sum over states includes an integral over the continuum states.

C.2 The resolvent

We study here a formal treatment of the perturbation problem which is sometimes useful. It is used in §6.3.1 in the development of iterative symmetry-forcing perturbation methods. First consider the operator $P_k \equiv |k^0\rangle\langle k^0|$, where $|k^0\rangle$ is one of the normalized eigenstates of $\mathcal{H}^0$. If we apply this to a wavefunction $\sum_j c_j|j^0\rangle$, we obtain $\sum_j c_j|k^0\rangle\langle k^0|j^0\rangle = \sum_j c_j|k^0\rangle\delta_{jk} = c_k|k^0\rangle$. That is, P_k projects from a general wavefunction the $|k^0\rangle$ component. It is a *projection operator*. All projection operators satisfy $P_k^2 = P_k$, easily verified in this case. We can also define a complementary projection operator $Q_k = 1 - P_k$, which removes the $|k^0\rangle$ component from any wavefunction, leaving the rest of the wavefunction unchanged. It is easy to verify that $Q_k^2 = Q_k$ also.

Now consider the operator

$$R_n = \frac{Q_n}{(W_n^0 - \mathcal{H}^0)}.$$

Operating on state $|n^0\rangle$ this gives zero, because $Q_n|n^0\rangle = 0$. (This can be made rigorous by replacing W_n^0 by z in the denominator, and taking the limit $z \to W_n^0$ after applying the operator (McWeeny 1989).) Otherwise it behaves like an inverse of the operator $W_n^0 - \mathcal{H}^0$. It is called the *reduced resolvent*.

Apply this operator to eqns (C.1.8–C.1.10). The result is

$$\begin{aligned} |n'\rangle &= -R_n(\mathcal{H}' - W_n')|n^0\rangle \\ &= -R_n\mathcal{H}'|n^0\rangle, \end{aligned} \tag{C.2.1}$$

$$\begin{aligned} |n''\rangle &= -R_n(\mathcal{H}' - W_n')|n'\rangle + R_n W_n''|n^0\rangle \\ &= -R_n(\mathcal{H}' - W_n')|n'\rangle, \end{aligned} \tag{C.2.2}$$

$$\begin{aligned} |n'''\rangle &= -R_n(\mathcal{H}' - W_n')|n''\rangle + R_n W_n''|n'\rangle + R_n W_n'''|n^0\rangle \\ &= -R_n(\mathcal{H}' - W_n')|n''\rangle + R_n W_n''|n'\rangle, \end{aligned} \tag{C.2.3}$$

and in general

$$|n^{(m)}\rangle = -R_n(\mathcal{H}' - W_n')|n^{(m-1)}\rangle + R_n \sum_{t=1}^{m-2} W_n^{(m-t)}|n^{(t)}\rangle. \tag{C.2.4}$$

By expressing R_n in the form $R_n = \sum_k{}'|k^0\rangle(W_n^0 - W_k^0)^{-1}\langle k^0|$ (with the prime on the sum denoting as usual that the term $k = n$ is omitted from the sum) we can recover the formulae of Rayleigh–Schrödinger perturbation theory.

C.3 Degenerate perturbation theory

If the state $|n^0\rangle$ is degenerate (a set of components $|n\alpha^0\rangle$ all with energy W_n^0) then we cannot apply the Rayleigh–Schrödinger formulae (C.1.21) and (C.1.17) because some of the energy denominators vanish. Also, if we take a particular perturbed state and allow λ

to tend to zero, there is no reason to expect that the resulting zeroth-order state $|nm^0\rangle$ will coincide with one of the $|n\alpha^0\rangle$. Rather we expect to get a linear combination of them:

$$|nm^0\rangle = \sum_\alpha c_{\alpha m}|n\alpha^0\rangle. \tag{C.3.1}$$

The first-order perturbation equation (C.1.8) now becomes, for the perturbed state $|nm\rangle$,

$$(\mathcal{H}^0 - W_n^0)|nm'\rangle + (\mathcal{H}' - W'_{nm})\sum_\alpha c_{\alpha m}|n\alpha^0\rangle = 0. \tag{C.3.2}$$

If we multiply by $\langle n\beta^0|$, the first term disappears as before, because $(\mathcal{H}^0 - W_n^0)|n\beta^0\rangle = 0$, and we get a set of secular equations:

$$\langle n\beta^0|\mathcal{H}' - W'_{nm}|n\alpha^0\rangle c_{\alpha m} = 0. \tag{C.3.3}$$

Consequently we need to solve these secular equations to find the first-order energies W'_{nm} and the corresponding eigenvectors $c_{\alpha m}$. Alternatively, since $(\mathcal{H}^0 - W_n^0)|n\alpha^0\rangle = 0$, we can replace (C.3.3) by

$$\langle n\beta^0|\mathcal{H}^0 + \lambda\mathcal{H}' - W_n^0 - \lambda W'_{nm}|n\alpha^0\rangle c_{\alpha m} = 0. \tag{C.3.4}$$

That is,

$$\sum_\alpha (H_{n\beta,n\alpha} - W\delta_{\beta\alpha})c_{\alpha m} = 0, \tag{C.3.5}$$

where $H_{n\beta,n\alpha}$ is the matrix element of the complete Hamiltonian and W is the energy correct to first order. Note that (C.3.5) is the result we would get from a variational treatment using the trial function (C.3.1), and that consequently the same result applies in the 'nearly degenerate' case, where the energy separations among a set of unperturbed states, although nonzero, are not large compared with the matrix elements of the perturbation. In such a case the formulae don't blow up, but the perturbation series fails to converge and the degenerate version must be used.

Now we can follow the Rayleigh–Schrödinger procedure to find the first-order wavefunction as before:

$$|nm'\rangle = -\sum_{pq}{}' \frac{\mathcal{H}'_{pq,nm}}{\Delta_{pn}}|pq^0\rangle, \tag{C.3.6}$$

where the sum is taken over the components $|pq^0\rangle$ of the other unperturbed states, and $\Delta_{pn} = W_p^0 - W_n^0$ as before. Similarly

$$W''_{nm} = -\sum_{pq}{}' \frac{|\mathcal{H}'_{pq,nm}|^2}{\Delta_{pn}}. \tag{C.3.7}$$

C.4 Time-dependent perturbation theory

Here we consider a Hamiltonian $\mathcal{H} = \mathcal{H}^0 + \mathcal{H}'$ consisting of a time-independent part $\mathcal{H}^0$ and a time-dependent perturbation $\mathcal{H}'$. The unperturbed states are stationary:

$$\Psi_n = \psi_n \exp(-\mathrm{i}W_n t/\hbar) = |n\rangle \exp(-\mathrm{i}\omega_n t), \tag{C.4.1}$$

where the $|n\rangle$ are the eigenstates of the unperturbed Hamiltonian: $\mathcal{H}^0|n\rangle = W_n|n\rangle$. The perturbed problem does not have stationary states; the solution Ψ must satisfy Schrödinger's time-dependent equation:

$$\mathcal{H}\Psi = \mathrm{i}\hbar\frac{\partial}{\partial t}\Psi. \tag{C.4.2}$$

To solve this, we use Dirac's method of *variation of constants*: let

$$\Psi = \sum_k a_k(t)\Psi_k(t) = \sum_k a_k(t)\psi_k \exp(-\mathrm{i}\omega_k t). \tag{C.4.3}$$

If the Hamiltonian were time-independent the a_k would be constant (and their values would be arbitrary). In the presence of the small time-dependent perturbation they evolve slowly with time. Substituting (C.4.3) into (C.4.2) gives

$$\mathcal{H}\sum_k a_k(t)\Psi_k(t) = \mathrm{i}\hbar\frac{\partial}{\partial t}\sum_k a_k(t)\Psi_k(t), \tag{C.4.4}$$

or

$$\sum_k \left[a_k W_k \Psi_k + a_k \mathcal{H}'(t)\Psi_k\right] = \sum_k \left[\mathrm{i}\hbar\frac{\partial a_k}{\partial t}\Psi_k + a_k W_k \Psi_k\right]. \tag{C.4.5}$$

Multiply by $\Psi_p^\star = \langle p|\exp(+\mathrm{i}W_p t/\hbar)$ and integrate over variables other than time:

$$\sum_k a_k \langle p|\mathcal{H}'(t)|k\rangle \exp(\mathrm{i}\omega_{pk}t) = \mathrm{i}\hbar\frac{\partial a_p}{\partial t}, \tag{C.4.6}$$

where $\omega_{pk} = (W_p - W_k)/\hbar$. We can integrate this formally to obtain

$$a_p(t) - a_p(0) = -\frac{\mathrm{i}}{\hbar}\sum_k \int_0^t a_k(\tau)\langle p|\mathcal{H}'(\tau)|k\rangle \exp(\mathrm{i}\omega_{pk}\tau)\,\mathrm{d}\tau. \tag{C.4.7}$$

If $\mathcal{H}'$ takes the form $\mathcal{H}' = \hat{V} f(t)$, where $\hat{V}$ is an operator independent of time and $f(t)$ is a function only of the time, then

$$a_p(t) - a_p(0) = -\frac{\mathrm{i}}{\hbar}\sum_k V_{pk}\int_0^t a_k(\tau) f(\tau)\exp(\mathrm{i}\omega_{pk}\tau)\,\mathrm{d}\tau. \tag{C.4.8}$$

All of this is *exact*, but is rarely useful as it stands since the expression for $a_p(t)$ involves all the a_k (including a_p) at times from 0 to t. We must therefore approximate:

we assume that $\hat{V}$ is sufficiently small that the a_k will not change much over the interval from 0 to t, so that we can put $a_k(\tau) = a_k(0)$ in (C.4.8), giving

$$a_p(t) - a_p(0) = -\frac{\mathrm{i}}{\hbar}\sum_k V_{pk}a_k(0)\int_0^t f(\tau)\exp(\mathrm{i}\omega_{pk}\tau)\,\mathrm{d}\tau. \tag{C.4.9}$$

If necessary we could substitute this result back into the r.h.s. of (C.4.8) to obtain a second-order formula, but in practice (C.4.9) is usually adequate.

For simplicity we now consider the case where the system is definitely in the stationary state n at time 0, so that $a_k(0) = \delta_{kn}$. Then $a_p(t)$ is the probability amplitude for a transition to state p having occurred by time t:

$$a_p(t) = -\frac{\mathrm{i}}{\hbar}V_{pn}\int_0^t f(\tau)\exp(\mathrm{i}\omega_{pn}\tau)\,\mathrm{d}\tau, \quad p \neq n. \tag{C.4.10}$$

The development of the theory from this point depends on the form of the function $f(t)$ that describes the time-dependence of the perturbation. See, for example, the discussion in §2.5 (p. 26) of the response of a molecule to an oscillating electric field.

APPENDIX D

CONVERSION FACTORS

The conversion factors in this appendix have been derived using values of fundamental constants from CODATA Bulletin No. 63 (November 1986). The uncertainties in these values are generally a few parts in the last digit quoted.

D.1 Multipole moments

Dipole moment

The Debye is defined as 10^{-18} esu (electrostatic units). The esu of length is the centimetre, while the esu of charge is numerically equal to $10/c$ C, but with c expressed in $\mathrm{cm\,s^{-1}}$ (Mills 1993).

	$10^{-30}\,\mathrm{C\,m}$	Debye	a.u. (ea_0)	$e\,\text{Å}$
$10^{-30}\,\mathrm{C\,m}$	1	0.299792458	0.1179474	0.06241506
1 Debye	3.33564095	1	0.3934301	0.2081942
1 a.u. (ea_0)	8.478358	2.541748	1	0.52917725
$1\,e\,\text{Å}$	16.02177	4.803207	1.8897260	1

Quadrupole moment

The Buckingham unit of quadrupole moment was introduced by Debye and named after A. D. Buckingham. It is the same as the Debye Å = 10^{-26} esu.

	$10^{-40}\,\mathrm{C\,m^2}$	Debye Å	a.u. (ea_0^2)	$e\,\text{Å}^2$
$10^{-40}\,\mathrm{C\,m^2}$	1	0.299792458	0.2228882	0.06241506
1 B = 1 Debye Å	3.33564095	1	0.7434750	0.2081942
1 a.u. (ea_0^2)	4.486554	1.3450351	1	0.2800286
$1\,e\,\text{Å}^2$	16.02177	4.803207	3.5710643	1

Octopole moment

Octopole	$10^{-50}\,\mathrm{C\,m^3}$	a.u. (ea_0^3)	$e\,\text{Å}^3$
$10^{-50}\,\mathrm{C\,m^3}$	1	0.4211976	0.0624151
1 a.u. (ea_0^3)	2.374182	1	0.1481847
$1\,e\,\text{Å}^3$	16.02177	6.748333	1

D.2 Polarizabilities

The dipole–dipole polarizability has dimensions of $[4\pi\varepsilon_0] \times [\text{length}^3]$. In the electrostatic unit system, $4\pi\varepsilon_0$ is dimensionless and is equal to 1, so that the polarizability has units of volume, typically Å^3. In atomic units, $4\pi\varepsilon_0$ is numerically equal to 1, but it is not dimensionless; nevertheless, polarizabilities in atomic units are often quoted as being in units of a_0^3. When converting any of these to S.I., it is necessary to insert the factor of $4\pi\varepsilon_0$. The S.I. unit of permittivity, i.e., of $4\pi\varepsilon_0$, is $\mathrm{F\,m^{-1}}$, where $\mathrm{F = C\,V^{-1}}$ is the farad, so the S.I. unit of polarizability is $\mathrm{F\,m^2}$.

	$10^{-40}\,\mathrm{F\,m^2}$	a.u. $(4\pi\varepsilon_0 a_0^3)$	$\text{Å}^3 = 10^{-24}\,\mathrm{cm^3}$
$10^{-40}\,\mathrm{F\,m^2}$	1	6.065099	0.898755
1 a.u.	0.1648778	1	0.1481847
$1\,\text{Å}^3 = 10^{-24}\,\mathrm{cm^3}$	1.1126501	6.748333	1

In general, the S.I. unit of rank-l–rank-l' polarizability is $\mathrm{F\,m}^{l+l'}$, and the atomic unit is $4\pi\varepsilon_0 \times a_0^{l+l'+1}$. Conversion factors for these are easily derived from those for dipole–dipole polarizability by applying the appropriate power of $a_0 = 0.52917725 \times 10^{-10}\,\mathrm{m}$. Thus for dipole–quadrupole polarizability:

$$1\,\text{a.u.} = 4\pi\varepsilon_0 a_0^4 = 0.08725 \times 10^{-50}\,\mathrm{F\,m^3}.$$

D.3 The C_6 coefficient

C_6 coefficients have the form [energy]×[length]^6, and various length and energy units have been used. We give just the conversions between atomic units and S.I.:

$$E_h a_0^6 = 57.65261 \times 10^{-60}\,\mathrm{kJ\,mol^{-1}\,m^6} = 57.65261\,\mathrm{kJ\,mol^{-1}}\,\text{Å}^6;$$

$$\mathrm{kJ\,mol^{-1}}\,\text{Å}^6 = 0.0173453 E_h a_0^6.$$

APPENDIX E

CARTESIAN–SPHERICAL CONVERSION TABLES

Table E.1 *Regular spherical harmonics and multipole moments*

$R_{00} = 1$	$Q_{00} = q$
$R_{10} = z$	$Q_{10} = \mu_z$
$R_{11c} = x$	$Q_{11c} = \mu_x$
$R_{11s} = y$	$Q_{11s} = \mu_y$
$R_{20} = \frac{1}{2}(3z^2 - r^2)$	$Q_{20} = \Theta_{zz}$
$R_{21c} = \sqrt{3}xz$	$Q_{21c} = \frac{2}{\sqrt{3}}\Theta_{xz}$
$R_{21s} = \sqrt{3}yz$	$Q_{21s} = \frac{2}{\sqrt{3}}\Theta_{yz}$
$R_{22c} = \frac{1}{2}\sqrt{3}(x^2 - y^2)$	$Q_{22c} = \frac{1}{\sqrt{3}}(\Theta_{xx} - \Theta_{yy})$
$R_{22s} = \sqrt{3}xy$	$Q_{22s} = \frac{2}{\sqrt{3}}\Theta_{xy}$
$R_{30} = \frac{1}{2}(5z^3 - 3zr^2)$	$Q_{30} = \Omega_{zzz}$
$R_{31c} = \frac{1}{4}\sqrt{6}(4xz^2 - x^3 - xy^2)$	$Q_{31c} = \sqrt{\frac{3}{2}}\Omega_{xzz}$
$R_{31s} = \frac{1}{4}\sqrt{6}(4yz^2 - x^2y - y^3)$	$Q_{31s} = \sqrt{\frac{3}{2}}\Omega_{yzz}$
$R_{32c} = \frac{1}{2}\sqrt{15}z(x^2 - y^2)$	$Q_{32c} = \sqrt{\frac{3}{5}}(\Omega_{xxz} - \Omega_{yyz})$
$R_{32s} = \sqrt{15}xyz$	$Q_{32s} = 2\sqrt{\frac{3}{5}}\Omega_{xyz}$
$R_{33c} = \frac{1}{4}\sqrt{10}(x^3 - 3xy^2)$	$Q_{33c} = \sqrt{\frac{1}{10}}(\Omega_{xxx} - 3\Omega_{xyy})$
$R_{33s} = \frac{1}{4}\sqrt{10}(3x^2y - y^3)$	$Q_{33s} = \sqrt{\frac{1}{10}}(3\Omega_{xxy} - \Omega_{yyy})$

continued ...

Table E.1 *continued. Regular spherical harmonics and multipole moments.*

$R_{40} = \frac{1}{8}[8z^4 - 24(x^2+y^2)z^2 + 3(x^4+2x^2y^2+y^4)]$	$Q_{40} = \Phi_{zzzz}$
$R_{41c} = \frac{1}{4}\sqrt{10}[4xz^3 - 3xz(x^2+y^2)]$	$Q_{41c} = \sqrt{\frac{8}{5}}\Phi_{xzzz}$
$R_{41s} = \frac{1}{4}\sqrt{10}[4yz^3 - 3yz(x^2+y^2)]$	$Q_{41s} = \sqrt{\frac{8}{5}}\Phi_{yzzz}$
$R_{42c} = \frac{1}{4}\sqrt{5}(x^2-y^2)(6z^2-x^2-y^2)$	$Q_{42c} = 2\sqrt{\frac{1}{5}}(\Phi_{xxzz} - \Phi_{yyzz})$
$R_{42s} = \frac{1}{2}\sqrt{5}xy(6z^2-x^2-y^2)$	$Q_{42s} = 4\sqrt{\frac{1}{5}}\Phi_{xyzz}$
$R_{43c} = \frac{1}{4}\sqrt{70}z(x^3-3xy^2)$	$Q_{43c} = 2\sqrt{\frac{2}{35}}(\Phi_{xxxz} - 3\Phi_{xyyz})$
$R_{43s} = \frac{1}{4}\sqrt{70}z(3x^2y-y^3)$	$Q_{43s} = 2\sqrt{\frac{2}{35}}(3\Phi_{xxyz} - \Phi_{yyyz})$
$R_{44c} = \frac{1}{8}\sqrt{35}(x^4-6x^2y^2+y^4)$	$Q_{44c} = \sqrt{\frac{1}{35}}(\Phi_{xxxx} - 6\Phi_{xxyy} + \Phi_{yyyy})$
$R_{44s} = \frac{1}{2}\sqrt{35}xy(x^2-y^2)$	$Q_{44s} = 4\sqrt{\frac{1}{35}}(\Phi_{xxxy} - \Phi_{xyyy})$

Table E.2 *Conversion from cartesian multipole moments to spherical.*

$$\Theta_{xx} = -\tfrac{1}{2}Q_{20} + \tfrac{1}{2}\sqrt{3}Q_{22c}$$
$$\Theta_{yy} = -\tfrac{1}{2}Q_{20} - \tfrac{1}{2}\sqrt{3}Q_{22c}$$
$$\Theta_{zz} = Q_{20}$$
$$\Theta_{xy} = \tfrac{1}{2}\sqrt{3}Q_{22s}$$
$$\Theta_{xz} = \tfrac{1}{2}\sqrt{3}Q_{21c}$$
$$\Theta_{yz} = \tfrac{1}{2}\sqrt{3}Q_{21s}$$

$$\Omega_{xxx} = \sqrt{\tfrac{5}{8}}Q_{33c} - \sqrt{\tfrac{3}{8}}Q_{31c}$$
$$\Omega_{xxy} = \sqrt{\tfrac{5}{8}}Q_{33s} - \sqrt{\tfrac{1}{24}}Q_{31s}$$
$$\Omega_{xyy} = \sqrt{\tfrac{5}{8}}Q_{33c} - \sqrt{\tfrac{1}{24}}Q_{31c}$$
$$\Omega_{yyy} = \sqrt{\tfrac{5}{8}}Q_{33s} - \sqrt{\tfrac{3}{8}}Q_{31s}$$
$$\Omega_{xxz} = \sqrt{\tfrac{5}{12}}Q_{32c} - \tfrac{1}{2}Q_{30}$$
$$\Omega_{xyz} = \sqrt{\tfrac{5}{12}}Q_{32s}$$
$$\Omega_{yyz} = -\sqrt{\tfrac{5}{12}}Q_{32c} - \tfrac{1}{2}Q_{30}$$
$$\Omega_{xzz} = \sqrt{\tfrac{2}{3}}Q_{31c}$$
$$\Omega_{yzz} = \sqrt{\tfrac{2}{3}}Q_{31s}$$
$$\Omega_{zzz} = Q_{30}$$

$$\Phi_{xxxx} = \tfrac{3}{8}Q_{40} - \tfrac{1}{4}\sqrt{5}Q_{42c} + \tfrac{1}{32}\sqrt{35}Q_{44c}$$
$$\Phi_{xxxy} = \tfrac{1}{8}(-\sqrt{5}Q_{42s} + \sqrt{35}Q_{44s})$$
$$\Phi_{xxyy} = \tfrac{1}{8}Q_{40} - \tfrac{1}{32}\sqrt{35}Q_{44c}$$
$$\Phi_{xyyy} = -\tfrac{1}{8}(\sqrt{5}Q_{42s} + \sqrt{35}Q_{44s})$$
$$\Phi_{yyyy} = \tfrac{3}{8}Q_{40} + \tfrac{1}{4}\sqrt{5}Q_{42c} + \tfrac{1}{32}\sqrt{35}Q_{44c}$$
$$\Phi_{xxxz} = \tfrac{1}{16}(-\sqrt{10}Q_{41c} + \sqrt{70}Q_{43c})$$
$$\Phi_{xxyz} = \tfrac{1}{16}(-\sqrt{10}Q_{41s} + \sqrt{70}Q_{43s})$$
$$\Phi_{xyyz} = -\tfrac{1}{16}(\sqrt{10}Q_{41c} + \sqrt{70}Q_{43c})$$
$$\Phi_{yyyz} = -\tfrac{1}{16}(3\sqrt{10}Q_{41s} + \sqrt{70}Q_{43s})$$
$$\Phi_{xxzz} = -\tfrac{1}{2}Q_{40} + \tfrac{1}{4}\sqrt{5}Q_{42c}$$
$$\Phi_{xyzz} = \tfrac{1}{4}\sqrt{5}Q_{42s}$$
$$\Phi_{yyzz} = -\tfrac{1}{2}Q_{40} - \tfrac{1}{4}\sqrt{5}Q_{42c}$$
$$\Phi_{xzzz} = \sqrt{\tfrac{5}{8}}Q_{41c}$$
$$\Phi_{yzzz} = \sqrt{\tfrac{5}{8}}Q_{41s}$$
$$\Phi_{zzzz} = Q_{40}$$

Table E.3 *Conversion from cartesian dipole–quadrupole polarizabilities to spherical.*

$A_{\alpha,\beta\gamma}$	α		
$\beta\gamma$	x	y	z
xx	$-\tfrac{1}{2}\alpha_{11c,20} + \tfrac{1}{2}\sqrt{3}\alpha_{11c,22c}$	$-\tfrac{1}{2}\alpha_{11s,20} + \tfrac{1}{2}\sqrt{3}\alpha_{11s,22c}$	$-\tfrac{1}{2}\alpha_{10,20} + \tfrac{1}{2}\sqrt{3}\alpha_{10,22c}$
yy	$-\tfrac{1}{2}\alpha_{11c,20} - \tfrac{1}{2}\sqrt{3}\alpha_{11c,22c}$	$-\tfrac{1}{2}\alpha_{11s,20} - \tfrac{1}{2}\sqrt{3}\alpha_{11s,22c}$	$-\tfrac{1}{2}\alpha_{10,20} - \tfrac{1}{2}\sqrt{3}\alpha_{10,22c}$
zz	$\alpha_{11c,20}$	$\alpha_{11s,20}$	$\alpha_{10,20}$
xy	$\tfrac{1}{2}\sqrt{3}\alpha_{11c,22s}$	$\tfrac{1}{2}\sqrt{3}\alpha_{11s,22s}$	$\tfrac{1}{2}\sqrt{3}\alpha_{10,22s}$
xz	$\tfrac{1}{2}\sqrt{3}\alpha_{11c,21c}$	$\tfrac{1}{2}\sqrt{3}\alpha_{11s,21c}$	$\tfrac{1}{2}\sqrt{3}\alpha_{10,21c}$
yz	$\tfrac{1}{2}\sqrt{3}\alpha_{11c,21s}$	$\tfrac{1}{2}\sqrt{3}\alpha_{11s,21s}$	$\tfrac{1}{2}\sqrt{3}\alpha_{10,21s}$

Table E.4 *Conversion from spherical quadrupole–quadrupole polarizabilities to cartesian.*

$\alpha_{2\kappa,2\kappa'}$	κ'				
κ	0	1*c*	1*s*	2*c*	2*s*
0	$3C_{zz,zz}$	$2\sqrt{3}C_{zz,xz}$	$2\sqrt{3}C_{zz,yz}$	$\sqrt{3}(C_{zz,xx}-C_{zz,yy})$	$2\sqrt{3}C_{zz,xy}$
1*c*	$2\sqrt{3}C_{xz,zz}$	$4C_{xz,xz}$	$4C_{xz,yz}$	$2(C_{xz,xx}-C_{xz,yy})$	$4C_{xz,xy}$
1*s*	$2\sqrt{3}C_{yz,zz}$	$4C_{yz,xz}$	$4C_{yz,yz}$	$2(C_{yz,xx}-C_{yz,yy})$	$4C_{yz,xy}$
2*c*	$\sqrt{3}(C_{xx,zz}-C_{yy,zz})$	$2(C_{xx,xz}-C_{yy,xz})$	$2(C_{xx,yz}-C_{yy,yz})$	$(C_{xx,xx}-2C_{xx,yy}+C_{yy,yy})$	$2(C_{xx,xy}-C_{yy,xy})$
2*s*	$2\sqrt{3}C_{xy,zz}$	$4C_{xy,xz}$	$4C_{xy,yz}$	$2(C_{xy,xx}-C_{xy,yy})$	$4C_{xy,xy}$

Table E.5 *Conversion from cartesian quadrupole–quadrupole polarizabilities to spherical.*

$$
\begin{aligned}
12C_{xx,xx} &= \alpha_{20,20} - 2\sqrt{3}\alpha_{20,22c} + 3\alpha_{22c,22c} \\
12C_{xx,xy} &= -\sqrt{3}\alpha_{20,22s} + 3\alpha_{22c,22s} \\
12C_{xx,xz} &= -\sqrt{3}\alpha_{20,21c} + 3\alpha_{22c,21c} \\
12C_{xx,yy} &= \alpha_{20,20} - 3\alpha_{22c,22c} \\
12C_{xx,yz} &= -\sqrt{3}\alpha_{20,21s} + 3\alpha_{22c,21s} \\
12C_{xx,zz} &= -2\alpha_{20,20} + 2\sqrt{3}\alpha_{22c,20} \\
12C_{xy,xy} &= 3\alpha_{22s,22s} \\
12C_{xy,xz} &= 3\alpha_{22s,21c} \\
12C_{xy,yy} &= -\sqrt{3}\alpha_{22s,20} - 3\alpha_{22s,22c} \\
12C_{xy,yz} &= 3\alpha_{22s,21s} \\
12C_{xy,zz} &= 2\sqrt{3}\alpha_{22s,20} \\
12C_{xz,xz} &= 3\alpha_{21c,21c} \\
12C_{xz,yy} &= -\sqrt{3}\alpha_{21c,20} - 3\alpha_{21c,22c} \\
12C_{xz,yz} &= 3\alpha_{21c,21s} \\
12C_{xz,zz} &= 2\sqrt{3}\alpha_{21c,20} \\
12C_{yy,yy} &= \alpha_{20,20} + 2\sqrt{3}\alpha_{20,22c} + 3\alpha_{22c,22c} \\
12C_{yy,yz} &= -\sqrt{3}\alpha_{20,21s} - 3\alpha_{22c,21s} \\
12C_{yy,zz} &= -2\alpha_{20,20} - 2\sqrt{3}\alpha_{22c,20} \\
12C_{yz,yz} &= 3\alpha_{21s,21s} \\
12C_{yz,zz} &= 2\sqrt{3}\alpha_{21s,20} \\
12C_{zz,zz} &= 4\alpha_{20,20}
\end{aligned}
$$

APPENDIX F

INTERACTION FUNCTIONS

Here are listed the interaction functions T^{ab}_{tu} that arise in the spherical-tensor formulation of electrostatic interactions (see Chapter 3). The terms in R^{-n} up to $n = 5$ have been given previously by Stone (1991), but there were misprints in the formulae for $T_{21c,21c}$ and $T_{21s,21s}$: the coefficient $+6$ in the last term should have been -6 in each case. The formulae given here for these two functions and for $T_{21c,21s}$ are in a different and more symmetrical form given by Hättig and Hess (1993), but the two versions are equivalent. All the formulae for $n = 6$ were derived by Hättig and Hess.

The interaction function T^{ab}_{tu} for the electrostatic interaction between a multipole moment Q_t on site a and a moment Q_u on site b, both referred to local axes, is given in terms of the direction cosines r^a_α, r^b_β and $c_{\alpha\beta}$. If $\mathbf{e}^a_x$, $\mathbf{e}^a_y$ and $\mathbf{e}^a_z$ are the unit vectors defining the local axis system for site a, $\mathbf{e}^b_x$, $\mathbf{e}^b_y$ and $\mathbf{e}^b_z$ are the unit vectors for site b, and $\mathbf{e}_{ab}$ is a unit vector in the direction from a to b, then $r^a_\alpha = \mathbf{e}^a_\alpha \cdot \mathbf{e}_{ab}$, $r^b_\beta = \mathbf{e}^b_\beta \cdot \mathbf{e}_{ba} = -\mathbf{e}^b_\beta \cdot \mathbf{e}_{ab}$ (note the minus sign) and $c_{\alpha\beta} = \mathbf{e}^a_\alpha \cdot \mathbf{e}^b_\beta$. Thus r^a_α, $\alpha = x, y, z$, are the components of a unit vector in the direction from a to b, expressed in the local axis system of site a, and r^b_β, $\beta = x, y, z$, are the components of a unit vector in the direction from b to a, expressed in the local axis system of site b.

With this notation, it is possible to obtain $T^{ab}_{ut} = T^{ba}_{tu}$ from T^{ab}_{tu} merely by replacing a by b and *vice versa*, and exchanging the suffixes in every $c_{\alpha\beta}$.

The components of a dipole moment may be written as Q_{10}, Q_{11c} and Q_{11s} or as Q_{1z}, Q_{1x} and Q_{1y}. The latter notation is used here, with the usual cartesian-tensor convention of α to mean x, y or z.

Table F.1 *Interaction functions $T^{a\ b}_{l\kappa l'\kappa'}$ for $0 \leq l + l' \leq 5$, i.e. for all terms in R^{-n} up to $n = 6$.*

t	u	$4\pi\varepsilon_0 T^{ab}_{tu}$
00	00	R^{-1}
1α	00	$R^{-2}.r^a_\alpha$
20	00	$R^{-3}.\frac{1}{2}(3r^{a2}_z - 1)$
$21c$	00	$R^{-3}.\sqrt{3}r^a_x r^a_z$
$21s$	00	$R^{-3}.\sqrt{3}r^a_y r^a_z$
$22c$	00	$R^{-3}.\frac{1}{2}\sqrt{3}(r^{a2}_x - r^{a2}_y)$
$22s$	00	$R^{-3}.\sqrt{3}r^a_x r^a_y$
1α	1β	$R^{-3}.(3r^a_\alpha r^b_\beta + c_{\alpha\beta})$
30	00	$R^{-4}.\frac{1}{2}(5r^{a3}_z - 3r^a_z)$
$31c$	00	$R^{-4}.\frac{1}{4}\sqrt{6}r^a_x(5r^{a2}_z - 1)$
$31s$	00	$R^{-4}.\frac{1}{4}\sqrt{6}r^a_y(5r^{a2}_z - 1)$
$32c$	00	$R^{-4}.\frac{1}{2}\sqrt{15}r^a_z(r^{a2}_x - r^{a2}_y)$
$32s$	00	$R^{-4}.\sqrt{15}r^a_x r^a_y r^a_z$
$33c$	00	$R^{-4}.\frac{1}{4}\sqrt{10}r^a_x(r^{a2}_x - 3r^{a2}_y)$
$33s$	00	$R^{-4}.\frac{1}{4}\sqrt{10}r^a_y(3r^{a2}_x - r^{a2}_y)$
20	1β	$R^{-4}.\frac{1}{2}(15r^{a2}_z r^b_\beta + 6r^a_z c_{z\beta} - 3r^b_\beta)$
$21c$	1β	$R^{-4}.\sqrt{3}(r^a_x c_{z\beta} + c_{x\beta}r^a_z + 5r^a_x r^a_z r^b_\beta)$
$21s$	1β	$R^{-4}.\sqrt{3}(r^a_y c_{z\beta} + c_{y\beta}r^a_z + 5r^a_y r^a_z r^b_\beta)$
$22c$	1β	$R^{-4}.\frac{1}{2}\sqrt{3}(5(r^{a2}_x - r^{a2}_y)r^b_\beta + 2r^a_x c_{x\beta} - 2r^a_y c_{y\beta})$
$22s$	1β	$R^{-4}.\sqrt{3}(5r^a_x r^a_y r^b_\beta + r^a_x c_{y\beta} + r^a_y c_{x\beta})$

Table F.1 *continued*

t	u	$4\pi\varepsilon_0 T_{tu}^{ab}$
40	00	$R^{-5}.\frac{1}{8}(35r_z^{a^4} - 30r_z^{a^2} + 3)$
41c	00	$R^{-5}.\frac{1}{4}\sqrt{10}(7r_x^a r_z^{a^3} - 3r_x^a r_z^a)$
41s	00	$R^{-5}.\frac{1}{4}\sqrt{10}(7r_y^a r_z^{a^3} - 3r_y^a r_z^a)$
42c	00	$R^{-5}.\frac{1}{4}\sqrt{5}(7r_z^{a^2} - 1)(r_x^{a^2} - r_y^{a^2})$
42s	00	$R^{-5}.\frac{1}{2}\sqrt{5}(7r_z^{a^2} - 1)r_x^a r_y^a$
43c	00	$R^{-5}.\frac{1}{4}\sqrt{70}r_x^a r_z^a(r_x^{a^2} - 3r_y^{a^2})$
43s	00	$R^{-5}.\frac{1}{4}\sqrt{70}r_y^a r_z^a(3r_x^{a^2} - r_y^{a^2})$
44c	00	$R^{-5}.\frac{1}{8}\sqrt{35}(r_x^{a^4} - 6r_x^{a^2} r_y^{a^2} + r_y^{a^4})$
44s	00	$R^{-5}.\frac{1}{2}\sqrt{35}r_x^a r_y^a(r_x^{a^2} - r_y^{a^2})$
30	1β	$R^{-5}.\frac{1}{2}(35r_z^{a^3} r_\beta^b + 15r_z^{a^2} c_{z\beta} - 15r_z^a r_\beta^b - 3c_{z\beta})$
31c	1β	$R^{-5}.\frac{1}{4}\sqrt{6}(35r_x^a r_z^{a^2} r_\beta^b + 5r_z^{a^2} c_{x\beta} + 10r_x^a r_z^a c_{z\beta} - 5r_x^a r_\beta^b - c_{x\beta})$
31s	1β	$R^{-5}.\frac{1}{4}\sqrt{6}(35r_y^a r_z^{a^2} r_\beta^b + 5r_z^{a^2} c_{y\beta} + 10r_y^a r_z^a c_{z\beta} - 5r_y^a r_\beta^b - c_{y\beta})$
32c	1β	$R^{-5}.\frac{1}{2}\sqrt{15}((r_x^{a^2} - r_y^{a^2})(7r_z^a r_\beta^b + c_{z\beta}) + 2r_z^a(r_x^a c_{x\beta} - r_y^a c_{y\beta}))$
32s	1β	$R^{-5}.\sqrt{15}(r_x^a r_y^a(7r_z^a r_\beta^b + c_{z\beta}) + r_z^a(r_x^a c_{y\beta} + r_y^a c_{x\beta}))$
33c	1β	$R^{-5}.\frac{1}{4}\sqrt{10}(7r_x^{a^3} r_\beta^b + 3(r_x^{a^2} - r_y^{a^2})c_{x\beta} - 21r_x^a r_y^{a^2} r_\beta^b - 6r_x^a r_y^a c_{y\beta})$
33s	1β	$R^{-5}.\frac{1}{4}\sqrt{10}(-7r_y^{a^3} r_\beta^b + 3(r_x^{a^2} - r_y^{a^2})c_{y\beta} + 21r_x^{a^2} r_y^a r_\beta^b + 6r_x^a r_y^a c_{x\beta})$
20	20	$R^{-5}.\frac{3}{4}(35r_z^{a^2} r_z^{b^2} - 5r_z^{a^2} - 5r_z^{b^2} + 20r_z^a r_z^b c_{zz} + 2c_{zz}^2 + 1)$
20	21c	$R^{-5}.\frac{1}{2}\sqrt{3}(35r_z^{a^2} r_x^b r_z^b - 5r_x^b r_z^b + 10r_z^a r_x^b c_{zz} + 10r_z^a r_z^b c_{zx}$ $+ 2c_{zx}c_{zz})$
20	21s	$R^{-5}.\frac{1}{2}\sqrt{3}(35r_z^{a^2} r_y^b r_z^b - 5r_y^b r_z^b + 10r_z^a r_y^b c_{zz} + 10r_z^a r_z^b c_{zy}$ $+ 2c_{zy}c_{zz})$

Table F.1 *continued*

t	u	$4\pi\varepsilon_0 T_{tu}^{ab}$
20	22c	$R^{-5}.\frac{1}{4}\sqrt{3}(35r_z^{a2}r_x^{b2} - 35r_z^{a2}r_y^{b2} - 5r_x^{b2} + 5r_y^{b2} + 20r_z^a r_x^b c_{zx}$ $- 20r_z^a r_y^b c_{zy} + 2c_{zx}^2 - 2c_{zy}^2)$
20	22s	$R^{-5}.\frac{1}{2}\sqrt{3}(35r_z^{a2}r_x^b r_y^b - 5r_x^b r_y^b + 10r_z^a r_x^b c_{zy} + 10r_z^a r_y^b c_{zx}$ $+ 2c_{zx}c_{zy})$
21c	21c	$R^{-5}.(35r_x^a r_z^a r_x^b r_z^b + 5r_x^a r_x^b c_{zz} + 5r_x^a r_z^b c_{zx} + 5r_z^a r_x^b c_{xz}$ $+ 5r_z^a r_z^b c_{xx} + c_{xx}c_{zz} + c_{xz}c_{zx})$
21c	21s	$R^{-5}.(35r_x^a r_z^a r_y^b r_z^b + 5r_x^a r_y^b c_{zz} + 5r_x^a r_z^b c_{zy} + 5r_z^a r_y^b c_{xz}$ $+ 5r_z^a r_z^b c_{xy} + c_{xy}c_{zz} + c_{xz}c_{zy})$
21c	22c	$R^{-5}.\frac{1}{2}(35r_x^a r_z^a r_x^{b2} - 35r_x^a r_z^a r_y^{b2} + 10r_x^a r_x^b c_{zx} - 10r_x^a r_y^b c_{zy}$ $+ 10r_z^a r_x^b c_{xx} - 10r_z^a r_y^b c_{xy} + 2c_{xx}c_{zx} - 2c_{xy}c_{zy})$
21c	22s	$R^{-5}.(35r_x^a r_z^a r_x^b r_y^b + 5r_x^a r_x^b c_{zy} + 5r_x^a r_y^b c_{zx} + 5r_z^a r_x^b c_{xy}$ $+ 5r_z^a r_y^b c_{xx} + c_{xx}c_{zy} + c_{xy}c_{zx})$
21s	21s	$R^{-5}.(35r_y^a r_z^a r_y^b r_z^b + 5r_y^a r_y^b c_{zz} + 5r_y^a r_z^b c_{zy} + 5r_z^a r_y^b c_{yz}$ $+ 5r_z^a r_z^b c_{yy} + c_{yy}c_{zz} + c_{yz}c_{zy})$
21s	22c	$R^{-5}.\frac{1}{2}(35r_y^a r_z^a r_x^{b2} - 35r_y^a r_z^a r_y^{b2} + 10r_y^a r_x^b c_{zx} - 10r_y^a r_y^b c_{zy}$ $+ 10r_z^a r_x^b c_{yx} - 10r_z^a r_y^b c_{yy} + 2c_{yx}c_{zx} - 2c_{yy}c_{zy})$
21s	22s	$R^{-5}.(35r_y^a r_z^a r_x^b r_y^b + 5r_y^a r_x^b c_{zy} + 5r_y^a r_y^b c_{zx} + 5r_z^a r_x^b c_{yy}$ $+ 5r_z^a r_y^b c_{yx} + c_{yx}c_{zy} + c_{yy}c_{zx})$
22c	22c	$R^{-5}.\frac{1}{4}(35r_x^{a2}r_x^{b2} - 35r_x^{a2}r_y^{b2} - 35r_y^{a2}r_x^{b2} + 35r_y^{a2}r_y^{b2}$ $+ 20r_x^a r_x^b c_{xx} - 20r_x^a r_y^b c_{xy} - 20r_y^a r_x^b c_{yx} + 20r_y^a r_y^b c_{yy}$ $+ 2c_{xx}^2 - 2c_{xy}^2 - 2c_{yx}^2 + 2c_{yy}^2)$
22c	22s	$R^{-5}.\frac{1}{2}(35r_x^{a2}r_x^b r_y^b - 35r_y^{a2}r_x^b r_y^b + 10r_x^a r_x^b c_{xy} + 10r_x^a r_y^b c_{xx}$ $- 10r_y^a r_x^b c_{yy} - 10r_y^a r_y^b c_{yx} + 2c_{xx}c_{xy} - 2c_{yx}c_{yy})$
22s	22s	$R^{-5}.(35r_x^a r_y^a r_x^b r_y^b + 5r_x^a r_x^b c_{yy} + 5r_x^a r_y^b c_{yx} + 5r_y^a r_x^b c_{xy}$ $+ 5r_y^a r_y^b c_{xx} + c_{xx}c_{yy} + c_{xy}c_{yx})$

Table F.1 *continued*

t	u	$4\pi\varepsilon_0 T^{ab}_{tu}$
50	00	$R^{-6}.\frac{1}{8}(63r_z^{a^5}-70r_z^{a^3}+15r_z^a)$
51c	00	$R^{-6}.\frac{1}{8}\sqrt{15}(21r_x^ar_z^{a^4}-14r_x^ar_z^{a^2}+r_x^a)$
51s	00	$R^{-6}.\frac{1}{8}\sqrt{15}(21r_y^ar_z^{a^4}-14r_y^ar_z^{a^2}+r_y^a)$
52c	00	$R^{-6}.\frac{1}{4}\sqrt{105}(3r_x^{a^2}r_z^{a^3}-3r_y^{a^2}r_z^{a^3}-r_x^{a^2}r_z^a+r_y^{a^2}r_z^a)$
52s	00	$R^{-6}.\frac{1}{2}\sqrt{105}(3r_x^ar_y^ar_z^{a^3}-r_x^ar_y^ar_z^a)$
53c	00	$R^{-6}.\frac{1}{16}\sqrt{70}(9r_x^{a^3}r_z^{a^2}-27r_x^ar_y^{a^2}r_z^{a^2}-r_x^{a^3}+3r_x^ar_y^{a^2})$
53s	00	$R^{-6}.\frac{1}{16}\sqrt{70}(27r_x^{a^2}r_y^ar_z^{a^2}-9r_y^{a^3}r_z^{a^2}-3r_x^{a^2}r_y^a+r_y^{a^3})$
54c	00	$R^{-6}.\frac{3}{8}\sqrt{35}(r_x^{a^4}r_z^a-6r_x^{a^2}r_y^{a^2}r_z^a+r_y^{a^4}r_z^a)$
54s	00	$R^{-6}.\frac{3}{2}\sqrt{35}(r_x^{a^3}r_y^ar_z^a-r_x^ar_y^{a^3}r_z^a)$
55c	00	$R^{-6}.\frac{3}{16}\sqrt{14}(r_x^{a^5}-10r_x^{a^3}r_y^{a^2}+5r_x^ar_y^{a^4})$
55s	00	$R^{-6}.\frac{3}{16}\sqrt{14}(5r_x^{a^4}r_y^a-10r_x^{a^2}r_y^{a^3}+r_y^{a^5})$
40	1β	$R^{-6}.\frac{5}{8}(63r_z^{a^4}r_\beta^b-42r_z^{a^2}r_\beta^b+28r_z^{a^3}c_{z\beta}+3r_\beta^b-12r_z^ac_{z\beta})$
41c	1β	$R^{-6}.\frac{1}{4}\sqrt{10}(63r_x^ar_z^{a^3}r_\beta^b-21r_x^ar_z^ar_\beta^b+21r_x^ar_z^{a^2}c_{z\beta}+7r_z^{a^3}c_{x\beta}$ $-3r_x^ac_{z\beta}-3r_z^ac_{x\beta})$
41s	1β	$R^{-6}.\frac{1}{4}\sqrt{10}(63r_y^ar_z^{a^3}r_\beta^b-21r_y^ar_z^ar_\beta^b+21r_y^ar_z^{a^2}c_{z\beta}+7r_z^{a^3}c_{y\beta}$ $-3r_y^ac_{z\beta}-3r_z^ac_{y\beta})$
42c	1β	$R^{-6}.\frac{1}{4}\sqrt{5}(63r_x^{a^2}r_z^{a^2}r_\beta^b-63r_y^{a^2}r_z^{a^2}r_\beta^b-7r_x^{a^2}r_\beta^b+7r_y^{a^2}r_\beta^b$ $+14r_x^{a^2}r_z^ac_{z\beta}+14r_x^ar_z^{a^2}c_{x\beta}-14r_y^{a^2}r_z^ac_{z\beta}-14r_y^ar_z^{a^2}c_{y\beta}$ $-2r_x^ac_{x\beta}+2r_y^ac_{y\beta})$
42s	1β	$R^{-6}.\frac{1}{2}\sqrt{5}(63r_x^ar_y^ar_z^{a^2}r_\beta^b-7r_x^ar_y^ar_\beta^b+14r_x^ar_y^ar_z^ac_{z\beta}+7r_x^ar_z^{a^2}c_{y\beta}$ $+7r_y^ar_z^{a^2}c_{x\beta}-r_x^ac_{y\beta}-r_y^ac_{x\beta})$
43c	1β	$R^{-6}.\frac{1}{4}\sqrt{70}(9r_x^{a^3}r_z^ar_\beta^b-27r_x^ar_y^{a^2}r_z^ar_\beta^b+r_x^{a^3}c_{z\beta}+3r_x^{a^2}r_z^ac_{x\beta}$ $-3r_x^ar_y^{a^2}c_{z\beta}-6r_x^ar_y^ar_z^ac_{y\beta}-3r_y^{a^2}r_z^ac_{x\beta})$
43s	1β	$R^{-6}.\frac{1}{4}\sqrt{70}(27r_x^{a^2}r_y^ar_z^ar_\beta^b-9r_y^{a^3}r_z^ar_\beta^b+3r_x^{a^2}r_y^ac_{z\beta}+3r_x^{a^2}r_z^ac_{y\beta}$ $+6r_x^ar_y^ar_z^ac_{x\beta}-r_y^{a^3}c_{z\beta}-3r_y^{a^2}r_z^ac_{y\beta})$
44c	1β	$R^{-6}.\frac{1}{8}\sqrt{35}(9r_x^{a^4}r_\beta^b-54r_x^{a^2}r_y^{a^2}r_\beta^b+9r_y^{a^4}r_\beta^b+4r_x^{a^3}c_{x\beta}$ $-12r_x^{a^2}r_y^ac_{y\beta}-12r_x^ar_y^{a^2}c_{x\beta}+4r_y^{a^3}c_{y\beta})$
44s	1β	$R^{-6}.\frac{1}{2}\sqrt{35}(9r_x^{a^3}r_y^ar_\beta^b-9r_x^ar_y^{a^3}r_\beta^b+r_x^{a^3}c_{y\beta}+3r_x^{a^2}r_y^ac_{x\beta}$ $-3r_x^ar_y^{a^2}c_{y\beta}-r_y^{a^3}c_{x\beta})$

Table F.1 *continued*

t	u	$4\pi\varepsilon_0 T_{tu}^{ab}$
30	20	$R^{-6}.\frac{5}{4}(63r_z^{a3}r_z^{b2} - 7r_z^{a3} - 21r_z^a r_z^{b2} + 42r_z^{a2}r_z^b c_{zz} + 3r_z^a - 6r_z^b c_{zz} + 6r_z^a c_{zz}^2)$
30	21c	$R^{-6}.\frac{5}{2}\sqrt{3}(21r_z^{a3}r_x^b r_z^b - 7r_z^a r_x^b r_z^b + 7r_z^{a2}r_x^b c_{zz} + 7r_z^{a2}r_z^b c_{zx} - r_x^b c_{zz} - r_z^b c_{zx} + 2r_z^a c_{zx}c_{zz})$
30	21s	$R^{-6}.\frac{5}{2}\sqrt{3}(21r_z^{a3}r_y^b r_z^b - 7r_z^a r_y^b r_z^b + 7r_z^{a2}r_y^b c_{zz} + 7r_z^{a2}r_z^b c_{zy} - r_y^b c_{zz} - r_z^b c_{zy} + 2r_z^a c_{zy}c_{zz})$
30	22c	$R^{-6}.\frac{5}{4}\sqrt{3}(21r_z^{a3}r_x^{b2} - 21r_z^{a3}r_y^{b2} - 7r_z^a r_x^{b2} + 7r_z^a r_y^{b2} + 14r_z^{a2}r_x^b c_{zx} - 14r_z^{a2}r_y^b c_{zy} - 2r_x^b c_{zx} + 2r_y^b c_{zy} + 2r_z^a c_{zx}^2 - 2r_z^a c_{zy}^2)$
30	22s	$R^{-6}.\frac{5}{2}\sqrt{3}(21r_z^{a3}r_x^b r_y^b - 7r_z^a r_x^b r_y^b + 7r_z^{a2}r_x^b c_{zy} + 7r_z^{a2}r_y^b c_{zx} - r_x^b c_{zy} - r_y^b c_{zx} + 2r_z^a c_{zx}c_{zy})$
31c	20	$R^{-6}.\frac{5}{8}\sqrt{6}(63r_x^a r_z^{a2} r_z^{b2} - 7r_x^a r_z^{a2} - 7r_x^a r_z^{b2} + 28r_x^a r_z^a r_z^b c_{zz} + 14r_z^{a2}r_z^b c_{xz} + r_x^a - 2r_z^b c_{xz} + 2r_x^a c_{zz}^2 + 4r_z^a c_{xz}c_{zz})$
31c	21c	$R^{-6}.\frac{5}{4}\sqrt{2}(63r_x^a r_z^{a2} r_x^b r_z^b - 7r_x^a r_x^b r_z^b + 14r_x^a r_z^a r_x^b c_{zz} + 14r_x^a r_z^a r_z^b c_{zx} + 7r_z^{a2} r_x^b c_{xz} + 7r_z^{a2} r_z^b c_{xx} - r_x^b c_{xz} - r_z^b c_{xx} + 2r_x^a c_{zx}c_{zz} + 2r_z^a c_{xx}c_{zz} + 2r_z^a c_{xz}c_{zx})$
31c	21s	$R^{-6}.\frac{5}{4}\sqrt{2}(63r_x^a r_z^{a2} r_y^b r_z^b - 7r_x^a r_y^b r_z^b + 14r_x^a r_z^a r_y^b c_{zz} + 14r_x^a r_z^a r_z^b c_{zy} + 7r_z^{a2} r_y^b c_{xz} + 7r_z^{a2} r_z^b c_{xy} - r_y^b c_{xz} - r_z^b c_{xy} + 2r_x^a c_{zy}c_{zz} + 2r_z^a c_{xy}c_{zz} + 2r_z^a c_{xz}c_{zy})$
31c	22c	$R^{-6}.\frac{5}{8}\sqrt{2}(63r_x^a r_z^{a2} r_x^{b2} - 63r_x^a r_z^{a2} r_y^{b2} - 7r_x^a r_x^{b2} + 7r_x^a r_y^{b2} + 28r_x^a r_z^a r_x^b c_{zx} - 28r_x^a r_z^a r_y^b c_{zy} + 14r_z^{a2} r_x^b c_{xx} - 14r_z^{a2} r_y^b c_{xy} - 2r_x^b c_{xx} + 2r_y^b c_{xy} + 2r_x^a c_{zx}^2 - 2r_x^a c_{zy}^2 + 4r_z^a c_{xx}c_{zx} - 4r_z^a c_{xy}c_{zy})$
31c	22s	$R^{-6}.\frac{5}{4}\sqrt{2}(63r_x^a r_z^{a2} r_x^b r_y^b - 7r_x^a r_x^b r_y^b + 14r_x^a r_z^a r_x^b c_{zy} + 14r_x^a r_z^a r_y^b c_{zx} + 7r_z^{a2} r_x^b c_{xy} + 7r_z^{a2} r_y^b c_{xx} - r_x^b c_{xy} - r_y^b c_{xx} + 2r_x^a c_{zx}c_{zy} + 2r_z^a c_{xx}c_{zy} + 2r_z^a c_{xy}c_{zx})$
31s	20	$R^{-6}.\frac{5}{8}\sqrt{6}(63r_y^a r_z^{a2} r_z^{b2} - 7r_y^a r_z^{a2} - 7r_y^a r_z^{b2} + 28r_y^a r_z^a r_z^b c_{zz} + 14r_z^{a2} r_z^b c_{yz} + r_y^a - 2r_z^b c_{yz} + 2r_y^a c_{zz}^2 + 4r_z^a c_{yz}c_{zz})$
31s	21c	$R^{-6}.\frac{5}{4}\sqrt{2}(63r_y^a r_z^{a2} r_x^b r_z^b - 7r_y^a r_x^b r_z^b + 14r_y^a r_z^a r_x^b c_{zz} + 14r_y^a r_z^a r_z^b c_{zx} + 7r_z^{a2} r_x^b c_{yz} + 7r_z^{a2} r_z^b c_{yx} - r_x^b c_{yz} - r_z^b c_{yx} + 2r_y^a c_{zx}c_{zz} + 2r_z^a c_{yx}c_{zz} + 2r_z^a c_{yz}c_{zx})$

Table F.1 *continued*

t	u	$4\pi\varepsilon_0 T^{ab}_{tu}$
31s	21s	$R^{-6}.\frac{5}{4}\sqrt{2}(63r^a_y r^{a^2}_z r^b_y r^b_z - 7r^a_y r^b_y r^b_z + 14r^a_y r^a_z r^b_y c_{zz}$
		$+ 14r^a_y r^a_z r^b_z c_{zy} + 7r^{a^2}_z r^b_y c_{yz} + 7r^{a^2}_z r^b_z c_{yy} - r^b_y c_{yz} - r^b_z c_{yy}$
		$+ 2r^a_y c_{zy} c_{zz} + 2r^a_z c_{yy} c_{zz} + 2r^a_z c_{yz} c_{zy})$
31s	22c	$R^{-6}.\frac{5}{8}\sqrt{2}(63r^a_y r^{a^2}_z r^{b^2}_x - 63r^a_y r^{a^2}_z r^{b^2}_y - 7r^a_y r^{b^2}_x + 7r^a_y r^{b^2}_y$
		$+ 28r^a_y r^a_z r^b_x c_{zx} - 28r^a_y r^a_z r^b_y c_{zy} + 14r^{a^2}_z r^b_x c_{yx} - 14r^{a^2}_z r^b_y c_{yy}$
		$- 2r^b_x c_{yx} + 2r^b_y c_{yy} + 2r^a_y c^2_{zx} - 2r^a_y c^2_{zy}$
		$+ 4r^a_z c_{yx} c_{zx} - 4r^a_z c_{yy} c_{zy})$
31s	22s	$R^{-6}.\frac{5}{4}\sqrt{2}(63r^a_y r^{a^2}_z r^b_x r^b_y - 7r^a_y r^b_x r^b_y + 14r^a_y r^a_z r^b_x c_{zy}$
		$+ 14r^a_y r^a_z r^b_y c_{zx} + 7r^{a^2}_z r^b_x c_{yy} + 7r^{a^2}_z r^b_y c_{yx} - r^b_x c_{yy} - r^b_y c_{yx}$
		$+ 2r^a_y c_{zx} c_{zy} + 2r^a_z c_{yx} c_{zy} + 2r^a_z c_{yy} c_{zx})$
32c	20	$R^{-6}.\frac{1}{4}\sqrt{15}(63r^{a^2}_x r^a_z r^{b^2}_z - 63r^{a^2}_y r^a_z r^{b^2}_z - 7r^{a^2}_x r^a_z + 7r^{a^2}_y r^a_z$
		$+ 14r^{a^2}_x r^b_z c_{zz} + 28r^a_x r^a_z r^b_z c_{xz} - 14r^{a^2}_y r^b_z c_{zz} - 28r^a_y r^a_z r^b_z c_{yz}$
		$+ 4r^a_x c_{xz} c_{zz} - 4r^a_y c_{yz} c_{zz} + 2r^a_z c^2_{xz} - 2r^a_z c^2_{yz})$
32c	21c	$R^{-6}.\frac{1}{2}\sqrt{5}(63r^{a^2}_x r^a_z r^b_x r^b_z - 63r^{a^2}_y r^a_z r^b_x r^b_z + 7r^{a^2}_x r^b_x c_{zz} + 7r^{a^2}_x r^b_z c_{zx}$
		$+ 14r^a_x r^a_z r^b_x c_{xz} + 14r^a_x r^a_z r^b_z c_{xx} - 7r^{a^2}_y r^b_x c_{zz} - 7r^{a^2}_y r^b_z c_{zx}$
		$- 14r^a_y r^a_z r^b_x c_{yz} - 14r^a_y r^a_z r^b_z c_{yx} + 2r^a_x c_{xx} c_{zz} + 2r^a_x c_{xz} c_{zx}$
		$- 2r^a_y c_{yx} c_{zz} - 2r^a_y c_{yz} c_{zx} + 2r^a_z c_{xx} c_{xz} - 2r^a_z c_{yx} c_{yz})$
32c	21s	$R^{-6}.\frac{1}{2}\sqrt{5}(63r^{a^2}_x r^a_z r^b_y r^b_z - 63r^{a^2}_y r^a_z r^b_y r^b_z + 7r^{a^2}_x r^b_y c_{zz} + 7r^{a^2}_x r^b_z c_{zy}$
		$+ 14r^a_x r^a_z r^b_y c_{xz} + 14r^a_x r^a_z r^b_z c_{xy} - 7r^{a^2}_y r^b_y c_{zz} - 7r^{a^2}_y r^b_z c_{zy}$
		$- 14r^a_y r^a_z r^b_y c_{yz} - 14r^a_y r^a_z r^b_z c_{yy} + 2r^a_x c_{xy} c_{zz} + 2r^a_x c_{xz} c_{zy}$
		$- 2r^a_y c_{yy} c_{zz} - 2r^a_y c_{yz} c_{zy} + 2r^a_z c_{xy} c_{xz} - 2r^a_z c_{yy} c_{yz})$
32c	22c	$R^{-6}.\frac{1}{4}\sqrt{5}(63r^{a^2}_x r^a_z r^{b^2}_x - 63r^{a^2}_x r^a_z r^{b^2}_y - 63r^{a^2}_y r^a_z r^{b^2}_x$
		$+ 63r^{a^2}_y r^a_z r^{b^2}_y + 14r^{a^2}_x r^b_x c_{zx} - 14r^{a^2}_x r^b_y c_{zy} + 28r^a_x r^a_z r^b_x c_{xx}$
		$- 28r^a_x r^a_z r^b_y c_{xy} - 14r^{a^2}_y r^b_x c_{zx} + 14r^{a^2}_y r^b_y c_{zy} - 28r^a_y r^a_z r^b_x c_{yx}$
		$+ 28r^a_y r^a_z r^b_y c_{yy} + 4r^a_x c_{xx} c_{zx} - 4r^a_x c_{xy} c_{zy} - 4r^a_y c_{yx} c_{zx}$
		$+ 4r^a_y c_{yy} c_{zy} + 2r^a_z c^2_{xx} - 2r^a_z c^2_{xy} - 2r^a_z c^2_{yx} + 2r^a_z c^2_{yy})$
32c	22s	$R^{-6}.\frac{1}{2}\sqrt{5}(63r^{a^2}_x r^a_z r^b_x r^b_y - 63r^{a^2}_y r^a_z r^b_x r^b_y + 7r^{a^2}_x r^b_x c_{zy} + 7r^{a^2}_x r^b_y c_{zx}$
		$+ 14r^a_x r^a_z r^b_x c_{xy} + 14r^a_x r^a_z r^b_y c_{xx} - 7r^{a^2}_y r^b_x c_{zy} - 7r^{a^2}_y r^b_y c_{zx}$
		$- 14r^a_y r^a_z r^b_x c_{yy} - 14r^a_y r^a_z r^b_y c_{yx} + 2r^a_x c_{xx} c_{zy} + 2r^a_x c_{xy} c_{zx}$
		$- 2r^a_y c_{yx} c_{zy} - 2r^a_y c_{yy} c_{zx} + 2r^a_z c_{xx} c_{xy} - 2r^a_z c_{yx} c_{yy})$

Table F.1 *continued*

t	u	$4\pi\varepsilon_0 T_{tu}^{ab}$
$32s$	20	$R^{-6}.\frac{1}{2}\sqrt{15}(63r_x^a r_y^a r_z^a r_z^{b^2} - 7r_x^a r_y^a r_z^a + 14r_x^a r_y^a r_z^b c_{zz} + 14r_x^a r_z^a r_z^b c_{yz} + 14r_y^a r_z^a r_z^b c_{xz} + 2r_x^a c_{yz}c_{zz} + 2r_y^a c_{xz}c_{zz} + 2r_z^a c_{xz}c_{yz})$
$32s$	$21c$	$R^{-6}.\sqrt{5}(63r_x^a r_y^a r_z^a r_x^b r_z^b + 7r_x^a r_y^a r_x^b c_{zz} + 7r_x^a r_y^a r_z^b c_{zx} + 7r_x^a r_z^a r_x^b c_{yz} + 7r_x^a r_z^a r_z^b c_{yx} + 7r_y^a r_z^a r_x^b c_{xz} + 7r_y^a r_z^a r_z^b c_{xx} + r_x^a c_{yx}c_{zz} + r_x^a c_{yz}c_{zx} + r_y^a c_{xx}c_{zz} + r_y^a c_{xz}c_{zx} + r_z^a c_{xx}c_{yz} + r_z^a c_{xz}c_{yx})$
$32s$	$21s$	$R^{-6}.\sqrt{5}(63r_x^a r_y^a r_z^a r_y^b r_z^b + 7r_x^a r_y^a r_y^b c_{zz} + 7r_x^a r_y^a r_z^b c_{zy} + 7r_x^a r_z^a r_y^b c_{yz} + 7r_x^a r_z^a r_z^b c_{yy} + 7r_y^a r_z^a r_y^b c_{xz} + 7r_y^a r_z^a r_z^b c_{xy} + r_x^a c_{yy}c_{zz} + r_x^a c_{yz}c_{zy} + r_y^a c_{xy}c_{zz} + r_y^a c_{xz}c_{zy} + r_z^a c_{xy}c_{yz} + r_z^a c_{xz}c_{yy})$
$32s$	$22c$	$R^{-6}.\frac{1}{2}\sqrt{5}(63r_x^a r_y^a r_z^a r_x^{b^2} - 63r_x^a r_y^a r_z^a r_y^{b^2} + 14r_x^a r_y^a r_x^b c_{zx} - 14r_x^a r_y^a r_y^b c_{zy} + 14r_x^a r_z^a r_x^b c_{yx} - 14r_x^a r_z^a r_y^b c_{yy} + 14r_y^a r_z^a r_x^b c_{xx} - 14r_y^a r_z^a r_y^b c_{xy} + 2r_x^a c_{yx}c_{zx} - 2r_x^a c_{yy}c_{zy} + 2r_y^a c_{xx}c_{zx} - 2r_y^a c_{xy}c_{zy} + 2r_z^a c_{xx}c_{yx} - 2r_z^a c_{xy}c_{yy})$
$32s$	$22s$	$R^{-6}.\sqrt{5}(63r_x^a r_y^a r_z^a r_x^b r_y^b + 7r_x^a r_y^a r_x^b c_{zy} + 7r_x^a r_y^a r_y^b c_{zx} + 7r_x^a r_z^a r_x^b c_{yy} + 7r_x^a r_z^a r_y^b c_{yx} + 7r_y^a r_z^a r_x^b c_{xy} + 7r_y^a r_z^a r_y^b c_{xx} + r_x^a c_{yx}c_{zy} + r_x^a c_{yy}c_{zx} + r_y^a c_{xx}c_{zy} + r_y^a c_{xy}c_{zx} + r_z^a c_{xx}c_{yy} + r_z^a c_{xy}c_{yx})$
$33c$	20	$R^{-6}.\frac{1}{8}\sqrt{10}(63r_x^{a^3} r_z^{b^2} - 189r_x^a r_y^{a^2} r_z^{b^2} - 7r_x^{a^3} + 21r_x^a r_y^{a^2} + 42r_x^{a^2} r_z^b c_{xz} - 84r_x^a r_y^a r_z^b c_{yz} - 42r_y^{a^2} r_z^b c_{xz} + 6r_x^a c_{xz}^2 - 6r_x^a c_{yz}^2 - 12r_y^a c_{xz}c_{yz})$
$33c$	$21c$	$R^{-6}.\frac{1}{4}\sqrt{30}(21r_x^{a^3} r_x^b r_z^b - 63r_x^a r_y^{a^2} r_x^b r_z^b + 7r_x^{a^2} r_x^b c_{xz} + 7r_x^{a^2} r_z^b c_{xx} - 14r_x^a r_y^a r_x^b c_{yz} - 14r_x^a r_y^a r_z^b c_{yx} - 7r_y^{a^2} r_x^b c_{xz} - 7r_y^{a^2} r_z^b c_{xx} + 2r_x^a c_{xx}c_{xz} - 2r_x^a c_{yx}c_{yz} - 2r_y^a c_{xx}c_{yz} - 2r_y^a c_{xz}c_{yx})$
$33c$	$21s$	$R^{-6}.\frac{1}{4}\sqrt{30}(21r_x^{a^3} r_y^b r_z^b - 63r_x^a r_y^{a^2} r_y^b r_z^b + 7r_x^{a^2} r_y^b c_{xz} + 7r_x^{a^2} r_z^b c_{xy} - 14r_x^a r_y^a r_y^b c_{yz} - 14r_x^a r_y^a r_z^b c_{yy} - 7r_y^{a^2} r_y^b c_{xz} - 7r_y^{a^2} r_z^b c_{xy} + 2r_x^a c_{xy}c_{xz} - 2r_x^a c_{yy}c_{yz} - 2r_y^a c_{xy}c_{yz} - 2r_y^a c_{xz}c_{yy})$
$33c$	$22c$	$R^{-6}.\frac{1}{8}\sqrt{30}(21r_x^{a^3} r_x^{b^2} - 21r_x^{a^3} r_y^{b^2} - 63r_x^a r_y^{a^2} r_x^{b^2} + 63r_x^a r_y^{a^2} r_y^{b^2} + 14r_x^{a^2} r_x^b c_{xx} - 14r_x^{a^2} r_y^b c_{xy} - 28r_x^a r_y^a r_x^b c_{yx} + 28r_x^a r_y^a r_y^b c_{yy} - 14r_y^{a^2} r_x^b c_{xx} + 14r_y^{a^2} r_y^b c_{xy} + 2r_x^a c_{xx}^2 - 2r_x^a c_{xy}^2 - 2r_x^a c_{yx}^2 + 2r_x^a c_{yy}^2 - 4r_y^a c_{xx}c_{yx} + 4r_y^a c_{xy}c_{yy})$

Table F.1 *continued*

t	u	$4\pi\varepsilon_0 T_{tu}^{ab}$
$33c$	$22s$	$R^{-6}.\frac{1}{4}\sqrt{30}(21r_x^{a^3}r_x^b r_y^b - 63r_x^a r_y^{a^2} r_x^b r_y^b + 7r_x^{a^2} r_x^b c_{xy} + 7r_x^{a^2} r_y^b c_{xx}$ $- 14r_x^a r_y^a r_x^b c_{yy} - 14r_x^a r_y^a r_y^b c_{yx} - 7r_y^{a^2} r_x^b c_{xy} - 7r_y^{a^2} r_y^b c_{xx}$ $+ 2r_x^a c_{xx}c_{xy} - 2r_x^a c_{yx}c_{yy} - 2r_y^a c_{xx}c_{yy} - 2r_y^a c_{xy}c_{yx})$
$33s$	20	$R^{-6}.\frac{1}{8}\sqrt{10}(189r_x^{a^2} r_y^a r_z^{b^2} - 63r_y^{a^3} r_z^{b^2} - 21r_x^{a^2} r_y^a + 7r_y^{a^3}$ $+ 42r_x^{a^2} r_z^b c_{yz} + 84r_x^a r_y^a r_z^b c_{xz} - 42r_y^{a^2} r_z^b c_{yz} + 12r_x^a c_{xz}c_{yz}$ $+ 6r_y^a c_{xz}^2 - 6r_y^a c_{yz}^2)$
$33s$	$21c$	$R^{-6}.\frac{1}{4}\sqrt{30}(63r_x^{a^2} r_y^a r_x^b r_z^b - 21r_y^{a^3} r_x^b r_z^b + 7r_x^{a^2} r_x^b c_{yz} + 7r_x^{a^2} r_z^b c_{yx}$ $+ 14r_x^a r_y^a r_x^b c_{xz} + 14r_x^a r_y^a r_z^b c_{xx} - 7r_y^{a^2} r_x^b c_{yz} - 7r_y^{a^2} r_z^b c_{yx}$ $+ 2r_x^a c_{xx}c_{yz} + 2r_x^a c_{xz}c_{yx} + 2r_y^a c_{xx}c_{xz} - 2r_y^a c_{yx}c_{yz})$
$33s$	$21s$	$R^{-6}.\frac{1}{4}\sqrt{30}(63r_x^{a^2} r_y^a r_y^b r_z^b - 21r_y^{a^3} r_y^b r_z^b + 7r_x^{a^2} r_y^b c_{yz} + 7r_x^{a^2} r_z^b c_{yy}$ $+ 14r_x^a r_y^a r_y^b c_{xz} + 14r_x^a r_y^a r_z^b c_{xy} - 7r_y^{a^2} r_y^b c_{yz} - 7r_y^{a^2} r_z^b c_{yy}$ $+ 2r_x^a c_{xy}c_{yz} + 2r_x^a c_{xz}c_{yy} + 2r_y^a c_{xy}c_{xz} - 2r_y^a c_{yy}c_{yz})$
$33s$	$22c$	$R^{-6}.\frac{1}{8}\sqrt{30}(63r_x^{a^2} r_y^a r_x^{b^2} - 63r_x^{a^2} r_y^a r_y^{b^2} - 21r_y^{a^3} r_x^{b^2} + 21r_y^{a^3} r_y^{b^2}$ $+ 14r_x^{a^2} r_x^b c_{yx} - 14r_x^{a^2} r_y^b c_{yy} + 28r_x^a r_y^a r_x^b c_{xx} - 28r_x^a r_y^a r_y^b c_{xy}$ $- 14r_y^{a^2} r_x^b c_{yx} + 14r_y^{a^2} r_y^b c_{yy} + 4r_x^a c_{xx}c_{yx} - 4r_x^a c_{xy}c_{yy}$ $+ 2r_y^a c_{xx}^2 - 2r_y^a c_{xy}^2 - 2r_y^a c_{yx}^2 + 2r_y^a c_{yy}^2)$
$33s$	$22s$	$R^{-6}.\frac{1}{4}\sqrt{30}(63r_x^{a^2} r_y^a r_x^b r_y^b - 21r_y^{a^3} r_x^b r_y^b + 7r_x^{a^2} r_x^b c_{yy} + 7r_x^{a^2} r_y^b c_{yx}$ $+ 14r_x^a r_y^a r_x^b c_{xy} + 14r_x^a r_y^a r_y^b c_{xx} - 7r_y^{a^2} r_x^b c_{yy} - 7r_y^{a^2} r_y^b c_{yx}$ $+ 2r_x^a c_{xx}c_{yy} + 2r_x^a c_{xy}c_{yx} + 2r_y^a c_{xx}c_{xy} - 2r_y^a c_{yx}c_{yy})$

REFERENCES

Abramowitz, M. and Stegun, I. A. (1965), *Handbook of Mathematical Functions*, Dover, New York.

Adams, W. H. (1994), 'The polarization approximation and the Amos–Musher intermolecular perturbation theories compared to infinite order at finite separation', *Chem. Phys. Lett.* **229**, 472–480.

Ahlrichs, R. (1976), 'Convergence properties of the intermolecular force series ($1/r$ expansion)', *Theor. Chim. Acta* **41**, 7–15.

Ahlrichs, R., Penco, P. and Scoles, G. (1977), 'Intermolecular forces in simple systems', *Chem. Phys.* **19**, 119–130.

Alemán, C., Orozco, M. and Luque, F. J. (1994), 'Multicentric charges for the accurate representation of electrostatic interactions in force-field calculations for small molecules', *Chem. Phys.* **189**, 573–584.

Allen, M. P. and Tildesley, D. J. (1987), *Computer Simulation of Liquids*, Clarendon Press, Oxford.

Allinger, N. L. (1976), 'Calculation of molecular structure and energy by force-field methods', *Adv. Phys. Org. Chem.* **13**, 1–82.

Allinger, N. L. (1977), 'Conformational analysis. 130. MM2. A hydrocarbon force field utilizing V_1 and V_2 torsional terms', *J. Amer. Chem. Soc.* **99**, 8127–8134.

Allinger, N. L., Yuh, Y. H. and Lii, J.-H. (1989), 'Molecular mechanics. The MM3 force field for hydrocarbons. I.', *J. Amer. Chem. Soc.* **111**, 8551–8566.

Althorpe, S. C. and Clary, D. C. (1994), 'Calculation of the intermolecular bound states for water dimer', *J. Chem. Phys.* **101**, 3603–3609.

Althorpe, S. C. and Clary, D. C. (1995), 'A new method for calculating the rovibrational states of polyatomics with application to water dimer', *J. Chem. Phys.* **102**, 4390–4399.

Altman, R. S., Marshall, M. D. and Klemperer, W. (1982), *Disc. Faraday Soc.* **73**, 116.

Altman, R. S., Marshall, M. D. and Klemperer, W. (1983*a*), 'Electric dipole moment and quadrupole hyperfine structure of OC – HCl and OC – DCl.', *J. Chem. Phys.* **79**, 52–56.

Altman, R. S., Marshall, M. D. and Klemperer, W. (1983*b*), 'Microwave spectrum and molecular structure of N_2 – HCl', *J. Chem. Phys.* **79**, 57–64.

Amos, A. T. (1970), 'The derivation of symmetry-adapted perturbation theories', *Chem. Phys. Lett.* **5**, 587–590.

Amos, A. T. and Crispin, R. J. (1976*a*), 'Calculations of intermolecular interaction energies', *Theoretical Chemistry: Advances and Perspectives* **2**, 1–66. Note that eqn (A2) is incorrect.

Amos, A. T. and Crispin, R. J. (1976*b*), 'Intermolecular forces between large molecules', *Molec. Phys.* **31**, 159–176.

Amos, R. D. (1985), 'Multipole moments of N_2 and F_2 using SCF and Møller–Plesset calculation', *Chem. Phys. Lett.* **113**, 19–22.

Anderson, J. B. (1975), 'A random-walk simulation of the Schrödinger equation: H_3^+', *J. Chem. Phys.* **63**, 1499–1503.

Anderson, J. B. (1976), 'Quantum chemistry by random walk', *J. Chem. Phys.* **65**, 4121.

Anderson, J. B. (1980), 'Quantum chemistry by random walk: higher accuracy', *J. Chem. Phys.* **73**, 3897–3899.

Anderson, J. B. (1995), 'Fixed-node quantum Monte Carlo', *Int. Rev. Phys. Chem.* **14**, 85–112.

Ángyán, J. G., Jansen, G., Loos, M., Hättig, C. and Hess, B. A. (1994), 'Distributed polarizabilities using the topological theory of atoms in molecules', *Chem. Phys. Lett.* **219**, 267–273.

Applequist, J. (1983), 'Cartesian polytensors', *J. Math. Phys.* **24**, 736–741.

Applequist, J. (1985), 'A multipole interaction theory of electric polarization of atomic and molecular assemblies.', *J. Chem. Phys.* **83**, 809–826.

Applequist, J., Carl, J. R. and Fung, K.-K. (1972), 'An atom dipole interaction model for molecular polarizability. Application to polyatomic molecules and determination of atom polarizabilities', *J. Amer. Chem. Soc.* **94**, 2952–2960.

Axilrod, P. M. and Teller, E. (1943), 'Interaction of the Van der Waals type between three atoms', *J. Chem. Phys.* **11**, 299–300.

Aziz, R. A. (1984), *in* M. L. Klein, ed., 'Inert Gases, Potentials, Dynamics, and Energy Transfer in Doped Crystals', Springer, Berlin, chapter 2.

Bačić, Z. and Light, J. C. (1989), 'Theoretical methods for rovibrational spectra of floppy molecules', *Ann. Rev. Phys. Chem.* **40**, 469–498.

Bader, R. (1990), *Atoms in Molecules*, Clarendon Press, Oxford.

Barker, J. A. (1953), 'Statistical mechanics of interacting dipoles', *Proc. Roy. Soc. A* **219**, 367–372.

Barker, J. A. (1986), 'Many-body interactions in rare gases', *Molec. Phys.* **57**, 755–760.

Barker, J. A., Fisher, R. A. and Watts, R. O. (1971), 'Liquid argon: Monte Carlo and molecular dynamics calculations', *Molec. Phys.* **21**, 657–673.

Barnes, P., Finney, J. L., Nicholas, J. D. and Quinn, J. E. (1979), 'Cooperative effects in simulated water', *Nature* **282**, 459–464.

Basilevsky, M. V. and Berenfeld, M. M. (1972*a*), 'Intermolecular interactions in the region of small overlap', *Int. J. Quantum Chem.* **6**, 23–45.

Basilevsky, M. V. and Berenfeld, M. M. (1972*b*), 'SCF perturbation theory and intermolecular interactions', *Int. J. Quantum Chem.* **6**, 555–574.

Bell, R. J. (1970), 'Multipolar expansion for the non-additive third-order interaction energy of three atoms', *J. Phys. B* **3**, 751–62.

Ben-Naim, A. (1980), *Hydrophobic Interactions*, Plenum Press, New York.

Berendsen, H. J. C., Postma, J. P. M., von Gunsteren, W. F. and Hermans, J. (1981), 'Interaction models for water in relation to protein hydration', *in* B. Pullman, ed., 'Intermolecular Forces', D. Reidel, Dordrecht, Holland, pp. 331–342.

Bernal, J. D. and Fowler, R. H. (1933), 'A theory of water and ionic solutions', *J. Chem. Phys.* **1**, 515–548.

Berne, B. J. and Pechukas, P. (1972), 'Gaussian model potentials for molecular interactions', *J. Chem. Phys.* **56**, 4213–4216.

Berweger, C. D., van Gunsteren, W. F. and Müller-Plathe, F. (1995), 'Force field parametrisation by weak coupling: re-engineering SPC water', *Chem. Phys. Lett.* **232**, 429–436.

Birge, R. R. (1980), 'Calculation of molecular polarizabilities using an anisotropic atom point dipole interaction model which includes the effect of electron repulsion', *J. Chem. Phys.* **72**, 5312–5319.

Birnbaum, G. and Cohen, E. R. (1975), 'Far infrared collision-induced absorption in gaseous methane. II. Determination of the octopole and hexadecapole moments', *J. Chem. Phys.* **62**, 3807–3812.

Böhm, H.-J. and Ahlrichs, R. (1982), 'A study of short-range repulsions', *J. Chem. Phys.* **77**, 2028–2034.

Bondi, A. (1964), 'Van der Waals volumes and radii', *J. Phys. Chem.* **68**, 441–451.

Bone, R. G. A. and Handy, N. C. (1990), '*Ab initio* studies of internal rotation barriers and vibrational frequencies of $(C_2H_2)_2$, $(CO_2)_2$ and $C_2H_2\cdots CO_2$', *Theor. Chim. Acta* **78**, 133–163.

Bone, R. G. A., Rowlands, T. W., Handy, N. C. and Stone, A. J. (1991), 'Transition states from molecular symmetry groups: analysis of non-rigid acetylene trimer', *Molec. Phys.* **72**, 33–73.

Born, M. and Mayer, J. E. (1932), *Z. Phys.* **75**, 1.

Böttcher, C. J. F., van Belle, O. C., Bordewijk, P. and Rip, A. (1972), *Theory of Electric Polarization*, Elsevier.

Boys, S. F. (1950), 'Electronic wave functions. I. A general method of calculation for the stationary states of any molecular system', *Proc. Roy. Soc. A* **200**, 542–554.

Boys, S. F. and Bernardi, F. (1970), 'The calculation of small molecular interactions by the difference of separate total energies. Some procedures with reduced errors', *Molec. Phys.* **19**, 553–566.

Brink, D. M. and Satchler, G. R. (1993), *Angular Momentum*, 3rd edn, Clarendon Press, Oxford.

Brooks, B. R., Bruccoleri, R. E., Olafson, B. D., States, D. J., Swaminathan, S. and Karplus, M. (1983), 'CHARMM: A program for macromolecular energy, minimization and dynamics calculations', *J. Comput. Chem.* **4**, 187–217.

Brooks, F. C. (1952), 'Convergence of intermolecular force series', *Phys. Rev.* **86**, 92–97.

Buch, V. (1992), 'Treatment of rigid bodies by diffusion Monte Carlo: application to the para–$H_2\cdots H_2O$ and ortho–$H_2\cdots H_2O$ clusters', *J. Chem. Phys.* **97**, 726–729.

Buck, U., Huisken, F., Kohlhase, A., Otten, D. and Schaefer, J. (1983), 'State resolved rotational excitations in $D_2\cdots H_2$ collisions', *J. Chem. Phys.* **78**, 4439–4450.

Buckingham, A. D. (1960), 'Solvent effects in vibrational spectroscopy', *Trans. Faraday Soc.* **56**, 753–760.

Buckingham, A. D. (1967), 'Permanent and induced molecular moments and long-range intermolecular forces', *Adv. Chem. Phys.* **12**, 107–143.

Buckingham, A. D. (1978), 'Basic theory of intermolecular forces: applications to small molecules', *in* B. Pullman, ed., 'Intermolecular Interactions from Diatomics to Biopolymers', Wiley, Chichester, pp. 1–67.

Buckingham, A. D. and Disch, R. L. (1963), 'The quadrupole moment of carbon dioxide', *Proc. Roy. Soc. A* **273**, 275–289.

Buckingham, A. D. and Fowler, P. W. (1983), 'Do electrostatic interactions predict structures of Van der Waals molecules?', *J. Chem. Phys.* **79**, 6426–6428.

Buckingham, A. D. and Fowler, P. W. (1985), 'A model for the geometries of Van der Waals complexes', *Canad. J. Chem.* **63**, 2018–2025.

Buckingham, A. D., Fowler, P. W. and Stone, A. J. (1986), 'Electrostatic predictions of shapes and properties of Van der Waals molecules', *Internat. Rev. Phys. Chem.* **5**, 107–114.

Buckingham, A. D., Graham, C. and Williams, J. H. (1983), 'Electric field-gradient-induced birefringence in N_2, C_2H_6, C_3H_6, Cl_2, N_2O and CH_3F', *Molec. Phys.* **49**, 703–710.

Buckingham, R. A. and Corner, J. (1947), 'Tables of second virial and low-pressure Joule–Thompson coefficients for intermolecular potentials with exponential repulsion', *Proc. Roy. Soc.A* **189**, 118–129.

Cammi, R., Bonaccorsi, R. and Tomasi, J. (1985), 'Counterpoise corrections to the interaction energy components in bimolecular complexes', *Theor. Chim. Acta* **68**, 271–283.

Car, R. and Parrinello, M. (1985), 'Unified approach for molecular dynamics and density functional theory', *Phys. Rev. Lett.* **55**, 2471–2474.

Casimir, H. B. G. and Polder, D. (1948), 'The influence of retardation on the London–Van der Waals forces', *Phys. Rev.* **73**, 360–372.

Cerjan, C. J. and Miller, W. H. (1981), 'On finding transition states', *J. Chem. Phys.* **75**, 2800–2806.

Chemical Reviews (1994), 'Van der Waals molecules', *Chem. Rev.* **94**(7), 1721–2160.

Child, M. S. (1991), *Semiclassical Mechanics with Molecular Applications*, Clarendon Press, Oxford.

Chipot, C., Angyan, J. G., Maigret, A. and Scheraga, H. A. (1994), 'Modelling amino-acid side-chains. 3. Influence of intramolecular and intermolecular environment on point charges', *J. Phys. Chem.* **98**, 1518.

Čižek, J. and Paldus, J. (1980), 'The coupled-cluster approach', *Physica Scripta* **21**, 251.

Claverie, P. (1971), 'Theory of intermolecular forces. I. On the inadequacy of the usual Rayleigh–Schrödinger perturbation method for the treatment of intermolecular forces', *Int. J. Quantum Chem.* **5**, 273–296.

Claverie, P. (1978), 'Elaboration of approximate formulas for the interactions between large molecules: applications in organic chemistry', *in* B. Pullman, ed., 'Intermolecular Interactions: from Diatomics to Biopolymers', Wiley, pp. 69–305.

Cochran, W. (1959), 'Theory of the lattice vibrations of germanium', *Proc. Roy. Soc.* **A 253**, 260–276.

Cohen, R. C. and Saykally, R. J. (1990), 'Extending the collocation method to multidimensional molecular dynamics: direct determination of the intermolecular potential of Ar – H_2O from tunable far infrared laser spectroscopy', *J. Phys. Chem.* **94**, 7991–8000.

Cohen, R. C. and Saykally, R. J. (1993), 'Determination of an improved global potential energy surface for Ar – H_2O from vibration–rotation–tunnelling spectroscopy', *J. Chem. Phys.* **98**, 6007–6030.

Cooper, A. R. and Hutson, J. M. (1993), 'Non-additive intermolecular forces from the spectroscopy of Van der Waals trimers: calculations on $Ar_2\cdots HCl$', *J. Chem. Phys.* **98**, 5337–5351.

Cooper, D. L. and Stutchbury, N. C. J. (1985), 'Distributed multipole analysis from charge partitioning by zero-flux surfaces: the structure of HF complexes', *Chem. Phys. Lett.* **120**, 167–172.

Corner, J. (1948), 'The second virial coefficient of a gas of non-spherical molecules', *Proc. Roy. Soc.* **A 192**, 275–292.

Costas, M., Kronberg, B. and Silveston, R. (1994), 'General thermodynamic analysis of the dissolution of nonpolar molecules into water—origin of hydrophobicity', *J. Chem. Soc. Faraday Trans.* **90**, 1513–1522.

Cox, S. R. and Williams, D. E. (1981), 'Representation of the molecular electrostatic potential by a net atomic charge model', *J. Comput. Chem.* **2**, 304–323.

Cozzi, F., Cinquini, M., Annunziata, R. and Siegel, J. S. (1993), 'Dominance of polar/π over charge transfer effects: stacked phenyl interactions', *J. Amer. Chem. Soc.* **115**, 5330–5331.

Craig, S. L. and Stone, A. J. (1994), 'Stereoselectivity and regioselectivity in Diels–Alder reactions studied by intermolecular perturbation theory', *J. Chem. Soc. Faraday Trans.* **90**, 1663–1668.

Cwiok, T., Jeziorski, B., Kołos, W., Moszynski, R., Rychlewski, J. and Szalewicz, K. (1992), 'Convergence properties and large-order behavior of the polarization expansion for the interaction energy of hydrogen atoms', *Chem. Phys. Lett.* **195**, 67–76.

Cwiok, T., Jeziorski, B., Kołos, W., Moszynski, R. and Szalewicz, K. (1994), 'Symmetry-adapted perturbation theory of potential energy surfaces for weakly bound molecular complexes', *J. Mol. Struct. (Theochem)* **307**, 135–151.

Cybulski, S. M. and Scheiner, S. (1990), 'Comparison of Morokuma and perturbation-theory approaches to decomposition of interaction energy: $NH_4^+ \cdots NH_3$', *Chem. Phys. Lett.* **166**, 57–64.

Dalgarno, A. (1967), 'New methods for calculating long-range intermolecular forces', *Adv. Chem. Phys.* **12**, 143–166.

Dalgarno, A. and Lynn, N. (1957*a*), 'An exact calculation of second order long range forces', *Proc. Phys. Soc. (London)* **A70**, 223–225.

Dalgarno, A. and Lynn, N. (1957*b*), 'Properties of the helium atom', *Proc. Phys. Soc. (London)* **A70**, 802–808.

Davidson, E. R. and Chakravorty, S. J. (1994), 'A possible definition of basis set superposition error', *Chem. Phys. Lett.* **217**, 48–54.

Davydov, A. S. (1962), *Theory of Molecular Excitons*, McGraw-Hill, New York. translated by M. Kasha and M. Oppenheimer, Jr. from the Russian edition of 1951.

del Bene, J. E., Person, W. B. and Szczepaniak, K. (1995), 'Properties of hydrogen-bonded complexes obtained from the B3LYP functional with 6–31G(d,p) and 6–31+G(d,p) basis sets', *J. Phys. Chem.* **99**, 10705–10707.

Dewar, M. J. S. and Thompson, C. C. (1966), 'π–molecular complexes. III. A critique of charge transfer, and stability constants for some TCNE–hydrocarbon complexes', *Tetrahedron* **S7**, 97–114.

Dewar, M. J. S., Zoebisch, E. G., Healy, E. F. and Stewart, J. J. P. (1985), 'AM1: a new general-purpose quantum mechanical molecular model', *J. Amer. Chem. Soc.* **107**, 3902–3909.

Dham, A. K., Allnatt, A. R., Meath, W. J. and Aziz, R. A. (1989), 'The Kr–Kr potential energy curve and related physical properties: the XC and HFD-B potential models', *Molec. Phys.* **67**, 1291–1307.

Dick, B. G. and Overhauser, A. W. (1958), 'Theory of the dielectric constants of alkali halide crystals', *Phys. Rev.* **112**, 90–103.

Doran, M. B. and Zucker, I. J. (1971), 'Higher-order multipole three-body Van der Waals interactions and stability of rare gas solids', *J. Phys. C* **4**, 307–312.

Douketis, C., Scoles, G., Marchetti, S., Zen, M. and Thakkar, A. J. (1982), 'Intermolecular forces *via* hybrid Hartree–Fock-SCF plus damped dispersion (HFD) energy calculations: an improved spherical model', *J. Chem. Phys.* **76**, 3057–3063.

Du, Q., Freysz, E. and Shen, Y. R. (1994), 'Surface vibrational spectroscopic studies of hydrogen bonding and hydrophobicity', *Science* **264**, 826–828.

Dulmage, W. J. and Lipscomb, W. N. (1951), 'The crystal structures of hydrogen cyanide, HCN', *Acta. Cryst.* **4**, 330–334.

Dyke, T. R., Howard, B. J. and Klemperer, W. (1972), 'Radiofrequency and microwave spectrum of HF dimer', *J. Chem. Phys.* **56**, 2442–2454.

Eberly, J. H. (1989), 'Quantum optics at very high laser intensities', *Adv. Chem. Phys.* **73**, 801–822.

Eggenberger, R., Gerber, S., Huber, H. and Searles, D. (1991), 'Basis set superposition errors in intermolecular structures and force constants', *Chem. Phys. Lett.* **183**, 223–226.

Eisenschitz, L. and London, F. (1930), *Z. Phys.* **60**, 491.

Elliott, J. P. and Dawber, P. G. (1979), *Symmetry in Physics*, MacMillan, London.

Elrod, M. J. and Saykally, R. J. (1994), 'Many-body effects in intermolecular forces', *Chem. Rev.* **94**, 1975–1997.

Elrod, M. J., Steyert, D. W. and Saykally, R. J. (1991), 'Tunable far-infrared laser spectroscopy of a ternary Van der Waals cluster Ar_2HCl: a sensitive probe of three-body forces', *J. Chem. Phys.* **94**, 58–66.

Elrod, M. J., Host, B. C., Steyert, D. W. and Saykally, R. J. (1993*a*), 'Far-infrared vibration-rotation-tunnelling spectroscopy of ArDCl. A critical test of the H6(4,3,0) potential surface', *Molec. Phys.* **79**, 245–251.

Elrod, M. J., Loeser, J. G. and Saykally, R. J. (1993*b*), 'An investigation of three-body effects in intermolecular forces. III. Far infrared laser vibration–rotation–tunnelling spectra of the lowest internal rotor state of Ar_2HCl', *J. Chem. Phys.* **98**, 5352–5361.

Epstein, S. T. and Johnson, R. E. (1968), 'The application of perturbation theories for exchange forces to a simple model', *Chem. Phys. Lett.* **1**, 602–604.

Ernesti, A. and Hutson, J. M. (1994), 'Non-additive intermolecular forces from the spectroscopy of Van der Waals trimers: the effect of monomer vibrational excitation in $Ar_2\cdots HF$ and $Ar_2\cdots HCl$', *J. Chem. Soc. Faraday Disc.* **97**, 119–129.

Etters, R. D. and Danilowicz, R. (1979), 'Three-body interactions in small rare-gas clusters', *J. Chem. Phys.* **71**, 4767–4768.

Ewald, P. (1921), *Ann. Phys.* **64**, 253–287.

Ewing, J. J., Tigelaar, H. L. and Flygare, W. H. (1972), 'Molecular Zeeman effect, magnetic properties and electric quadrupole moments in ClF, BrF, ClCN, BrCN and ICN', *J. Chem. Phys.* pp. 1957–1966.

Faraday (1994), 'Structure and dynamics of Van der Waals complexes', *Faraday Disc.* **97**, 1–461.

Feller, D. (1992), 'Application of systematic sequences of wavefunctions to the water dimer', *J. Chem. Phys.* **96**, 6104–6114.

Ferenczy, G. G. (1991), 'Charges derived from distributed multipole series', *J. Comp. Chem.* **12**, 913–917.

Figari, G. and Magnasco, V. (1985), 'On the evaluation of the cofactors occuring in the matrix elements between multiply-excited determinantal wavefunctions of non-orthogonal orbitals', *Molec. Phys.* **55**, 319–330.

Filippini, G. and Gavezzotti, A. (1993), 'Empirical intermolecular potentials for organic crystals: the '6-exp' approximation revisited', *Acta Cryst.* **B49**, 868–880.

Fleming, I. (1976), *Frontier Orbitals and Organic Chemical Reactions*, Wiley, London, New York.

Forster, T. (1955), *Z. Electrochem.* **59**, 976.

Fowler, P. W. and Madden, P. A. (1984), 'In-crystal polarizabilities of alkali and halide ions', *Phys. Rev. B* **29**, 1035–1042.

Fowler, P. W. and Madden, P. A. (1985), 'In-crystal polarizability of O^{2-}', *J. Phys. Chem.* **89**, 2581–2585.

Fowler, P. W. and Stone, A. J. (1987), 'Induced dipole moments of Van der Waals complexes', *J. Phys. Chem.* **91**, 509–511.

Francl, M. M., Pietro, W. J., Hehre, W. J., Binkley, J. S., Gordon, M. S., DeFrees, D. J. and Pople, J. A. (1982), 'SCF MO methods. 23. A polarization-type basis set for second-row elements', *J. Chem. Phys.* **77**, 3654–3665.

Fraser, G. T., Suenram, R. D., Lovas, F. J., Pine, A. S., Hougen, J. T., Lafferty, W. J. and Muenter, J. S. (1988), 'Infrared and microwave investigations of interconversion tunnelling in the acetylene dimer', *J. Chem. Phys.* **89**, 6028–6045.

Frey, R. F. and Davidson, E. R. (1989), 'Energy partitioning of the self-consistent-field interaction energy of ScCO', *J. Chem. Phys.* **90**, 5555–5562.

Freyriafava, C., Dovesi, F., Saunders, V. R., Leslie, M. and Roetti, C. (1993), 'Ca and Be substitution in bulk MgO: ab initio Hartree–Fock and ionic model supercell calculations', *J. Phys. Condensed Matter* **5**, 4793–4804.

Frisch, H. L. and Helfand, E. (1960), 'Conditions imposed by gross properties on the intermolecular potential', *J. Chem. Phys.* **32**, 269–270.

Frisch, M. J., del Bene, J. E., Binkley, J. S. and Schaefer, H. F. (1986), 'Extensive theoretical studies of the hydrogen-bonded complexes $(H_2O)_2$, $(H_2O)_2H^+$, $(HF)_2$, F_2H^- and $(NH_3)_2$', *J. Chem. Phys.* **84**, 2279–2289.

Fukui, K. and Fujimoto, H. (1968), 'An MO-theoretical interpretation of the nature of chemical reaction. I. Partitioning analysis of the interaction energy', *Bull. Chem. Soc. Japan* **41**, 1989–1997.

Gavezzotti, A. and Filippini, G. (1994), 'Geometry of the intermolecular X–H···Y (X, Y = N, O) hydrogen bond and the calibration of empirical hydrogen-bond potentials', *J. Phys. Chem.* **98**, 4831–4837.

Gerratt, J. and Mills, I. M. (1968), 'Force constants and dipole moment derivatives of molecules from perturbed Hartree–Fock calculations', *J. Chem. Phys.* **49**, 1719–1729.

Gouyet, J. F. (1973), 'Use of biorthogonal orbitals in calculation by perturbation of intermolecular interactions', *J. Chem. Phys.* **59**, 4637–4641.

Gray, C. G. and Gubbins, K. E. (1984), *Theory of Molecular Fluids*, Vol. 1: Fundamentals, Oxford University Press, Oxford.

Gregory, J. K. and Clary, D. C. (1994), 'A comparison of conventional and rigid body diffusion Monte Carlo techniques. Application to water dimer.', *Chem. Phys. Lett.* **228**, 547–554.

Gregory, J. K. and Clary, D. C. (1995), 'Calculations of the tunnelling splittings in water dimer and trimer using diffusion Monte Carlo', *J. Chem. Phys.* **102**, 7817–7829.

Gregory, J. K. and Clary, D. C. (1996), 'Three-body effects on molecular properties in the water trimer', *J. Chem. Phys.* **103**, 8924–8930.

Gussoni, M., Castiglioni, C. and Zerbi, G. (1986), 'Molecular point charges as derived from infrared intensities and from *ab initio* calculations', *Theochem (J. Mol. Struct.)* **31**, 203–212.

Gutowski, M. and Piela, L. (1988), 'Interpretation of the Hartree–Fock interaction energy between closed-shell systems', *Molec. Phys.* **64**, 337–355.

Gutowski, M., Van Duijneveldt, F. B., Chałasiński, G. and Piela, L. (1987), 'Proper correction for the basis set superposition error in SCF calculations of intermolecular interactions', *Molec. Phys.* **61**, 233–247.

Gutowski, M., van Duijneveldt-van de Rijdt, J. G. C. M., van Lenthe, J. H. and van Duijneveldt, F. B. (1993), 'Accuracy of the Boys and Bernardi function counterpoise method', *J. Chem. Phys.* **98**, 4728–4737.

Hall, D. and Pavitt, N. (1984), 'An appraisal of molecular force fields for the representation of polypeptides', *J. Comput. Chem.* **5**, 441–450.

Handy, N. C. (1994), 'Density functional theory: an alternative to quantum chemistry?', *in* R. Broer, P. J. C. Aerts and P. S. Bagus, eds, 'New Challenges in Computational Quantum Chemistry', University of Groningen, pp. 59–70.

Hariharan, P. C. and Pople, J. A. (1973), 'Self-consistent-field molecular orbital methods. XII. Further extension of Gaussian-type basis sets for use in molecular-orbital studies of organic molecules', *Theor. Chim. Acta* **28**, 213–222.

Hättig, C. and Heß, B. A. (1994), 'Calculation of orientation dependent double-tensor moments for Coulomb-type molecular interactions', *Molec. Phys.* **81**, 813–824.

Havekort, J. E. M., Bass, F. and Beenakker, J. J. M. (1983), 'Measurements of depolarization ratios of linear chain molecules: a test of the principle of additivity of bond polarizabilities', *Chem. Phys.* **79**, 105–109.

Hayes, I. C. and Stone, A. J. (1984*a*), 'An intermolecular perturbation theory for the region of moderate overlap', *Molec. Phys.* **53**, 83–105.

Hayes, I. C. and Stone, A. J. (1984*b*), 'Matrix elements between determinantal wavefunctions of non-orthogonal orbitals', *Molec. Phys.* **53**, 69–82.

Hehre, W. J., Ditchfield, R. and Pople, J. A. (1971), 'Self-consistent-field molecular-orbital methods. XII. Further extensions of Gaussian-type basis sets for use in molecular orbital studies of organic molecules', *J. Chem. Phys.* **56**, 2257–2261.

Hepburn, J., Scoles, G. and Penco, R. (1975), 'A simple but reliable method for prediction of intermolecular potentials', *Chem. Phys. Lett.* **36**, 451–456.

Hermans, J., Berendsen, H. J. C., van Gunsteren, W. F. and Postma, J. P. M. (1984), 'A consistent empirical potential for water–protein interactions', *Biopolymers* **23**, 1513–1518.

Hess, O., Caffarel, M., Huiszoon, C. and Claverie, P. (1990), 'Second-order exchange effects in intermolecular interactions. The water dimer', *J. Chem. Phys.* **92**, 6049–6060.

Hirschfelder, J. O. (1967), 'Perturbation theory for exchange forces', *Chem. Phys. Lett.* **1**, 325–329, 363–368.

Hirschfelder, J. O., Curtiss, C. F. and Bird, R. B. (1954), *Molecular Theory of Liquids and Gases*, Wiley, New York and London.

Hirschfelder, J. O. and Silbey, R. (1966), 'New type of molecular perturbation treatment', *J. Chem. Phys.* **45**, 2188–2192.

Hohenberg, P. and Kohn, W. (1964), 'Inhomogeneous electron gas', *Phys. Rev.* **136**, B864–B871.

Holmgren, S. L., Waldman, M. and Klemperer, W. (1977), 'Internal dynamics of Van der Waals complexes. I. Born–Oppenheimer separation of radial and angular motion', *J. Chem. Phys.* **67**, 4414–4422.

Huang, Z. S. and Miller, R. E. (1989), 'The structure of CO_2–HCCH from near infrared spectroscopy', *Chem. Phys.* **132**, 185–196.

Hunter, C. A. (1993), 'Sequence-dependent DNA structure: the role of base stacking interactions', *J. Molec. Biol.* **230**, 1025–1054.

Hunter, C. A. (1994), 'The role of aromatic interactions in molecular recognition', *Chem. Soc. Rev.* **23**, 101–109.

Hunter, C. A. and Sanders, J. K. M. (1990), 'The nature of $\pi - \pi$ interactions', *J. Amer. Chem. Soc.* **112**, 5525–5534.

Hunter, C. A., Sanders, J. K. M. and Stone, A. J. (1989), 'Exciton coupling in porphyrin dimers', *Chem. Phys.* **133**, 395–404.

Hunter, C. A., Singh, J. and Thornton, J. M. (1991), '$\pi - \pi$ interactions: the geometry and energetics of phenylalanine–phenylalanine interactions in proteins', *J. Molec. Biol.* **218**, 837–846.

Huot, J. and Bose, T. K. (1991), 'Determination of the quadrupole moment of nitrogen from the dielectric second virial coefficient', *J. Chem. Phys.* **94**, 3849–3854.

Hutson, J. M. (1989*a*), 'The intermolecular potential of Ne–HCl: determination from high-resolution spectroscopy', *J. Chem. Phys.* **91**, 4448–4454.

Hutson, J. M. (1989*b*), 'Anisotropic intermolecular forces. III. Rare gas–hydrogen bromide systems', *J. Chem. Phys.* **91**, 4455–4461.

Hutson, J. M. (1990*a*), 'BOUND', A computer program distributed by EPSRC Collaborative Computational Project No. 6 on Heavy Particle Dynamics.

Hutson, J. M. (1990*b*), 'Dynamics of Van der Waals complexes: beyond atom-diatom systems', *in* N. Halberstadt and K. C. Janda, eds, 'Dynamics of Polyatomic Van der Waals Complexes', NATO ASI, Plenum, pp. 67–79.

Hutson, J. M. (1992), 'Vibrational dependence of the anisotropic intermolecular potential of Ar–HCl', *J. Phys. Chem.* **96**, 4237–4247.

Hutson, J. M. and Howard, B. J. (1980), 'Spectroscopic properties and potential surfaces for atom–diatom Van der Waals molecules', *Molec. Phys.* **41**, 1123–1141.

Jalink, H., Parker, D. H. and Stolte, S. (1987), 'Experimental verification of the sign of the electric dipole moment of N_2O.', *J. Mol. Spectr.* **121**, 236–237.

Janda, K. C., Klemperer, W. and Novick, S. E. (1976), 'Measurement of the sign of the dipole moment of ClF', *J. Chem. Phys.* **64**, 2698–2699.

Jansen, L. (1957), 'Interactions between permanent multipole moments', *Physica* **23**, 599–604.

Jansen, L. (1958), 'Tensor formalism for Coulomb interaction and asymptotic properties of multipole expansions', *Phys. Rev.* **110**, 661–669.

Jansen, L. (1967), 'Schrödinger perturbation formalism for exchange interactions between atoms or molecules', *Phys. Rev.* **162**, 63–68.

Jarque, C. and Buckingham, A. D. (1989), 'Ion–ion interaction in a polarizable lattice', *Chem. Phys. Lett.* **164**, 485–490.

Jarque, C. and Buckingham, A. D. (1992), 'Ion–ion interaction in a polarizable medium', *in* J. J. C. Teixeira-Dias, ed., 'Molecular Liquids: New Perspectives in Physics and Chemistry', Kluwer, Dordrecht, pp. 253–265.

Jeffreys, H. (1931), *Cartesian Tensors*, Cambridge University Press.

Jeziorski, B. and Kołos, W. (1977), 'On symmetry forcing in the perturbation theory of weak intermolecular interactions', *Int. J. Quantum Chem.* **12, Suppl. 1**, 91–117.

Jeziorski, B., Moszynski, R., Ratkiewicz, A., Rybak, S., Szalewicz, K. and Williams, H. L. (1993), 'SAPT: a program for many-body symmetry-adapted perturbation theory calculations of intermolecular interaction energies', *in* E. Clementi, ed., 'Methods and Techniques in Computational Chemistry: METECC94', Vol. B, STEF, Cagliari, p. 79.

Jeziorski, B., Moszynski, R. and Szalewicz, K. (1994), 'Perturbation theory approach to intermolecular potential energy surfaces of Van der Waals complexes', *Chem. Rev.* **94**, 1887–1930.

Jeziorski, B., Szalewicz, K. and Chałasinski, G. (1978), 'Symmetry forcing and convergence properties of perturbation expansions for molecular interaction energies', *Int. J. Quantum Chem.* **14**, 271–287.

Jhanwar, B. L. and Meath, W. J. (1982), 'Dipole oscillator strength distributions, sums and dispersion energy coefficients for CO and CO_2', *Chem. Phys.* **67**, 185–199.

Jhanwar, B. L., Meath, W. J. and MacDonald, J. C. F. (1981), 'Dipole oscillator strength distributions and sums for C_2H_6, C_3H_8, n-C_4H_{10}, n-C_5H_{12}, n-C_6H_{14}, n-C_7H_{16}, and n-C_8H_{18}', *Canad. J. Phys.* **59**, 185–197.

Johnson, B. R. (1978), 'The renormalized Numerov method applied to calculations of bound states of the coupled-channel Schrödinger equation', *J. Chem. Phys.* **69**, 4678–4688.

Johnson, R. E. and Epstein, S. T. (1968), 'Connection between several perturbation theories of intermolecular forces', *Chem. Phys. Lett.* **1**, 599–601.

Jorgensen, W. L., Chandrasekhar, J., Madura, J. D., Impey, R. W. and Klein, M. L. (1983), 'Comparison of simple model potentials for simulating liquid water', *J. Chem. Phys.* **79**, 926–935.

Jorgensen, W. L. and Tirado-Rives, J. (1988), 'The OPLS potential function for proteins. Energy minimization for crystals of cyclic peptides and crambin', *J. Amer. Chem. Soc.* **110**, 1657–1666.

Joslin, C. G., Gray, C. G. and Singh, S. (1985), 'Far infrared absorption in gaseous CH_4 and CF_4. A theoretical study.', *Molec. Phys.* **54**, 1469–1489.

Juanós i Timoneda, J. and Hunt, K. L. C. (1986), 'Label-free exchange perturbation approximation for the collision-induced dipole of He···H', *J. Chem. Phys.* **84**, 3954–3962.

Karlström, G. (1982), 'Local polarizabilities in molecules, based on *ab initio* Hartree–Fock calculations', *Theor. Chim. Acta* **60**, 535–541.

Kell, G. S., McLaurin, G. E. and Whalley, E. (1968), 'PVT properties of water. Virial coefficients in the range 150°–450°C', *J. Chem. Phys.* **48**, 3805–3813.

Kell, G. S., McLaurin, G. E. and Whalley, E. (1989), 'PVT properties of water. VII. Vapour densities of light and heavy water from 150 to 500°C', *Proc. Roy. Soc. A* **425**, 49–71.

Keller, J. B. and Zumino, B. (1959), 'Determination of intermolecular potentials from thermodynamic data and the law of corresponding states', *J. Chem. Phys.* **30**, 1351–1353.

Kihara, T. (1978), *Intermolecular Forces*, Wiley.

Kim, Y. S., Kim, S. K. and Lee, W. D. (1981), 'Dependence of the closed-shell repulsive interaction on the overlap of the electron densities', *Chem. Phys. Lett.* **80**, 574–575.

King, B. F. and Weinhold, F. (1995), 'Structure and spectroscopy of $(HCN)_n$ clusters: cooperative and electronic delocalization effects in C–H···N hydrogen bonding', *J. Chem. Phys.* **103**, 333–347.

Kita, S., Noda, K. and Inouye, H. (1976), 'Repulsion potentials for Cl^-–R and Br^-–R (R = He, Ne and Ar) derived from beam experiments', *J. Chem. Phys.* **64**, 3446–3449.

Kitaura, K. and Morokuma, K. (1976), 'A new energy decomposition scheme for molecular interactions within the Hartree–Fock approximation', *Int. J. Quantum Chem.* **10**, 325–340.

Kittel, C. (1987), *Quantum theory of solids*, Wiley, New York & Chichester.

Klopman, G. (1968), 'Chemical reactivity and the concept of charge- and frontier-controlled reactions', *J. Amer. Chem. Soc.* **90**, 223–234.

Klopman, G. and Hudson, R. F. (1967), 'Polyelectronic perturbation theory of chemical reactivity', *Theor. Chim. Acta* **8**, 165 174.

Knowles, P. J. and Meath, W. J. (1986*a*), 'Non-expanded dispersion and induction energies, and damping functions, for molecular interactions with application to HF···He', *Molec. Phys.* **59**, 965–984.

Knowles, P. J. and Meath, W. J. (1986*b*), 'Non-expanded dispersion energies and damping functions for Ar_2 and Li_2', *Chem. Phys. Lett.* **124**, 164–171.

Knowles, P. J. and Meath, W. J. (1987), 'A separable method for the calculation of dispersion and induction energy damping functions with applications to the dimers arising from He, Ne and HF', *Molec. Phys.* **60**, 1143–1158.

Koch, U., Popelier, P. L. A. and Stone, A. J. (1995), 'Conformational dependence of atomic multipole moments', *Chem. Phys. Lett.* **238**, 253–260.

Kochanski, E. and Gouyet, J. F. (1975*a*), '*Ab initio* calculation of the first order term of the intermolecular energy near the Van der Waals minimum', *Theor. Chim. Acta* **39**, 329–337.

Kochanski, E. and Gouyet, J. F. (1975*b*), '*Ab initio* studies of the intermolecular interactions between two hydrogen molecules near the Van der Waals minimum from a perturbation procedure using biorthogonal orbitals', *Molec. Phys.* **29**, 693–701.

Koide, A., Meath, W. J. and Allnatt, A. R. (1981), 'Second-order charge overlap effects and damping functions for isotropic atomic and molecular interactions', *Chem. Phys.* **58**, 105–119.

Kołos, W. and Wolniewicz, L. (1974), 'Variational calculations of the long-range interaction between two ground state hydrogen atoms', *Chem. Phys. Lett.* **24**, 457–460.

Kreek, H. and Meath, W. J. (1969), 'Charge-overlap effects. Dispersion and induction forces', *J. Chem. Phys.* **50**, 2289–2302.

Kuharsky, R. A. and Rossky, P. J. (1985), 'A quantum mechanical study of structure in liquid H_2O and D_2O', *J. Chem. Phys.* **82**, 5164–5177.

Kumar, A. and Meath, W. J. (1984), 'Pseudo-spectral dipole oscillator-strength distributions for SO_2, CS_2 and OCS and values of some related dipole–dipole and triple-dipole dispersion energy constants', *Chem. Phys.* **91**, 411–418.

Kumar, A. and Meath, W. J. (1985), 'Pseudo-spectral dipole oscillator strengths and dipole–dipole and triple-dipole dispersion energy coefficients for HF, HCl, HBr, He, Ne, Ar, Kr and Xe', *Molec. Phys.* **54**, 823–833.

Kumar, A. and Meath, W. J. (1992), 'Dipole oscillator strength properties and dispersion energies for acetylene and benzene', *Molec. Phys.* **75**, 311–324.

Kumar, A. and Meath, W. J. (1994), 'Reliable isotropic and anisotropic dipole properties, and dipolar dispersion energy coefficients, for CO evaluated using constrained dipole oscillator strength techniques', *Chem. Phys.* **189**, 467–477.

Kutzelnigg, W. (1980), 'The 'primitive' wavefunction in the theory of intermolecular interactions', *J. Chem. Phys.* **73**, 343–359.

Kutzelnigg, W. (1992), 'Does the polarization approximation converge for large *r* to a primitive or a symmetry-adapted function?', *Chem. Phys. Lett.* **195**, 77–84.

Kvasnicka, V., Laurinc, H. and Hubac, I. (1974), 'Many-body perturbation theory of intermolecular interactions', *Phys. Rev. A* **10**, 2016–2026.

Laasonen, K., Sprik, M., Parrinello, M. and Car, R. (1993), '"*Ab initio*" liquid water', *J. Chem. Phys.* **99**, 9080–9089.

Le Fèvre, R. J. W. (1965), 'Molecular polarizability and refractivity', *Adv. Phys. Org. Chem.* **3**, 1–90.

Le Roy, R. J. and Van Kranendonk, J. (1974), 'Anisotropic intermolecular potentials from an analysis of spectra of H_2– and D_2–inert-gas complexes', *J. Chem. Phys.* **61**, 4750–4769.

Le Sueur, C. R. and Stone, A. J. (1993), 'Practical schemes for distributed polarizabilities', *Molec. Phys.* **78**, 1267–1291.

Le Sueur, C. R. and Stone, A. J. (1994), 'Localization methods for distributed polarizabilities', *Molec. Phys.* **83**, 293–308.

Le Sueur, C. R., Stone, A. J. and Fowler, P. W. (1991), 'Induced dipole moments in acetylene complexes', *J. Phys. Chem.* **95**, 3519–3522.

Leavitt, R. P. (1980), 'An irreducible tensor method of deriving the long-range anisotropic interactions between molecules of arbitrary symmetry', *J. Chem. Phys.* **72**, 3472–3482.

Leforestier, C. (1994), 'Grid method for the Wigner functions. Application to the Van der Waals system Ar–H_2O', *J. Chem. Phys.* **101**, 7357–7363.

Legon, A. C. and Millen, D. J. (1982), 'Determination of properties of hydrogen-bonded dimers by rotational spectroscopy and a classification of dimer geometries', *Faraday Disc. Chem. Soc.* **73**, 71–87, 127, 128.

Legon, A. C., Millen, D. J. and Mjöberg, P. J. (1977), 'The hydrogen cyanide dimer: Identification and structure from microwave spectroscopy', *Chem. Phys. Lett.* **47**, 589–591.

Leighton, P., Cowan, J. A., Abraham, R. J. and Sanders, J. K. M. (1988), 'Geometry of porphyrin–porphyrin interactions', *J. Org. Chem.* **53**, 733–740.

Leslie, M. (1983), 'A symmetry-adapted method for the determination of the lattice energy and properties of ionic crystals', *Solid State Ionics* **8**, 243–246.

Lighthill, M. J. (1958), *Fourier Analysis and Generalized Functions*, Cambridge University Press, Cambridge.

Lii, J. H. and Allinger, N. L. (1991), 'The MM3 force field for amides, polypeptides and proteins', *J. Comput. Chem.* **12**, 186–199.

London, F. (1930*a*), *Z. Phys.* **63**, 245.

London, F. (1930*b*), *Z. Physik. Chem. B* **11**, 222–251.

London, F. (1937), 'The general theory of molecular forces', *Trans. Faraday Soc.* **33**, 8–26.

Longuet-Higgins, H. C. (1956), 'The electronic states of composite systems', *Proc. Roy. Soc. A* **235**, 537–543.

Loudon, R. (1973), *The Quantum Theory of Light*, Clarendon Press, Oxford.

Magnasco, V. and Figari, G. (1986), 'Epstein–Nesbet calculation of interatomic interactions in the Van der Waals region', *Molec. Phys.* **59**, 689–705.

Maitland, G. C., Rigby, M., Smith, E. B. and Wakeham, W. A. (1981), *Intermolecular Forces: Their Origin and Determination*, Clarendon Press, Oxford.

Mancera, R. L. and Buckingham, A. D. (1995*a*), 'Temperature effects on the hydrophobic hydration of ethane', *J. Phys. Chem.* **99**, 14632–14640.

Mancera, R. L. and Buckingham, A. D. (1995*b*), 'Further evidence for a temperature-dependent hydrophobic interaction—the aggregation of ethane in aqueous solutions', *Chem. Phys. Lett.* **234**, 296–303.

Manolopoulos, D. E. (1988), Close Coupled Equations, PhD thesis, Cambridge University.

Margenau, H. (1939), 'Van der Waals forces', *Rev. Mod. Phys.* **11**, 1–35.

Mason, E. A. (1957), 'Scattering of low-velocity molecular beams in gases', *J. Chem. Phys.* **26**, 667–677.

Matsuoka, O., Clementi, E. and Yoshimine, M. (1976), 'Configuration interaction study of the water dimer potential surface', *J. Chem. Phys.* **64**, 1351–1367.

Mayer, I. and Surjan, P. (1993), 'Handling overlap as a perturbation', *Croatica Chem. Acta* **66**, 161–165.

McClellan, A. L. (1963), *Tables of Experimental Dipole Moments*, Vol. 1, Freeman.

McClellan, A. L. (1974), *Tables of Experimental Dipole Moments*, Vol. 2, Rahara Enterprises, El Cerrito.

McClellan, A. L. (1989), *Tables of Experimental Dipole Moments*, Vol. 3, Rahara Enterprises, El Cerrito.

McDowell, S. A. C., Le Sueur, C. R., Buckingham, A. D. and Stone, A. J. (1992), 'Using monomer properties to obtain integrated intensities for vibrational transitions of Van der Waals complexes', *Molec. Phys.* **77**, 823–835.

McGurk, J., Norris, C. L., Tigelaar, H. L. and Flygare, W. H. (1973), 'Molecular magnetic properties of FCl', *J. Chem. Phys.* **58**, 3118–3120.

McIlroy, A., Lascola, R., Lovejoy, C. M. and Nesbitt, D. J. (1991), 'Structural dependence of HF vibrational red shifts in Ar_nHF, $n = 1$–4, via high-resolution slit jet infrared spectroscopy', *J. Phys. Chem.* **95**, 2636–2644.

McIlroy, A. and Nesbitt, D. J. (1992), 'Intermolecular motion in Ar_nHF micromatrices', *J. Chem. Phys.* **97**, 6044–6056.

McKellar, A. R. W. (1994), 'Long-path equilibrium IR spectra of weakly bound complexes at low temperatures', *Faraday Disc.* **97**, 69–80.

McQuarrie, D. A. (1976), *Statistical Mechanics*, Harper & Row, New York.

McWeeny, R. (1984), 'Weak interactions between molecules', *Croatica Chem. Acta* **57**, 865–878.

McWeeny, R. (1989), *Methods of Molecular Quantum Mechanics*, Academic Press, London.

Meath, W. J. and Aziz, R. A. (1984), 'On the importance and problems in the construction of many-body potentials.', *Molec. Phys.* **52**, 225–243.

Meath, W. J. and Koulis, M. (1991), 'On the construction and use of reliable two-body and many-body interatomic and intermolecular potentials', *Theochem. (J. Mol. Struct.)* **72**, 1–37.

Meath, W. J. and Kumar, A. (1990), 'Reliable isotropic and anisotropic dipole dispersion energies, evaluated using constrained dipole oscillator strength techniques, with application to interactions involving H_2, N_2, and the rare gases', *Int. J. Quantum Chem.* **S24**, 501–520.

Meath, W. J., Margoliash, D. J., Jhanwar, B. L., Koide, A. and Zeiss, G. D. (1981), 'Accurate molecular properties, their additivity, and their use in constructing intermolecular potentials', *in* B. Pullman, ed., 'Intermolecular forces', Reidel, Dordrecht, pp. 101–115.

Miller, K. J. (1990*a*), 'Additivity methods in molecular polarizability', *J. Amer. Chem. Soc.* **112**, 8533–8542.

Miller, K. J. (1990*b*), 'Calculation of the molecular polarizability tensor', *J. Amer. Chem. Soc.* **112**, 8543–8551.

Millot, C. and Stone, A. J. (1992), 'Towards an accurate intermolecular potential for water', *Molec. Phys.* **77**, 439–462.

Mills, I. M., ed. (1993), *Quantities, Units and Symbols in Physical Chemistry*, 2nd edn, Blackwell Scientific Publications, Oxford.

Milonni, P. W. and Eberly, J. H. (1988), *Lasers*, Wiley, New York.

Mirsky, K. (1978), 'The determination of the intermolecular interaction energy by empirical methods', *in* R. Schenk, R. Olthof-Hazenkamp, H. van Koningsveld and G. C. Bassi, eds, 'Computing in Crystallography', Delft University Press, p. 169.

Momany, F. A. (1978), 'Determination of partial atomic charge from *ab initio* molecular electrostatic potentials. Application to formamide, methanol and formic acid', *J. Phys. Chem.* **82**, 592.

Morgan, III, J. D. and Simon, B. (1980), 'Behaviour of molecular potential energy curves for large molecular separation', *Int. J. Quantum Chem.* **17**, 1143–1166.

Morokuma, K. (1971), 'Molecular orbital studies of hydrogen bonds. III. C=O···H–O hydrogen bond in $H_2CO\cdots H_2O$ and $H_2CO\cdots 2H_2O$', *J. Chem. Phys.* **55**, 1236–1244.

Morokuma, K. and Kitaura, K. (1981), 'Energy decomposition analysis of molecular interactions', *in* P. Politzer and D. G. Truhlar, eds, 'Chemical Applications of Atomic and Molecular Electrostatic Potentials', Plenum Press, New York, pp. 215–242.

Moszynski, R., Wormer, P. E. S. and Van der Avoird, A. (1995), 'Ab initio potential-energy surface and near infrared spectrum of the He–C_2H_2 complex', *J. Chem. Phys.* **102**, 8385–8397.

Muenter, J. S. (1989), 'Radio-frequency and microwave spectroscopy of the HCCH–CO_2 and DCCD–CO_2 Van der Waals complexes', *J. Chem. Phys.* **90**, 4048–4053.

Mulliken, R. S. (1952), 'Molecular compounds and their spectra', *J. Amer. Chem. Soc.* **74**, 811–824.

Murrell, J. N. and Bosanac, S. D. (1989), *Introduction to the Theory of Atomic and Molecular Collisions*, Wiley.

Murrell, J. N. and Laidler, K. J. (1968), 'Symmetries of activated complexes', *Trans. Faraday Soc.* **64**, 371–377.

Murrell, J. N. and Shaw, G. (1967), 'Intermolecular forces in the region of small orbital overlap', *J. Chem. Phys.* **46**, 1768–1772.

Murthy, C. S., O'Shea, S. F. and McDonald, I. R. (1983), 'Electrostatic interactions in molecular crystals: lattice dynamics of solid nitrogen and carbon dioxide', *Molec. Phys.* **50**, 531–541.

Musher, I. J. and Amos, A. T. (1967), 'Theory of weak atomic and molecular interactions', *Phys. Rev.* **164**, 31–43.

Muto, Y. (1943), *Proc. Phys.-Math. Soc. Japan* **17**, 629–631.

Némethy, G., Pottle, M. S. and Scheraga, H. A. (1983), 'Energy parameters in polypeptides. 9. Updating of geometrical parameters, nonbonded interactions and hydrogen-bond interactions for the naturally occurring amino acids', *J. Phys. Chem.* **87**, 1883–1887.

Nesbitt, D. J. (1994), 'Probing potential energy surfaces *via* high-resolution IR laser spectroscopy', *Faraday Discuss. Chem. Soc.* **97**, 1–18.

Nesbitt, D. J. and Child, M. S. (1993), 'Rotational RKR inversion of intermolecular stretching potentials: extension to linear hydrogen-bonded complexes', *J. Chem. Phys.* **98**, 478–486.

Nesbitt, D. J., Child, M. S. and Clary, D. C. (1989), 'Rydberg–Klein–Rees inversion of high-resolution Van der Waals infrared spectra: an intermolecular potential energy surface for Ar···HF ($v = 1$)', *J. Chem. Phys.* **90**, 4855–4864.

Novick, S. E., Janda, K. C. and Klemperer, W. (1976), 'HF–ClF: structure and bonding', *J. Chem. Phys.* **65**, 5115–5121.

Novoa, J. J., Planas, M. and Whangbo, M.-H. (1994), 'A numerical evaluation of the counterpoise method on hydrogen bond complexes using near complete basis sets', *Chem. Phys. Lett.* **225**, 240–246.

Nyburg, S. C. and Faerman, C. H. (1985), 'A revision of Van der Waals radii for molecular crystals: N, O, F, S, Cl, Se, Br and I bonded to carbon', *Acta Cryst. B: Struct. Sci.* **B41**, 274–279.

Ohshima, Y., Masumoto, Y., Takami, M. and Kuchitsu, K. (1988), 'The structure and tunneling motion of acetylene dimer studied by free-jet infrared absorption spectroscopy in the 14μm region', *Chem. Phys. Lett.* **147**, 1–6.

Pack, R. T. (1978), 'Anisotropic potentials and the damping of rainbow and diffraction oscillations in differential cross-sections', *Chem. Phys. Lett.* **55**, 197–201.

Pack, R. T., Piper, E., Pfeffer, G. A. and Toennies, J. P. (1984), 'Multiproperty empirical anisotropic intermolecular potentials. II. He···SF_6 and Ne···SF_6', *J. Chem. Phys.* **80**, 4940–4950.

Pack, R. T., Valentini, J. J. and Cross, J. D. (1982), 'Multiproperty empirical anisotropic intermolecular potentials. I. Ar···SF_6 and Kr···SF_6', *J. Chem. Phys.* **77**, 5486–5499.

Pauling, L. (1928), 'The shared-electron chemical bond', *Proc. Nat. Acad. Sci.* **14**, 359–362.

Pauling, L. (1960), *The Nature of the Chemical Bond*, 3rd edn, Cornell University Press.

Peet, A. C. and Yang, W. (1989), 'An adapted form of the collocation method for calculating energy levels of rotating atom–diatom complexes', *J. Chem. Phys.* **91**, 6598–6603.

Pérez-Jordá, J. M. and Becke, A. D. (1995), 'A density-functional study of Van der Waals forces: rare gas diatomics', *Chem. Phys. Lett.* **233**, 134–137.

Pitzer, K. S. (1959), 'Inter- and intramolecular forces and molecular polarizabilities', *Adv. Chem. Phys.* **11**, 59–83.

Poll, J. D. and Hunt, J. L. (1981), 'Analysis of the far infrared spectrum of gaseous N_2', *Canad. J. Phys.* **59**, 1448–1458.

Polymeropoulos, E. E., Brickmann, J., Jansen, L. and Block, R. (1984), 'Analysis of 3-body potentials in systems of rare-gas atoms—Axilrod–Teller versus 3-atom exchange interactions', *Phys. Rev. A* **30**, 1593–1599.

Popelier, P. L. A. and Stone, A. J. (1994), 'Formulae for the first and second derivatives of anisotropic potentials with respect to geometrical parameters', *Molec. Phys.* **82**, 411–425. Erratum, Molec. Phys. (1995) **84**, 811.

Pople, J. A., Schneider, W. G. and Bernstein, H. J. (1959), *High-Resolution Nuclear Magnetic Resonance*, McGraw-Hill, New York.

Price, S. L. and Stone, A. J. (1980), 'Evaluation of anisotropic model intermolecular pair potentials using an *ab initio* SCF-CI surface', *Molec. Phys.* **40**, 805–822.

Price, S. L. and Stone, A. J. (1982), 'The anisotropy of the $Cl_2 \cdots Cl_2$ pair potential as shown by the crystal structure—evidence for intermolecular bonding or lone-pair effects?', *Molec. Phys.* **47**, 1457–1470.

Price, S. L. and Stone, A. J. (1983), 'A distributed multipole analysis of the charge densities of the azabenzene molecules', *Chem. Phys. Lett.* **98**, 419–423.

Price, S. L. and Stone, A. J. (1984), 'A six-site intermolecular potential scheme for the azabenzene molecules, derived by crystal structure analysis', *Molec. Phys.* **51**, 569–583.

Price, S. L. and Stone, A. J. (1987), 'The electrostatic interactions in Van der Waals complexes involving aromatic molecules', *J. Chem. Phys.* **86**, 2859–2868.

Price, S. L., Stone, A. J. and Alderton, M. (1984), 'Explicit formulae for the electrostatic energy, forces and torques between a pair of molecules of arbitrary symmetry', *Molec. Phys.* **52**, 987–1001.

Price, S. L., Stone, A. J., Lucas, J., Rowland, R. S. and Thornley, A. (1994), 'On the nature of $-Cl \cdots Cl-$ intermolecular interactions', *J. Amer. Chem. Soc.* **116**, 4910–4918.

Prichard, J. S., Nandi, R. N., Muenter, J. S. and Howard, B. J. (1988), 'Vibration–rotation spectrum of the carbon-dioxide acetylene Van der Waals complex in the 3μ region', *J. Chem. Phys.* **89**, 1245–1250.

Rauk, A., Allen, L. C. and Clementi, E. (1970), 'Electronic structure and inversion barrier of ammonia', *J. Chem. Phys.* **52**, 4133–4144.

Reed, A. E., Curtiss, L. A. and Weinhold, F. (1988), 'Intermolecular interactions from a natural bond orbital, donor–acceptor viewpoint', *Chem. Rev.* **88**, 899–926.

Reed, A. E., Weinhold, F., Curtiss, L. A. and Pochatko, D. J. (1986), 'Natural bond orbital analysis of molecular interactions: theoretical studies of binary complexes of HF, H_2O, NH_3, N_2, O_2, F_2, CO and CO_2 with HF, H_2O and NH_3', *J. Chem. Phys.* **84**, 5687–5705.

Reimers, J. R., Watts, R. O. and Klein, M. L. (1982), 'Intermolecular potential functions and the properties of water', *Chem. Phys.* **64**, 95–114.

Rein, R. (1973), 'Physical properties and interactions of polyatomic molecules: with applications to molecular recognition in biology', *Adv. Quantum Chem.* **7**, 335–396.

Remler, D. K. and Madden, P. A. (1990), 'Molecular dynamics without effective potentials via the Car-Parrinello approach', *Molec. Phys.* **70**, 921–966.

Reynolds, C. A., Essex, J. W. and Richards, W. G. (1992*a*), 'Atomic charges for variable molecular-conformations', *J. Amer. Chem. Soc.* **114**, 9075–9079.

Reynolds, C. A., Essex, J. W. and Richards, W. G. (1992*b*), 'Errors in free energy perturbation calculations due to neglecting the conformational variation of atomic charges', *Chem. Phys. Letters* **199**, 257–260.

Rijks, W., Gerritsen, M. and Wormer, P. E. S. (1989), 'Computation of the short range repulsion energy from correlated monomer wavefunctions in Van der Waals dimers containing He, Ne and N_2', *Molec. Phys.* **66**, 929–953.

Rijks, W. and Wormer, P. E. S. (1989), 'Correlated Van der Waals coefficients. II. Dimers consisting of CO, HF, H_2O and NH_3', *J. Chem. Phys.* **90**, 6507–6519. Note subsequent Erratum, correcting many of the tabulated values, in *J. Chem. Phys.* (1990) **92**, 5754.

Rijks, W. and Wormer, P. E. S. (1990), 'Erratum: Correlated van der Waals coefficients. II. Dimers consisting of CO, HF, H_2O and NH_3', *J. Chem. Phys.* **92**, 5754.

Rodger, P. M., Stone, A. J. and Tildesley, D. J. (1987), 'Atomic anisotropy and the structure of liquid chlorine', *J. Chem. Soc. Faraday II* **83**, 1689–1702.

Rodger, P. M., Stone, A. J. and Tildesley, D. J. (1988*a*), 'The intermolecular potential of chlorine—a three-phase study', *Molec. Phys.* **63**, 173–188.

Rodger, P. M., Stone, A. J. and Tildesley, D. J. (1988*b*), 'Intermolecular interactions in halogens: bromine and iodine', *Chem. Phys. Lett.* **145**, 365–370.

Rodger, P. M., Stone, A. J. and Tildesley, D. J. (1992), 'Anisotropic site-site potentials in molecular dynamics', *Molec. Simulation* **8**, 145–164.

Roterman, I. K., Gibson, K. D. and Scheraga, H. A. (1989), 'A comparison of the CHARMm, AMBER and ECEPP potentials for peptides. 1. Conformational predictions for the tandemly repeated peptide (Asn–Ala–Asn–Pro)$_9$', *J. Biomol. Struct. Dyn.* **7**, 391–419.

Roterman, I. K., Lambert, M. H., Gibson, K. D. and Scheraga, H. A. (1989), 'A comparison of the CHARMm, AMBER and ECEPP potentials for peptides. 2. Phi-psi maps for normal-acetyl alanine N′-methyl amide—comparisons, contrasts and simple experimental tests', *J. Biomol. Struct. Dyn.* **7**, 421–453.

Rowlinson, J. S. (1949), 'The second virial coefficient of polar gases', *Trans. Faraday Soc.* **45**, 974.

Rowlinson, J. S. (1951), 'The lattice energy of ice and the second virial coefficient of water vapour', *Trans. Faraday Soc.* **47**, 120.

Rybak, S., Jeziorski, B. and Szalewicz, K. (1991), 'Many-body symmetry-adapted perturbation theory of intermolecular interactions—H_2O and HF dimers', *J. Chem. Phys.* **95**, 6576–6601.

Rybak, S., Szalewicz, K., Jeziorski, B. and Corongiu, G. (1992), 'Symmetry-adapted perturbation-theory calculations of uracil water interaction energy', *Chem. Phys. Lett.* **199**, 567–573.

Salem, L. (1968), 'Intermolecular orbital theory of the interaction between conjugated systems. I. General theory', *J. Amer. Chem. Soc.* **90**, 543–552.

Sandorfy, C. (1976), *Anharmonicity and Hydrogen Bonding*, North-Holland, Amsterdam. pp. 613–654.

Saykally, R. J. and Blake, G. A. (1993), 'Molecular interactions and hydrogen-bond tunneling dynamics—some new perspectives', *Science* **259**, 1570–1575.

Schmuttenmaer, C. A., Cohen, R. C. and Saykally, R. J. (1994), 'Spectroscopic determination of the intermolecular potential energy surface for Ar···NH_3', *J. Chem. Phys.* **101**, 146–173.

Schwenke, D. W. and Truhlar, D. G. (1988), 'Systematic study of basis set superposition errors in the calculated interaction energy of two HF molecules', *J. Chem. Phys.* **82**, 2418–2426.

Sikora, P. T. (1970), 'Combining rules for spherically symmetrical intermolecular potentials', *J. Phys. B* **3**, 1475–1482.

Silberstein, L. (1917), 'Molecular refractivity and atomic interaction', *Phil. Mag.* **33**, 92–128.

Singh, U. C. and Kollman, P. A. (1984), 'An approach to computing electrostatic charges for molecules', *J. Comput. Chem.* **5**, 129–145.

Sippl, M., Némethy, G. and Scheraga, H. A. (1984), 'Intermolecular potentials from crystal data. 6. Determination of empirical potentials for OH···O=C hydrogen bond from packing considerations', *J. Phys. Chem.* **88**, 6231–6233.

Skipper, N. T. (1993), 'Computer simulation of methane–water solutions. Evidence for a temperature-dependent hydrophobic attraction', *Chem. Phys. Lett.* **207**, 424–429.

Smith, W. (1982*a*), 'Point multipoles in the Ewald summation', *CCP5 Quarterly* **4**, 13–25.

Smith, W. (1982*b*), The program MDMULP, CCP5 program library, Daresbury Laboratory.

Sokalski, W. A., Hariharan, P. C. and Kaufman, J. J. (1983), 'A self-consistent field interaction energy decomposition study of twelve hydrogen-bonded dimers', *J. Phys. Chem.* **87**, 2803–2810.

Sokalski, W. A. and Poirier, R. A. (1983), 'Cumulative atomic multipole representation of the molecular charge distribution and its basis set dependence', *Chem. Phys. Lett.* **98**, 86–92.

Sokalski, W. A., Roszak, S., Hariharan, P. C. and Kaufman, J. J. (1983), 'Improved SCF interaction energy decomposition scheme corrected for basis set superposition effect', *Int. J. Quantum Chem.* **23**, 847–854.

Sokalski, W. A. and Sawaryn, A. (1992), 'Cumulative multicenter multipole moment databases and their applications', *J. Mol. Struct.* **256**, 91–112.

Spackman, M. A. (1992), 'Molecular electric moments from X-ray diffraction data', *Chemical Reviews* **92**, 1769–1797.

Sprik, M. (1991), 'Hydrogen bonding and the static dielectric constant in liquid water', *J. Chem. Phys.* **95**, 6762–6769.

Sprik, M. and Klein, M. L. (1988), 'A polarizable model for water using distributed charge sites', *J. Chem. Phys.* **89**, 7556–7560.

Stevens, R. M., Pitzer, R. and Lipscomb, W. N. (1963), 'Perturbed Hartree-Fock calculations. I. Magnetic susceptibility and shielding in the HF molecule', *J. Chem. Phys.* **38**, 550–560.

Stewart, J. J. P. (1989), 'Optimization of parameters for semi-empirical methods. I. Method. II. Applications', *J. Comput. Chem.* pp. 209–264.

Stillinger, F. H. and Rahman, A. (1974), 'Improved simulation of liquid water by molecular dynamics', *J. Chem. Phys.* **60**, 1545.

Stogryn, D. E. (1971), 'Higher order interaction energies for systems of asymmetric molecules', *Molec. Phys.* **22**, 81–103.

Stone, A. J. (1978), 'Theories of organic reactions', *Chem. Soc. Spec. Periodical Rep.* **3**, 39–69.

Stone, A. J. (1979), 'Intermolecular forces', *in* G. R. Luckhurst and G. W. Gray, eds, 'The Molecular Physics of Liquid Crystals', Academic Press, pp. 31–50.

Stone, A. J. (1981), 'Distributed multipole analysis; or how to describe a molecular charge distribution', *Chem. Phys. Lett.* **83**, 233–239.

Stone, A. J. (1985), 'Distributed polarizabilities', *Molec. Phys.* **56**, 1065–1082.

Stone, A. J. (1991), 'Classical electrostatics in molecular interactions', *in* Z. B. Maksić, ed., 'Theoretical Models of Chemical Bonding, vol. 4', Springer-Verlag, pp. 103–131.

Stone, A. J. (1993), 'Computation of charge-transfer energies by perturbation theory', *Chem. Phys. Lett.* **211**, 101–109.

Stone, A. J. and Alderton, M. (1985), 'Distributed multipole analysis—methods and applications', *Molec. Phys.* **56**, 1047–1064.

Stone, A. J. and Erskine, R. W. (1980), 'Intermolecular self-consistent-field perturbation theory for organic reactions. I. Theory and implementation; nucleophilic attack on carbonyl compounds', *J. Amer. Chem. Soc.* **102**, 7185–7192.

Stone, A. J. and Tong, C. S. (1989), 'Local and non-local dispersion models', *Chem. Phys.* **137**, 121–135.

Stone, A. J. and Tong, C.-S. (1994), 'Anisotropy of atom–atom repulsions', *J. Comput. Chem.* **15**, 1377–1392.

Stone, A. J. and Tough, R. J. A. (1984), 'Spherical tensor theory of long-range intermolecular forces', *Chem. Phys. Lett.* **110**, 123–129.

Storer, J. W., Giesen, D. J., Cramer, C. J. and Truhlar, D. G. (1995), 'Class IV charge models: a new semiempirical approach in quantum chemistry', *J. Comp.-Aided Molec. Design* **9**, 87–110.

Suhm, M. A. and Watts, R. O. (1991), 'Quantum Monte–Carlo studies of vibrational states in molecules and clusters', *Phys. Reports* **204**, 293–329.

Surjan, P. R. and Mayer, I. (1991), 'Intermolecular interactions: biorthogonal perturbation theory revisited', *Theochem* **72**, 47–58.

Surjan, P. R., Mayer, I. and Lukovits, I. (1985), 'Second-quantization-based perturbation theory for intermolecular interactions without basis set superposition error', *Chem. Phys. Lett.* **119**, 538–542.

Surjan, P. R. and Poirier, R. A. (1986), 'Intermolecular interactions using small basis sets: perturbation theory calculations avoiding basis set superposition error', *Chem. Phys. Lett.* **128**, 358–362.

Szabo, A. and Ostlund, N. S. (1989), *Modern Quantum Chemistry*, McGraw-Hill.

Szalewicz, K., Cole, S. J., Kolos, W. and Bartlett, R. J. (1988), 'A theoretical study of the water dimer interaction', *J. Chem. Phys.* **89**, 3662–3673.

Tang, K. T. (1969), 'Dynamic polarizabilities and Van der Waals coefficients', *Phys. Rev.* **177**, 108–114.

Tang, K. T. and Toennies, J. P. (1978), 'A simple model of the Van der Waals potential at intermediate distances. II. Anisotropic potential of He···H_2 and Ne···H_2', *J. Chem. Phys.* **68**, 5501–5517.

Tang, K. T. and Toennies, J. P. (1984), 'An improved simple model for the Van der Waals potential based on universal damping functions for the dispersion coefficients', *J. Chem. Phys.* **80**, 3726–3741.

Thakkar, A. J. (1988), 'Higher dispersion coefficients: accurate values for the hydrogen atom and simple estimates for other systems', *J. Chem. Phys.* **89**, 2092–2098.

Thakkar, A. J., Hettema, H. and Wormer, P. E. S. (1992), '*Ab initio* dispersion coefficients for interactions involving rare-gas atoms', *J. Chem. Phys.* **97**, 3252–3264.

Thole, B. T. (1981), 'Molecular polarizabilities calculated with a modified dipole interaction', *Chem. Phys.* **59**, 341–350.

Thomas, G. F. and Meath, W. J. (1977), 'Dipole spectrum, sums and properties of ground-state methane and their relation to the molar refractivity and dispersion energy constant', *Molec. Phys.* **34**, 113–125.

Thornley, A. E. and Hutson, J. M. (1992), 'The intermolecular potential of Ar–acetylene. Information from infrared and microwave spectroscopy', *Chem. Phys. Lett.* **198**, 1–8.

Tinkham, M. (1964), *Group Theory and Quantum Mechanics*, McGraw-Hill, New York.

Tough, R. J. A. and Stone, A. J. (1977), 'Properties of the regular and irregular solid harmonics', *J. Phys. A* **10**, 1261–1269.

Townes, C. H., Dousmanis, G. C., White, R. L. and Schwarz, R. F. (1955), 'Connections between molecular structure and certain magnetic effects in molecules', *Disc. Faraday Soc.* **19**, 56–64.

Unsöld (1927), *Z. Physik.* **43**, 563–574.

Van Bladel, J. W. I., Van der Avoird, A., Wormer, P. E. S. and Saykally, R. J. (1992), 'Computational exploration of the six-dimensional vibration–rotation–tunnelling dynamics of $(NH_3)_2$', *J. Chem. Phys.* **97**, 4750–4763.

Van der Avoird, A. (1967*a*), 'Intermolecular interactions by perturbation theory including exchange effects', *Chem. Phys. Lett.* **1**, 24–27.

Van der Avoird, A. (1967*b*), 'Note on a perturbation theory for intermolecular interactions in the wave operator formalism', *Chem. Phys. Lett.* **1**, 411–412.

Van der Avoird, A. (1967*c*), 'A perturbation theory for intermolecular interactions in the wave-operator formalism', *J. Chem. Phys.* **47**, 3649–3653.

Van der Avoird, A., Olthof, E. H. T. and Wormer, P. E. S. (1994), 'Is the $NH_3 \cdots NH_3$ riddle solved?', *Faraday Discussions* **97**, 43–55.

Varshalovich, D. A., Moskalev, A. N. and Khersonskii, V. K. (1988), *Quantum Theory of Angular Momentum*, World Scientific, Singapore.

Vigné-Maeder, F. and Claverie, P. (1988), 'The exact multicentre multipolar part of a molecular charge distribution and its simplified representations', *J. Chem. Phys.* **88**, 4934–4948.

Voisin, C. and Cartier, A. (1993), 'Determination of distributed polarizabilities to be used for peptide modelling', *Theochem (J. Mol. Struct.)* **105**, 35–45.

Waldman, M. and Hagler, A. T. (1993), 'New combining rules for rare gas Van der Waals parameters', *J. Comput. Chem.* **14**, 1077–1084.

Wales, D. J. (1991), 'Theoretical study of some small Van der Waals complexes containing inert gas atoms', *Molec. Phys.* **74**, 1–25.

Wallqvist, A., Ahlström, P. and Karlström, G. (1990), 'A new intermolecular energy calculation scheme: applications to potential surface and liquid properties of water', *J. Phys. Chem.* **94**, 1649–1656. Erratum J. Phys. Chem. (1991) 95, 4922.

Warshel, A. and Parson, W. W. (1987), 'Spectroscopic properties of photosynthetic reaction centres', *J. Amer. Chem. Soc.* **109**, 6143–6163.

Watanabe, A. and Welsh, H. L. (1964), 'Direct spectroscopic evidence of bound states of $(H_2)_2$ complexes at low temperature', *Phys. Rev. Lett.* **13**, 810–812.

Weiner, S. J., Kollman, P. A., Case, D. A., Singh, U. C., Ghio, C., Alagona, G., Profeta, S. and Weiner, P. (1984), 'A new force field for molecular mechanical simulation of nucleic acids and proteins', *J. Amer. Chem. Soc.* **106**, 765–784.

Weiner, S. J., Kollman, P. A., Nguyen, D. T. and Case, D. A. (1986), 'An all atom force field for simulations of proteins and nucleic acids', *J. Comput. Chem.* **7**, 230–252.

Wells, A. F. (1975), *Structural Inorganic Chemistry*, 4th edn, Clarendon Press, Oxford.

Wells, B. H. (1985), 'The differential Green's function Monte Carlo method. The dipole moment of LiH', *Chem. Phys. Lett.* **115**, 89–94.

Wells, B. H. and Wilson, S. (1986), 'Van der Waals interaction potentials: many-body effects in rare gas mixtures', *Molec. Phys.* **57**, 421–426.

Wells, B. H. and Wilson, S. (1989*a*), 'Van der Waals interaction potentials. Many-body effects in Ne_4', *Molec. Phys.* **66**, 457–464.

Wells, B. H. and Wilson, S. (1989*b*), 'Van der Waals potentials: convergence of the many-body expansion', *Molec. Phys.* **65**, 1363–1376.

Werner, H.-J. and Meyer, W. (1976), 'PNO–CI and PNO–CEPA studies of electron correlation effects. V. Static dipole polarizabilities of small molecules', *Molec. Phys.* **31**, 855–872.

Wheatley, R. J. (1993*a*), 'Gaussian multipole functions for describing molecular charge distributions', *Molec. Phys.* **79**, 597–610.

Wheatley, R. J. (1993*b*), 'A new distributed multipole procedure for linear molecules', *Chem. Phys. Lett.* **208**, 159–166.

Wheatley, R. J. and Meath, W. J. (1993*a*), 'Dispersion energy damping functions, and their relative scale with interatomic separation, for (H, He, Li)–(H, He, Li) interactions', *Molec. Phys.* **80**, 25–54.

Wheatley, R. J. and Meath, W. J. (1993*b*), 'On the relationship between first-order exchange and coulomb interaction energies for closed shell atoms and molecules', *Molec. Phys.* **79**, 253–275.

Wheatley, R. J. and Mitchell, J. B. O. (1994), 'Gaussian multipoles in practice: electrostatic energies for intermolecular potentials', *J. Comput. Chem.* **15**, 1187–1198.

Wheatley, R. J. and Price, S. L. (1990*a*), 'An overlap model for estimating the anisotropy of repulsion', *Molec. Phys.* **69**, 507–533.

Wheatley, R. J. and Price, S. L. (1990*b*), 'A systematic intermolecular potential method applied to chlorine', *Molec. Phys.* **71**, 1381–1404.

Wiberg, K. B. and Rablen, P. R. (1993), 'Comparison of atomic charges derived by different procedures', *J. Comput. Chem.* **14**, 1504–1518.

Williams, D. E. (1965), 'Non-bonded potential parameters derived from crystalline aromatic hydrocarbons', *J. Chem. Phys.* **45**, 3770–3778.

Williams, D. E. (1967), 'Non-bonded potential parameters derived from crystalline hydrocarbons', *J. Chem. Phys.* **47**, 4680–4684.

Williams, D. E. (1993), 'Net atomic charge and multipole models for the *ab initio* molecular electrostatic potential', *Rev. Comput. Chem.* **2**, 219–271.

Williams, H. L., Szalewicz, K., Jeziorski, B., Moszynski, R. and Rybak, S. (1993), 'Symmetry-adapted perturbation theory calculation of the Ar$\cdots$$H_2$ intermolecular potential energy surface', *J. Chem. Phys.* **98**, 1279–1292.

Willock, D. J., Leslie, M., Price, S. L. and Catlow, C. R. A. (1993), 'The need for realistic electrostatic models to predict the crystal structures of NLO molecules', *Mol. Cryst. Liq. Cryst.* **234**, 499–506.

Willock, D. J., Price, S. L., Leslie, M. and Catlow, C. R. A. (1995), 'The relaxation of molecular crystal structures using a distributed multipole electrostatic model.', *J. Comput. Chem.* **16**, 628–647.

Wilson, E. B. (1968), 'Some remarks on quantum chemistry', *in* A. Rich and N. Davidson, eds, 'Structural Chemistry and Molecular Biology', W. H. Freeman, San Francisco, pp. 753–760.

Wilson, E. B., J., Decius, J. C. and Cross, P. C. (1955), *Molecular Vibrations*, McGraw-Hill, New York.

Wilson, M. and Madden, P. A. (1994), 'Anion polarization and the stability of layered structures in MX_2 systems', *J. Phys. Condensed Matter* **6**, 159–170.

Woodward, R. B. and Hoffmann, R. (1970), *The Conservation of Orbital Symmetry*, Verlag Chemie, Weinheim.

Wormer, P. E. S. and Hettema, H. (1992), 'Many-body perturbation theory of frequency-dependent polarizabilities and Van der Waals coefficients: application to $H_2O$$\cdots$$H_2O$ and Ar$\cdots$$NH_3$', *J. Chem. Phys.* **97**, 5592–5606.

Xantheas, S. S. (1995), '*Ab initio* studies of cyclic water clusters $(H_2O)_n$, $n = 1$–6. III. Comparison of density functional with MP2 results', *J. Chem. Phys.* **102**, 4505–4517.

Yang, M. B. and Watts, R. O. (1994), 'The anisotropic potential energy surfaces of H_2, N_2, and Ar with C_2H_2 from total differential scattering experiments', *J. Chem. Phys.* **100**, 3582–3593.

Yashonath, S., Price, S. L. and McDonald, I. R. (1988), 'A six-site anisotropic atom–atom potential model for the condensed phases of benzene', *Molec. Phys.* **64**, 361–376.

Zare, R. N. (1988), *Angular Momentum*, Wiley Interscience.

INDEX

Page numbers in italic followed by *n*, *f* or *t* refer to footnotes, figures and tables respectively.

Printed in the United Kingdom
by Lightning Source UK Ltd.
108102UKS00001B/106

9 780198 558835